4

Springer Series in Chemical Physics

Edited by Fritz Peter Schäfer

Springer Series in Chemical Physics

Editors: V. I. Goldanskii R. Gomer F. P. Schäfer J. P. Toennies

Volume 1 **Atomic Spectra and Radiative Transitions**
By I. I. Sobelman

Volume 2 **Surface Crystallography by LEED,** Theory, Computation, and Structural Results
By M. A. Van Hove, S. Y. Tong

Volume 3 **Advances in Laser Chemistry**
Editor: A. H. Zewail

Volume 4 **Picosecond Phenomena**
Editors: C. V. Shank, E. P. Ippen, S. L. Shapiro

Volume 5 **Laser Spectroscopy,** Fundamentals and Techniques
By W. Demtröder

Volume 6 **Laser induced Processes in Molecules**
Editors: K. Kompa, D. Smith

Picosecond Phenomena

Proceedings of the First International Conference
on Picosecond Phenomena
Hilton Head, South Carolina, USA
May 24–26, 1978

Editors
C. V. Shank E. P. Ippen S. L. Shapiro

With 222 Figures

Springer-Verlag Berlin Heidelberg New York 1978

Series Editors

Professor Vitalii I. Goldanskii

Institute of Chemical Physics
Academy of Sciences
Vorobyevskoye Chaussee 2-b
Moscow V-334, USSR

Professor Dr. Fritz Peter Schäfer

Max-Planck-Institut für
Biophysikalische Chemie
D-3400 Göttingen-Nikolausberg
Fed. Rep. of Germany

Professor Robert Gomer

The James Franck Institute
The University of Chicago
5640 Ellis Avenue
Chicago, IL 60637, USA

Professor Dr. J. Peter Toennies

Max-Planck-Institut für Strömungsforschung
Böttingerstraße 6–8
D-3400 Göttingen
Fed. Rep. of Germany

Conference Chairman and Editor
Dr. Stanley L. Shapiro, University of California,
Los Alamos Scientific Laboratory, Los Alamos, NM 87545, USA

Program Co-Chairmen and Editors
Dr. Erich P. Ippen and Dr. Charles V. Shank, Bell Telephone Laboratories,
Holmdel, NJ 07733, USA

Program Committee
D. H. Auston D. J. Bradley J. Ducuing M. A. Duguay K. B. Eisenthal S. E. Harris
R. M. Hochstrasser W. Kaiser H. Mahr J. F. Reintjes M. C. Richardson M. O. Scully
M. W. Windsor E. Yablonovitch

Sponsored by
The Optical Society of America

Supported by Grants from
National Science Foundation
Office of Naval Research
U.S. Army Research Office
Spectra Physics
Coherent Radiation
GEAR
Hamamatsu

ISBN 3-540-09054-1 Springer-Verlag Berlin Heidelberg New York
ISBN 0-387-09054-1 Springer-Verlag New York Heidelberg Berlin

Offset printing: Zechnersche Buchdruckerei, Speyer.
Bookbinding: J. Schäffer OHG, Grünstadt.
2153/3130-543210

Preface

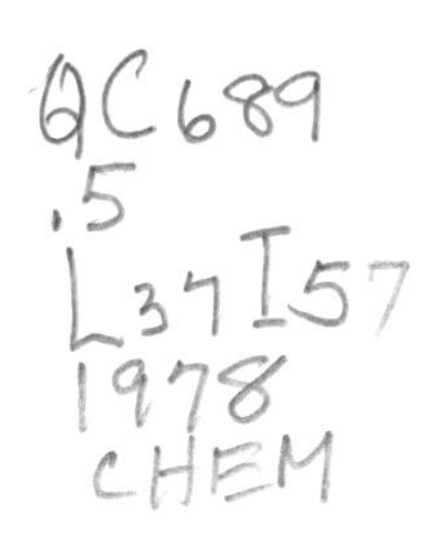

The first international conference devoted to Picosecond Phenomena was held May 24-26, 1978 in Hilton Head, South Carolina. After more than a decade of active research, this conference brought together scientists from widely varying disciplines who shared a common interest in studying ultrafast processes. It was organized as a Topical Meeting of the Optical Society of America and attracted 187 registered participants.

The conference reviewed current progress in experimental and theoretical understanding of phenomena that occur on a picosecond timescale. A recent upsurge of interest has developed because of the sudden availability of sophisticated and more powerful techniques. Consequently, the conference attracted chemists, physicists, biologists and engineers who had the opportunity to learn of the new and significant results in rapidly advancing areas. The enthusiasm of the scientists, the high quality of the research presented, and the attractive surroundings combined to produce a thoroughly successful and enjoyable conference.

Many people helped to make the conference so successful. Special thanks are due to Drs. Jarus Quinn and Jon Hagen, and their colleagues at the Optical Society of America for their vital assistance in smoothly implementing the meeting arrangements, and to the program committee for their advice and efforts. We gratefully acknowledge the financial support from our sponsors that allowed the conference to be more productive and more complete.

Holmdel, New Jersey

Los Alamos, New Mexico
August 1978

C.V. Shank
E.P. Ippen
S.L. Shapiro

Contents

I. Interactions in Liquids and Molecules

II. Poster Session

III. Sources and Techniques

I. Interactions in Liquids and Molecules

Dynamic Spectroscopy of Polyatomic Molecules with Tunable Picosecond Pulses

W. Kaiser, A. Seilmeier and A. Laubereau

Physik Department der Technischen Universität München
D-8000 München, Fed. Rep. of Germany

Light pulses of picosecond duration have opened a wide field of physical investigations. The numerous papers presented at this conference give clear evidence of this statement.

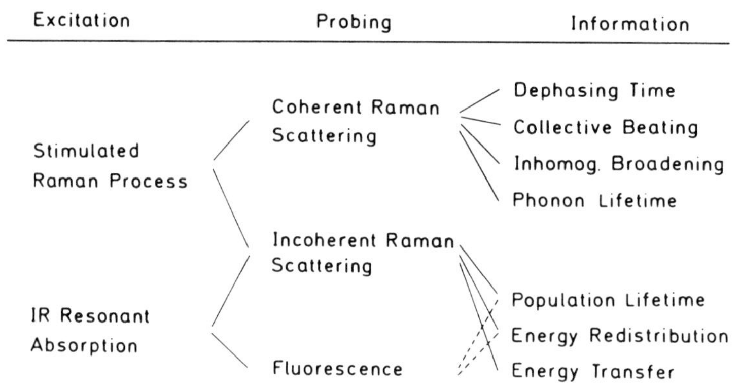

Fig. 1
Ultrashort excitation and probing processes used in our Laboratories for time resolved spectroscopic studies

For a number of years we were concerned with vibrational modes of polyatomic molecules in the electronic ground state. In the condensed phase - especially in liquids at room temperature - the molecular vibrational interactions proceed very rapidly, on the time scale of picoseconds. In Fig. 1 we recall two excitation processes, three probing techniques and various information which we have reported in a series of papers. By stimulated Raman scattering the medium is coherently excited allowing the investigation of coherent optical phenomena. With direct IR excitation a substantial excess population of selected vibrational modes is readily achieved with existing pulse intensities. Probing of the instantaneous vibrational population using anti-Stokes Raman scattering has provided valuable information of vibrational lifetimes, of decay routes and relaxation mechanisms and of energy transfer processes. Unfortunatly, this technique is experimentally difficult on account of the small Raman scattering cross sections. In a different more sensitive method we first excite the molecules by an infrared pulse of frequency ν_1 and promote the still excited molecules with a second pulse at ν_2 to a fluorescent level of the first excited singlet state S_1. Proper time delay between the two pulses allows us to study lifetimes of the intermediate states. Recently, we found that this two-pulse technique gives us interesting possibilities for spectroscopic investigations. In particular, large polyatomic molecules, which exhibit broad featureless absorption and fluorescence bands by standard spectroscopic techniques, show characteristic structure in the two-pulse spectra. Fig. 2 shows schematically the three types of investigations we have made. First (Fig. 2, left) we tuned the infrared frequency ν_1 holding the frequency ν_2 of the second visible

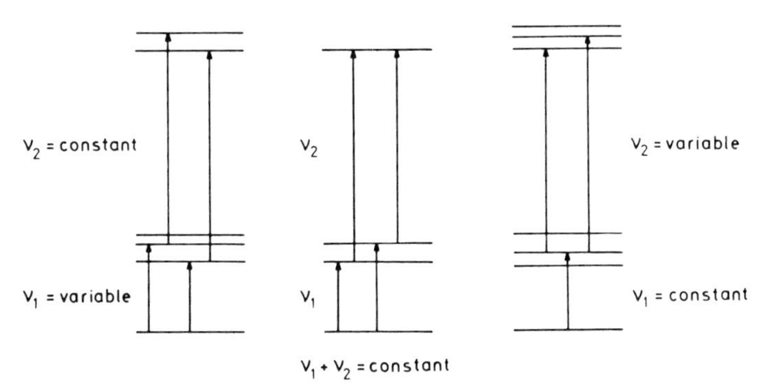

Fig. 2
Schematic of three types of investigations using the two-pulse technique

pulse constant. In this case we connect different states in S_o with different vibronic levels in S_1. Next (Fig. 2, middle) we tune both frequencies ν_1 and ν_2 keeping the sum $\nu_1+\nu_2$ fixed; i.e. the total photon energy supplied to the system is held constant. In a third experiment, the frequency ν_1 remains constant and ν_2 is varied. Before presenting spectra corresponding to the three investigations pictured in Fig. 2, the experimental system should be briefly introduced. Fig. 3 shows very schematically the mode-locked Nd:glass laser, the single pulse selector, the optical amplifier, and the two parametric generators which provide the two pulses of frequency ν_1 and ν_2. Both pulses are combined and travel through the sample at the same time. We emphasize the large tuning range, the favorable bandwidth and the high peak intensity of the light pulses (see contribution by Laubereau et al.) In Fig. 4 we present the two molecules coumarin 6 and nile-blue-A-oxazon which are investigated here. Both molecules consist of 42 atoms and have broad absorption bands extending over several thousand wave numbers in the visible range. Three absorption spectra obtained by the two-pulse technique are presented next. In Fig. 5a we tune the infrared frequency ν_1 over 600 cm^{-1} holding $\nu_2 = 2\nu_L$ at the second harmonic of the laser

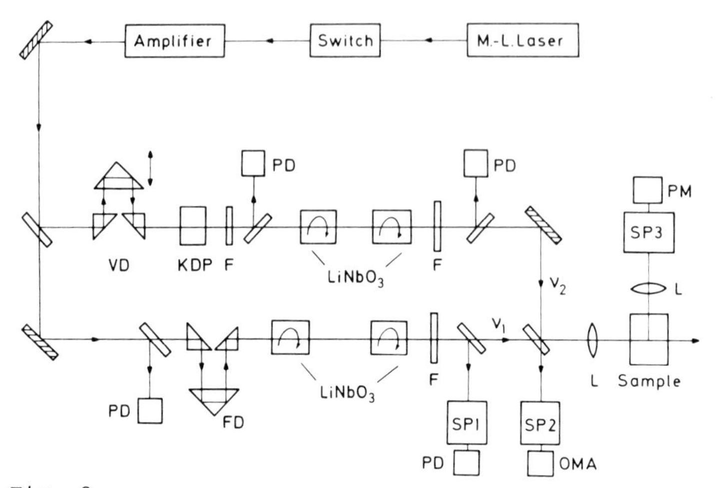

Fig. 3
Schematic of experimental set-up. A single pulse of a Nd:glass laser at ν_L = 9455 cm^{-1} generates a tunable infrared (ν_1) and a tunable visible (ν_2) pulse in two parametric generators consisting of two $LiNbO_3$ crystals each. The two pulses excite the sample selectively. The fluorescence is monitored with the help of spectrometer SP3

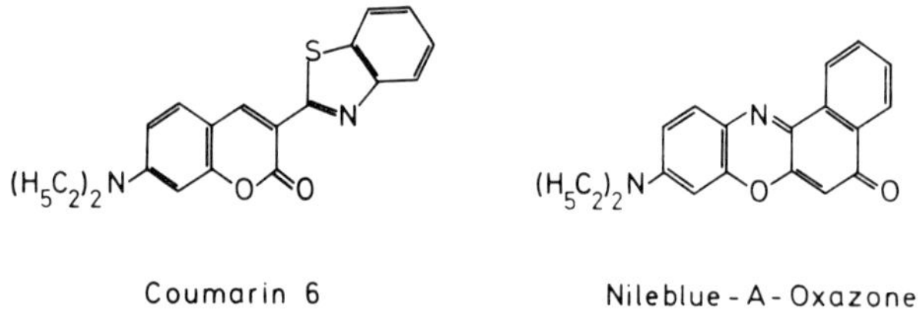

Coumarin 6

Nileblue - A - Oxazone

Fig. 4
The two molecules investigated. Coumarin 6 and nile-blue-A-oxazone

frequency. We find a distinct spectrum of more than ten maxima and shoulders. Comparison between Fig. 5a and the normal infrared spectrum of Fig. 5b shows little relation between the two pictures. The peaks in Fig. 5a result from favorable values of the product infrared transition moment times Franck-Condon factors while Fig. 5b is solely determined by infrared moments. In Fig.6a the two-pulse signal is plotted versus infrared frequency ν_1 for the case $\nu_1+\nu_2$ = 18,910 cm^{-1}. Again, we see a distinct spectrum differing substantially from the standard infrared spectrum of Fig. 6b. In Fig. 7b, finally, we present two-pulse data when ν_1 = 2935 cm^{-1} and ν_2 is varied between 15,500 cm^{-1} and 16,200 cm^{-1}. This spectrum differs drastically from the smooth absorption obtained from one-photon spectroscopy over the same total energy

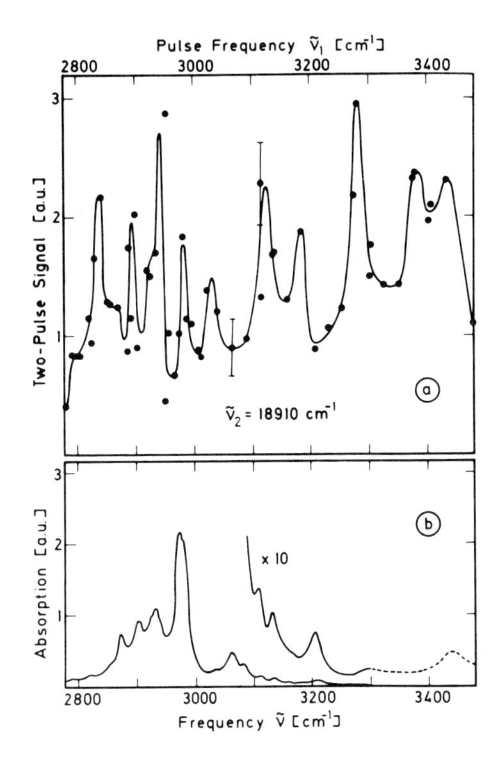

Fig. 5
a) Two-pulse spectrum versus frequency ν_1 of infrared pulse. Frequency ν_2=18,910 cm^{-1}; Coumarin 6 in CCl_4, 10^{-4} M
b) Conventional infrared spectrum in the same frequency range

range (Fig. 7a). The spectrum of Fig. 7b consists of three strong lines separated by 170 cm^{-1} which have satellites with a frequency interval of 25 cm^{-1}. We briefly summarize our tentative picture. In the two-pulse technique we excite molecules to the S_1 state via certain intermediate states.The total transition moment is proportional to infrared transition moments and Franck-Condon factors. The intermediate states which contribute predominantly to our spectra are thought to be mixed states made up of skeletal modes and low lying bending modes. The skeletal modes of the molecules are known to have good Franck-Condon factors since they couple well to the π-electrons of the chromophore. The low bending modes carry large dipole moments and enhance the infrared transition moment. Evidence for these bending modes comes from the distinct satellite structure of the spectrum in Fig. 7b. In a similar way, certain mixed intermediate states called combination states appear as peaks in the spectrum of

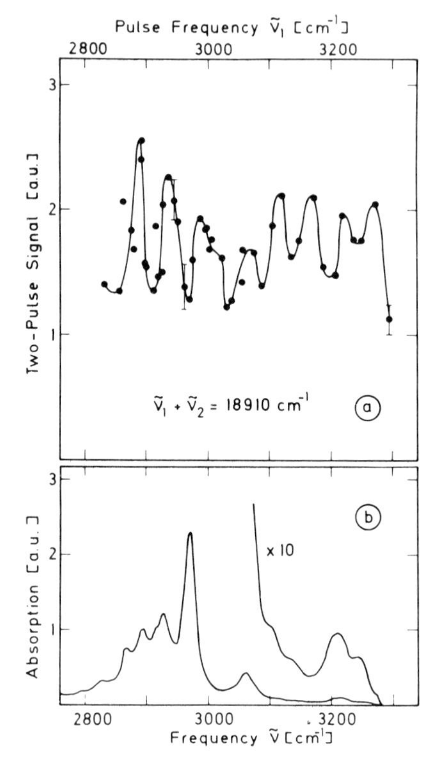

Fig. 6
a) Two-pulse spectrum versus frequency ν_1 of infrared pulse. The frequencies of ν_1 and ν_2 were adjusted to make the sum constant, $\nu_1+\nu_2$ = 18,910 cm^{-1}; nile-blue-A-oxazone in CCl_4, 10^{-4} M
b) Conventional infrared spectrum

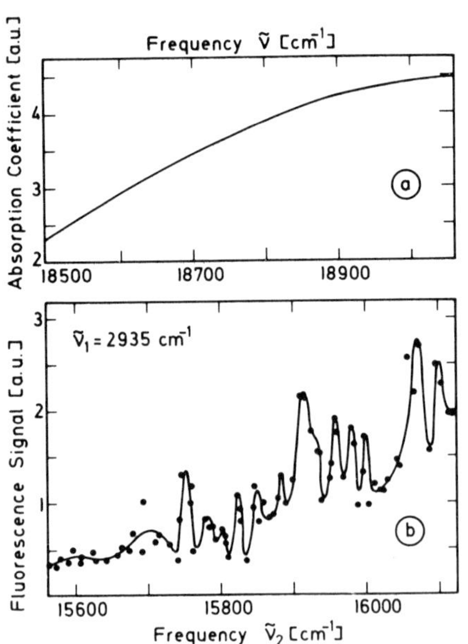

Fig. 7
a) Expanded part of the conventional absorption spectrum
b) Ultrafast two-pulse spectrum of nile-blue-A-oxazone (10^{-4}M in CCl_4). The observed fluorescence is plotted versus ν_2, the frequency of the red pulse; the frequency ν_1 of the infrared pulse is held constant at ν_1 = 2935 cm^{-1}

Fig. 6a. Certainly, more experimental data are necessary for a detailed assignment of the two-pulse spectra. We emphasize that the two-pulse spectra are characteristic of the investigated molecules; they should be of value for the identification and for interaction studies of large polyatomic molecules.

For further references of the authors:

A. Laubereau, A. Seilmeier and W. Kaiser, Chem. Phys. Lett. 36, 232 (1975)
J.P. Maier, A. Seilmeier, A. Laubereau and W. Kaiser, Chem. Phys. Lett. 46, 527 (1977)
A. Seilmeier, K. Spanner, A. Laubereau and W. Kaiser, Optics Commun. 24, 237 (1978
A. Seilmeier, W. Kaiser, A. Laubereau and S.F. Fischer, to be published.

A New Technique for Measurement of Raman Dephasing Dynamics and Recent Advances in cw Mode-Locked Dye Lasers

J.P. Heritage, C.P. Ausschnitt

Bell Telephone Laboratories, Holmdel, NJ 07733, USA

R.K. Jain

Department of Optical Physics
Hughes Research Laboratory, Malibu, CA 90265, USA

Recent advances in the development of synchronously mode-locked cw dye lasers have permitted the first low power measurement of the time resolved dephasing dynamics of a Raman active ground state molecular vibration. Excitation and delayed probing of a coherent ensemble of vibrators is accomplished with the help of synchronized picosecond pulse trains from two independently tunable mode-locked dye lasers. A new probe technique, based on a Raman gain messurement, is employed that provides a signal that scales linearly with the coherent vibrational amplitude. We demonstrate this technique with the first measurement of the dephasing of the ν_1 (656.5 cm^{-1}) of CS_2 in the liquid state. An exponential decay time of 20 ± 1 psec is obtained for the isotropic contribution to the coherent scattering.

The dephasing dynamics of certain Raman active molecular vibrations have previously been studied by excitation and time resolved probing with intense single picosecond optical pulses [1]. In this paper a new experimental technique for measurement of Raman dephasing dynamics is described. Our methods are suitable for use with low power cw mode-locked dye lasers and offer advantages stemming from the tunability of the lasers, the use of low peak power and from the linear scaling law obeyed by the detected signal. The gain (or loss) experienced by a delayed probe pulse-pair depends linearly upon the previously excited coherent vibrational amplitude and the interaction length. For thin samples, where the sample thickness is much less than the beam confocal parameter and for dilute samples, the signal to noise ratio will be larger than those obtained from a quadratic response [1] with the same peak power. Secondly, the use of low peak power (200 w) picosecond pulses helps avoid self-focusing in materials with a large nonlinear refractive index. Note also that this technique shares with other methods that employ synchronized tunable pulses [2], the ability to excite different Raman modes within the same sample. Even though the experimental techniques described in this paper differ from those reviewed in [1], the underlying physical phenomena are closely related. The stimulated Raman effect provides the coherent excitation. The mixing of applied probe optical fields with the coherent vibrations generates a polarization at well-defined frequencies that radiate and may be detected.

In the excitation process a coherent Raman vibrational amplitude grows efficiently during the presence of applied synchronized picosecond pulses provided that the difference between the optical frequencies is in resonance with the molecular vibration, and the pump pulse train and stokes pulse train overlap in time. Since the stimulated Raman gain obtained with a low power excitation pulse is small, the effect of the increase in the stokes field on the growth of the coherence may be neglected. In our experiments the gain experienced by the stokes pulse is calculated to be $g = \Delta Ps/Ps \leq 10^{-2}$ for CS_2 (ν_1). ΔPs is the change in stokes power after passing through the interaction region and Ps is the average stokes power presented to the sample; g is the gain. Very small stimulated Raman gains may be easily detected by chopping the cw laser pump with phase sensitive detection at the chop frequency of the change in the stokes average power [3]. Note that no temporal information about the Raman resonance is obtained when the pump laser pulse train is delayed with respect to the stokes laser pulse train. One obtains only a cross-correlation between the two laser pulse trains. For time resolved studies additional probe pulses are required.

Our probe technique involves introducing a delayed probe pair of coincident pump laser and stokes laser pulse trains subsequent to excitation with the coincident pump pair. The probe pair is obtained from the pump pair with beam splitters and adjustable delay lines. The probe pair mix with the Raman vibrators to produce, for example, a polarization at the stokes laser and pump laser frequencies. This polarization induces gain or loss on the stokes probe pulse train depending upon the relative phase of the probe pair and the material vibrations. This phase dependent gain appears in addition to the ordinary Raman gain that the probe-pair induce themselves. Experimentally we discriminate against this background by modulating the pump stokes beam with a chopper and phase sensitive detection of the probe stokes power. The decay of the coherence is monitored by carefully adjusting the optical phase of one beam with very small shifts of a delay line and finding the phase condition for peak gain (or loss). This procedure is repeated at each position of the probe-pair delay.

In a discussion of the experimental arrangement we start with the source of the synchronized, tunable, picosecond pulse trains. Recently, stable trains of broadly tunable picosecond pulses of high quality have been obtained from the mode-locked Argon ion laser synchronously pumped jet stream dye laser. Transform limited pulses ($\Delta\nu\Delta T \sim .5$) as short as 0.7 psec have been demonstrated [4,5]. The width of the pulse may be conveniently adjusted by controlling an intracavity filter bandwidth [6]. With a commercially available 3 plate Lyot filter and a 100 psec duration pump pulse, the dye laser pulse width is typically 3 to 4 psec. Outstanding dye laser performance may be obtained when a stable train of 80-100 psec pump pulses is available. The electronic oscillator that drives the loss modulator for mode-locking of the ion laser is the system clock and must be stable over a long term ($\sim$ 1 ppm) and exhibit very low phase noise. An HP 8640-B oscillator or a Rockland 5600 frequency synthesizer were found to be sufficiently stable. An ultrafast detector and sampling oscilloscope with system rise time of $\sim$ 50 psec is essential for diagnosis of the state of the Argon laser mode-locking. Twenty percent average power conversion from the pump to dye laser is obtained near the peak of the Rhodamine 6G band. This performance is obtained with about 600 mw pump power and 40 percent transmitting output couplers. Length tuning of the dye laser is critical, especially for sub-picosecond operation [7].

When two dye lasers are pumped from a single cw mode-locked pump, the output pulse trains are synchronized [8]. The degree of synchronization may be measured with a cross-correlation technique using sum frequency generation. Noise free cross-correlations as short as 5 psec have been obtained with two dye lasers tuned with 3 plate lyot filters. This result is obtained when the pump pulse is short (80 psec) and the two dye lasers operate with nearly the same threshold. A short pump pulse produces a fast rising gain which minimizes jitter from independent mechanical fluctuations in each laser. When the threshold of each laser occurs at the same point in the gain curve, the jitter arising from pump fluctuations is minimized.

A collinear pump pulse-pair and the delayed probe pulse-pair are derived from the output of the two synchronized lasers with the help of dichroic mirrors and optical delay lines. The two collinear pair of pump and probe beams cross in a thin sample cell (2 mm) after focusing with a 50 mm Achromat. The pump pair is blocked after passing through the sample and the probe pulse-pair is separated with brewster prisms. The change in the intensity of the probe stokes beam is monitored with a silicon photodiode and a lockin amplifier. Momentum matching is satisfied for collinear geometry and for crossed collinear pairs focused through crossing angles of a few degrees. The signal to noise ratio is limited by fluctuations in the probe laser. A commercially available electrooptic feedback loop provides a factor of 30 improvement over the unprocessed laser beam. Presently we are able to measure a gain of 10^{-5} with unit signal to noise ratio in one second of integration.

In Figure 1 is displayed the measured (peak) gain as a function of delay between the pump pair and the probe pair for (ν_1) 656.5 cm^{-1} mode of CS_2. A cross-correlation trace is also plotted which serves to identify zero relative delay and establish the time resolution. Two decay components are evident, one fast and one slow. The slow exponential decay with time constant $\tau = 20 \pm 1$ psec is due to the dephasing of the isotropic component of the Raman scattering tensor. The rapid decay near zero delay is due to molecular reorientation. The reorientational relaxation time has been previously measured to be $\sim$ 2 psec [9] and is not resolved here. The decay constant of 20 ± 1 psec compares well with the value of 21 psec calculated from the 0.5 cm^{-1} [10] linewidth of the isotropic component of the known spontaneous Raman linewidth assuming homogeneous broadening. In this measurement inhomogeneous as well as homogeneous contributions to the Raman linewidth are measured. Nonetheless, the good agreement with an exponential decay is evidence that the (ν_1) Raman mode is at most only weakly inhomogeneously broadened. The extent to which weak structure in the Raman band lying near the ν_1 fundamental interferes with the ν_1 decay awaits resolution with a frequency selective technique that will permit isolation of the single ν_1 vibration.

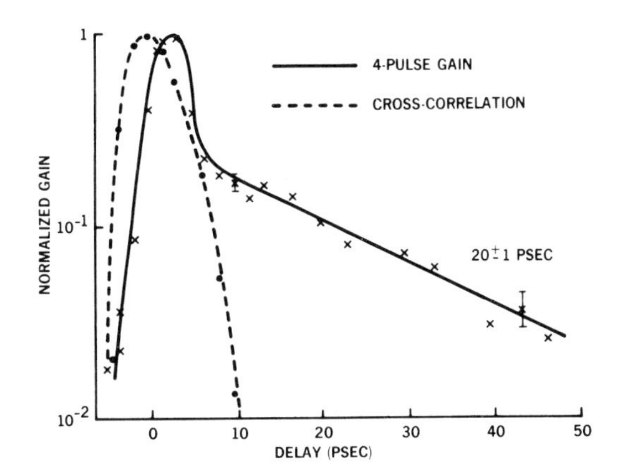

Fig. 1 Peak stokes gain versus delay time for CS_2.

In summary, we present a new technique for measurement of Raman dephasing times using cw mode-locked dye lasers. Several advantages of this method which stem from the use of low power, stable pulses and cw laser measurement techniques are presented and the technique is demonstrated with the first measurement of the dephasing of the ν_1 vibrational mode of CS_2.

References

1. A. Laubereau and W. Kaiser in Chemical and Biochemical Applications of Lasers, Vol. II, Ed. C. Bradely Moore (Academic Press, New York, 1977) p. 87.
2. Chi H. Lee and D. Ricard, Appl. Phys. Lett. 32, 168 (1978).
3. A. Owyoung, IEEE J. Quantum Electron. QE-14, 192 (1978).
4. J. P. Heritage and R. K. Jain, Appl. Phys. Lett. 32, 101 (1978).
5. R. K. Jain and C. P. Ausschnitt, Opt. Lett. 2, 117 (1978).
6. C. P. Ausschnitt and R. K. Jain, Appl. Phys. Lett. 32, 727 (1978).
7. C. P. Ausschnitt, J. P. Heritage and R. K. Jain (to be published.
8. R. K. Jain and J. P. Heritage, Appl. Phys. Lett. 32, 41 (1978).
9. E. P. Ippen, C. V. Shank, Appl. Phys. Lett. 26, 92 (1975).
10. W. R. L. Clements and B. P. Stoicheff, Appl. Phys. Lett. 12, 246 (1968).

Picosecond Vibrational and Electronic Relaxation Processes in Molecules

B.I. Greene, R.M. Hochstrasser, and R.B. Weisman
Department of Chemistry, and

Laboratory for Research on the Structure of Matter
University of Pennsylvania, Philadelphia, PA 19104, USA

1. Introduction

We describe in this paper an improved picosecond timescale spectrophotometer. As the accuracy of ultrashort kinetic techniques improves so can the sophistication of the models used to describe the responses of systems to picosecond pulses. No less than with traditional kinetic studies, it is vital in picosecond work to have analytical methods that characterize the species that are incorporated into the models. In our case the analytical method is spectrophotometry.

Time resolved absorption spectroscopy on the picosecond timescale arose from a discovery by ALFANO and SHAPIRO [1] that a continuum is coherently produced when a picosecond pulse is focused into various glasses. These authors later suggested [2] that the continuum could be used as a kinetic spectroscopy device. Subsequently numerous workers (see for example [3-5]) have utilized the picosecond generated continuum to observe transients in chemical and physical processes. Although there has also been considerable research to elucidate the nature of the continuum generation process, the detailed mechanism is not yet certain [6].

In the present paper we discuss first the expectations for fitting picosecond timescale measurements to simple kinetic models in order to illustrate the great need for accurate spectroscopic information. Then the picosecond spectrometer and its performance in some test cases is presented.

2. Kinetic Models for Picosecond Processes

Many systems that are studied by picosecond techniques are assumed to respond in accordance with the mechanism:

$$A \rightarrow A^* \quad ; \quad I_p$$

$$A^* \rightarrow X \quad ; \quad k_x$$

$$A^* \rightarrow B \quad ; \quad k_B$$

where I_p is the excitation pulse profile, k_B is the rate constant for decay into a particular channel B, and k_x represents the sum of rate constants for all other processes. The lifetime of the state A* is $k^{-1} = (k_x + k_B)^{-1}$. Measurements of the time evolution (rise times) of X or B, or measurements of the temporal decay of the population of A* should all give the same parameter k if the model is appropriate. The individual rate constants determine the yields into the various channels.

In typical applications A* might represent a singlet or other excited state created instantly by absorption of a photon. X may represent the result of a chemical dissociation, spontaneous emission, internal conversion to A, isomerization, or all of these. If B were to represent a triplet state, then k_B would be the rate constant for intersystem crossing. B could be any of those processes indicated by X, and in our notation B is singled out as the one that is actually measured in the experiment. The return of B to A might be very slow and would not need to be considered if B were to represent a triplet state.

We will assume that the optically thin (length ℓ) sample of A is pumped with a pulse $I_p(\tau)$ and probed with a pulse with a normalized intensity profile $I_{pr}(\tau)$ at wavelength λ. If a spatially uniform population of species A, A*, X, B is maintained throughout the length of the sample cell, and if A is considered not to be significantly depleted, then the induced transient optical density observed at time τ (delay between pump and probe) is given by

$$D(\lambda,\tau) = \ell \int_{-\infty}^{\infty} I_{pr}(\tau-\tau')\,d\tau' \left\{\sigma_{A^*}(\lambda)A^*(\tau') + k_B\sigma_B(\lambda)\int_{-\infty}^{\tau'} A^*(\tau'')d\tau''\right\} \quad (1)$$

where $A^*(\tau)$ is the number of A^* molecules per unit volume in the sample at time τ, whose integral from $-\infty$ to τ' is the amount of A* accumulated by τ'. The absorption cross-sections for wavelength λ are written as $\sigma(\lambda)$. A* is the following known function of time:

$$A^*(\tau') \approx \sigma_A A(o) \int_{-\infty}^{\tau'} I_p(\tau)\, e^{k(\tau-\tau')}\, d\tau \quad (2)$$

if the pulseshapes are known. Equation (1) contains two important terms within the framework of the approximation used here. The first term involves the contribution from excited singlet absorption (A* absorption) and the second term involves the absorption of B. In the approximation that the pulses of light are short compared with k^{-1}, equations (1) and (2) yield the well known result ($\tau \geq o$)

$$D(\lambda,\tau) \propto \sigma_{A^*}(\lambda)\, e^{-k\tau} + \varphi_B\sigma_B(\lambda)\,(1 - e^{-k\tau}) \quad (3)$$

the first term representing the decay of A* and the second representing the growth of B. φ_B is the quantum yield of formation of B.

A moderately high degree of precision is needed in order to extract a value of k from such an experiment unless $[\sigma_{A^*}(\lambda)/\sigma_B(\lambda)\varphi_B]$ is known from an independent experiment. Obviously it is necessary to obtain the spectra of A* and B separately in order to fit experimental values of

$D(\tau, \lambda)$ to the model except in certain limiting circumstances. In particular, the observation of a rapid risetime (pulse limited) is always suspect since if $\sigma_A*(\lambda) \approx \sigma_B(\lambda)\varphi_B$ the value of $D(\tau, \lambda)$ will rise instantly and then remain constant regardless of the value of k. Three calculated examples of the function $D(\tau)$ are given in Figure 1. Here the actual triplet state (or B) risetime is held constant (k = 25 ps) while the ratio of $[\sigma_A*/\sigma_B\varphi_B]$ is varied from zero (Figure 1a) through 0.5 (Figure 1b) to 1.0 (Figure 1c). The solid lines indicate the exact response $D(\tau)$ while the dashed lines correspond to a best fit of $D(\tau)$ to a single exponential growth constant convoluted with the laser pulse. The procedure of fitting picosecond absorption data to a single exponential growth constant has been widely used, mainly because the data quality has not merited fitting to more sophisticated models. In each of the three cases chosen, and for those in between, it is obvious that the function $D(\tau)$ can be fitted to a single exponential (k_f) growth function with a mismatch that is small compared with typical uncertainties of picosecond experimental data. Yet only in the first case (Figure 1a) does the best fit to a single exponential yield a valid result. In the others the computed values of the supposed exponential growth constant are meaningless in relation to the parameters of the model.

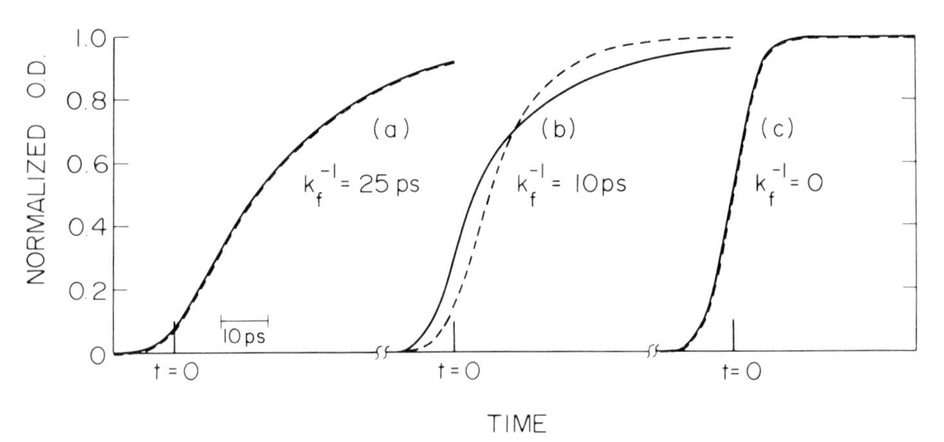

Fig. 1 Artifacts in risetime measurements. Solid lines: Calculated values of $D(\tau)$ for k^{-1} = 25 ps and 8 ps pulses. Dashed lines: Best fit of $D(\tau)$ to a single exponential growth having lifetime k_f^{-1}.

The conclusions of this standard kinetic analysis of the model are as follows:

(a) All temporal decays or growths depend only on k, the inverse lifetime of the state A* [7].

(b) If no analytical data are available regarding the amounts and identification of absorbing species at all times, the experimental growth constant is meaningful only if it is long compared with the pulse width.

(c) In experiments where a single wavelength is used to probe the system, and when no information is available regarding the spectra of intermediates, the measurement of a growth time comparable with the pulsewidth is not a convincing demonstration of the actual growth kinetics.

(d) Accurate time-resolved spectra are a necessary adjunct to temporal data at small delay times.

The presence of the two terms in equation (1) indicates that any small modifications in the absorption cross-sections σ_A and σ_B will also alter the apparent time evolution. Thus there are certain experimental variables such as choice of solvent to which the ratio σ_A/σ_B is certainly sensitive. These may determine the build-up times observed in particular experiments.

3. Transient Absorption Measurements

Our transient absorption spectrometer is of the excite and probe type in which a laser harmonic is used to create a sudden excitation of the sample and the absorption spectrum at a later, variable time is measured with a broad band picosecond continuum pulse. For a given delay time we obtain a full double-beam transient absorption spectrum from a single laser shot.

Both excitation and probe beams are derived from a single 8ps pulse extracted from the 1.06 μm output of a passively mode-locked Nd:glass laser. After amplification, this pulse carries 30 mJ, gives a second harmonic spectrum of 15 cm^{-1} width, and shows a TEM_{00} transverse mode structure. Once the second and third harmonics at 530 and 353 nm, respectively, have been generated, the third harmonic pulse is isolated by a dichroic beam splitter and filter, passed through a wave plate to reorient its polarization, and focused through a 320 μm diameter aperture into the sample. The residual first harmonic light is used to produce the probing continuum by focusing it into a 5 cm cell of H_2O. The emerging continuum pulse passes through a variable delay and scatter plate and is then split into two parallel beams which are focused into the sample cell through a dichroic beam combiner. One continuum beam passes through the aperture and excited volume collinearly with the excitation pulse while the second traverses an unexcited part of the sample. A lens focuses these two beams onto different positions along the length of the entrance slit of a low resolution spectrograph.

The result is two parallel dispersed spectra on the focal plane of the spectrograph, one representing $I_o(\lambda)$ the other $I(\lambda)$. Both of them are detected and recorded for each laser shot by a Princeton Applied Research Corp. model 1215/16/54 optical multichannel analyzer system. The two 500-channel tracks of spectral information are digitally processed in the following way: dark current spectra are subtracted from the raw data; one track is divided by the other; the result is divided by a corresponding ratio spectrum obtained with no excitation light present; and the logarithm of the new ratio spectrum is taken. This gives the change in absorbance in the excited sample volume as a function of wavelength for a single laser shot at the time delay determined by the setting of the optical delay line.

4. Results and Discussion

The three spectra shown in Figure 2 are typical of the spectral quality obtained by the apparatus described above. In previous experiments with benzophenone it was shown that the transient optical density at 530 nm was fitted to an exponential growth time of about 10 ps, whereas the growth of the xanthone transient was closer to being pulsewidth limited [5, 8]. It was known prior to these experiments that benzophenone has a triplet-triplet absorption maximum at 530 nm. [9]. It is clear from Figure 2(a) and 2(b) that even after about 10 ps the excited samples have not yet completely relaxed. The relaxed and unrelaxed spectra shown in the Figure are quite distinct with our present spectral accuracy, although with an earlier photographic technique these important differences were not apparent [5].

Similarly in the case of quinoxaline shown in Figure 2(c) an unrelaxed spectrum is clearly seen at short times. In this case the more relaxed spectrum is not fully developed until a few pulsewidths after excitation, consistent with the previously reported 41 ps fluorescence lifetime of singlet $n\pi^*$ excited quinoxaline [10], as cited in [5]. The two spectra are significantly more accurately observed in the present work than by a previous photographic method [5].

In all of these examples the gross features of the single wavelength kinetics should depend on the probing wavelength. It appears likely that the recently reported discrepancies in transient behavior of anthrone [11] may be attributed to these effects.

Our spectra raise serious doubts concerning the common view that only relaxed species are expected to be observed in solutions at 300 K in the decapicosecond regime. The early time spectra in Figure 2 are clearly unrelaxed. Whether the relaxation occurring is between thermalized electronic levels or whether it involves vibrational and electronic states is currently a question of great interest under study in this laboratory.

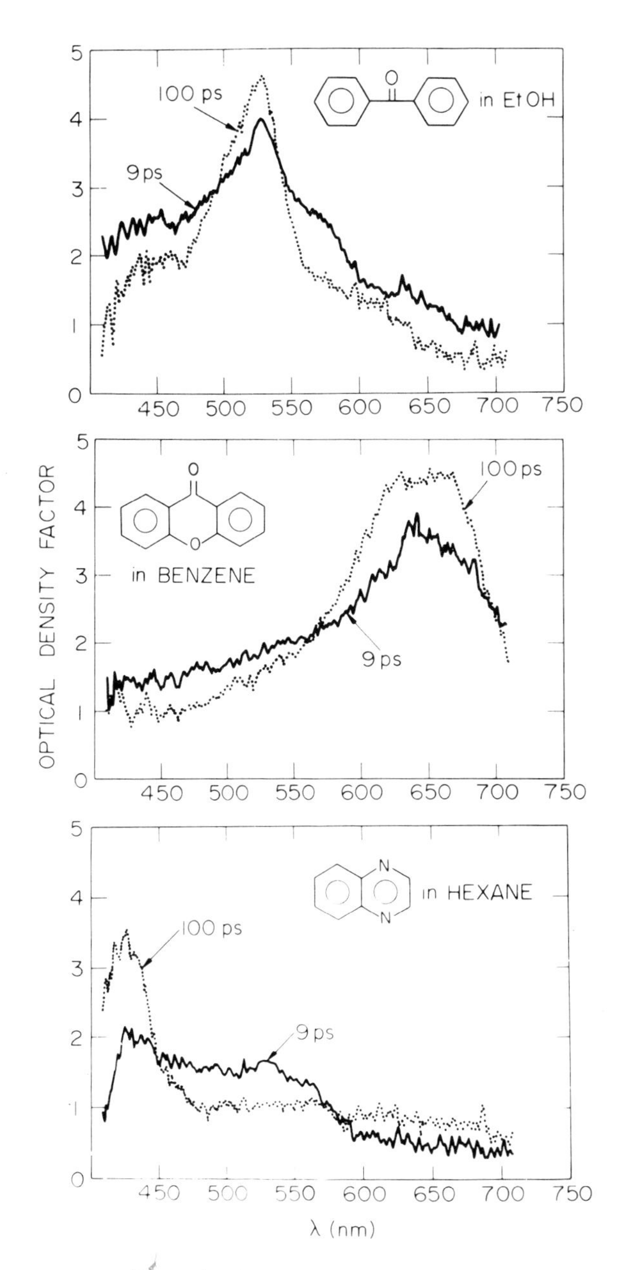

Fig. 2 Transient spectra of Molecules in Solution at 300 K. Top: Benzophenone in ethanol, Middle: Xanthone in benzene. Bottom: Quinoxaline in hexane. The early time spectra are shot at 9 ps (rather than considerably earlier) in order to avoid spectral distortion introduced by the frequency chirp of the probing continuum.

Acknowledgement

This research was supported by a grant from the NSF (GP 8442S) and in part by the MRL Program under Grant No. DMR76-80994. R. B. W. is grateful for the award of an N. S. F. National Needs Post-Doctoral Fellowship.

References

1. R. R. Alfano and S. L. Shapiro, Phys. Rev. Letts. 24, 584 (1970).
2. R. R. Alfano and S. L. Shapiro, Chem. Phys. Letts. 8, 631 (1971).
3. D. Magde and M. W. Windsor, Chem. Phys. Letts. 27, 31 (1974).
4. G. E. Busch, R. P. Jones and P. M. Rentzepis, Chem. Phys. Letts. 18, 178 (1973).
5. R. M. Hochstrasser and A. C. Nelson in *Lasers in Physical Chemistry and Biophysics*, Ed. J. Joussot-Dubien; Elsevier, 1975, p. 305.
6. D.H.Auston: *Ultrashort Light Pulses*, ed. by S.L.Shapiro, Topics in Appl. Phys., Vol.18 (Springer, Berlin, Heidelberg, New York 19
7. Although this is an elementary result there appear to be conceptual difficulties associated with it since erroneous applications of these principles still appear in the literature.
8. R. W. Anderson Jr., Ph.D. Dissertation, University of Pennsylvania (1976).
9. D. S. McClure and P. L. Hanst, J. Chem. Phys. 23, 1772 (1955).
10. Robin M. Hochstrasser and R. W. Anderson, unpublished results.
11. G. W. Scott and L. D. Talley, Chem. Phys. Letts., 52, 431 (1977).

The Excited State Absorption Kinetics of Quinoxaline in Room Temperature Solution[1]

G.W. Scott and L.D. Talley

Department of Chemistry, University of California
Riverside, CA 92521, USA

R.W. Anderson, Jr.

Xerox Corporation, Rochester, NY 14644, USA

1. Introduction

Aza-aromatic molecules have low energy $n\pi^*$ states that may couple efficiently by spin-orbit interaction with lower $\pi\pi^*$ states of different multiplicities. Quinoxaline is an example of such a molecule. Previous fluorescence decay measurements for quinoxaline in benzene solution gave a decay time of 40 ps [1]. In the present paper, we report the kinetics and spectra of excited state absorption of quinoxaline in room temperature solution.

2. Experimental

A single, third harmonic, 8 ps pulse (355 nm) from a Nd^{+3}:glass laser was used to excite the lowest $^1n\pi^*$ state of quinoxaline. Subsequent excited state absorption of quinoxaline was probed with a continuum pulse generated by the 1064 nm fundamental in a 2 cm CCl_4 cell. The continuum pulse was polarized nearly parallel to the excitation pulse. Spectra were recorded with a spectrograph on photographic plates utilizing a "double beam" geometry [2]. The absorption kinetics were obtained in separate experiments utilizing a suitably delayed, 4 ps pulse centered near the peak of the $T_n \leftarrow T_1$ absorption spectrum (λ_o = 422 nm, $\Delta\lambda$ ~ 8 nm). This probe pulse was obtained from a short cavity dye laser [3,4]. Both parallel and perpendicular probe polarizations were investigated, and multiple shots at each delay setting were normalized and averaged [5]. In the determinations of both the excited state absorption spectra and kinetics, the t=0 delay position, corresponding to maximum overlap of excitation and probe pulses, was obtained in a separate experiment using excited state absorption of coronene.

3. Results and Discussion

The room temperature transient absorption spectra of quinoxaline in benzene, obtained 3.3 ps and 1 ns after excitation at 355 nm, are shown in Fig.1. The overall features of both spectra are quite similar to a previously reported $T_n \leftarrow T_1$ absorption spectrum [6]. The weak, 3.3 ps spectrum appears to be somewhat

[1]This research was supported in part by the National Science Foundation, the Research Corporation, and the Committee on Research of the University of California, Riverside.

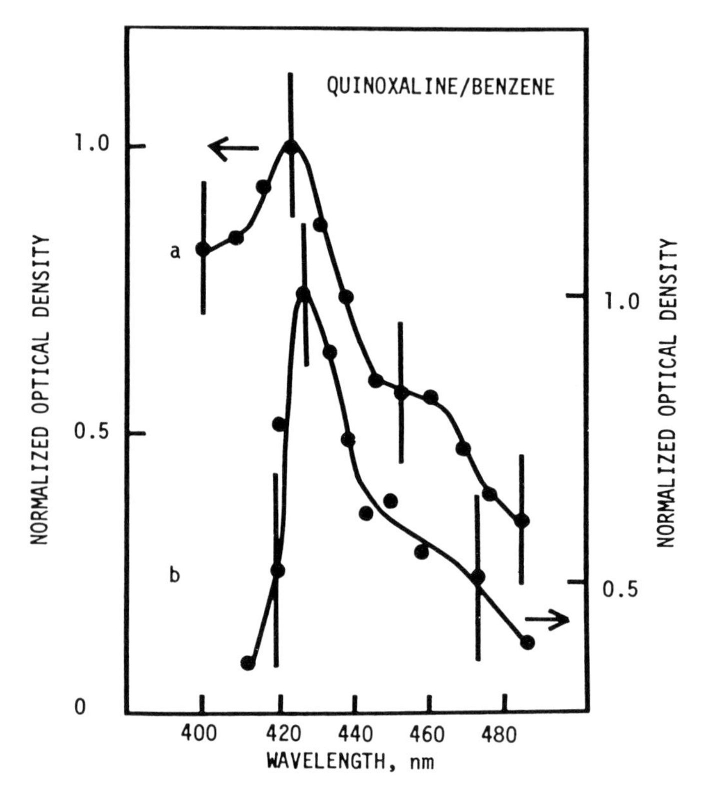

Fig.1 Normalized transient absorption spectrum of quinoxaline in benzene observed at (a) 1 ns (left ordinate) and (b) 3.3 ps (right ordinate) delays following excitation at 355 nm

sharper to the blue of the peak than the 1 ns spectrum. This broadening of the transient absorption spectrum with time could be due to sharp $S_n \leftarrow S_1$ or $T_n \leftarrow T_1{}^*$ contributions at short time. (Here $T_1{}^*$ stands for a vibrationally excited T_1 state or possibly the T_2 state.) However, the weakness of the 3.3 ps absorption as well as the relative weakness of the available continuum probe pulse below ~420 nm require that any conclusions based on spectral changes in this wavelength region remain tentative. An additional weak absorption tail from 500 to ~620 nm is easily observed. A check for transient absorption at 355 nm at both long and short times revealed no absorption with an optical density comparable to the 425 nm peak. Although the similarity of both the 3.3 ps and 1 ns spectra to the known $T_n \leftarrow T_1$ absorption spectrum of quinoxaline [6] suggests the same assignment for each, data discussed below indicate a different interpretation.

Typical kinetics data at 422 nm are shown in Fig.2 for both parallel and perpendicular probe polarizations, relative to the 355 nm excitation. The smooth curves shown for comparison in Fig.2 each represent exponential buildup times of 5 ps, taking

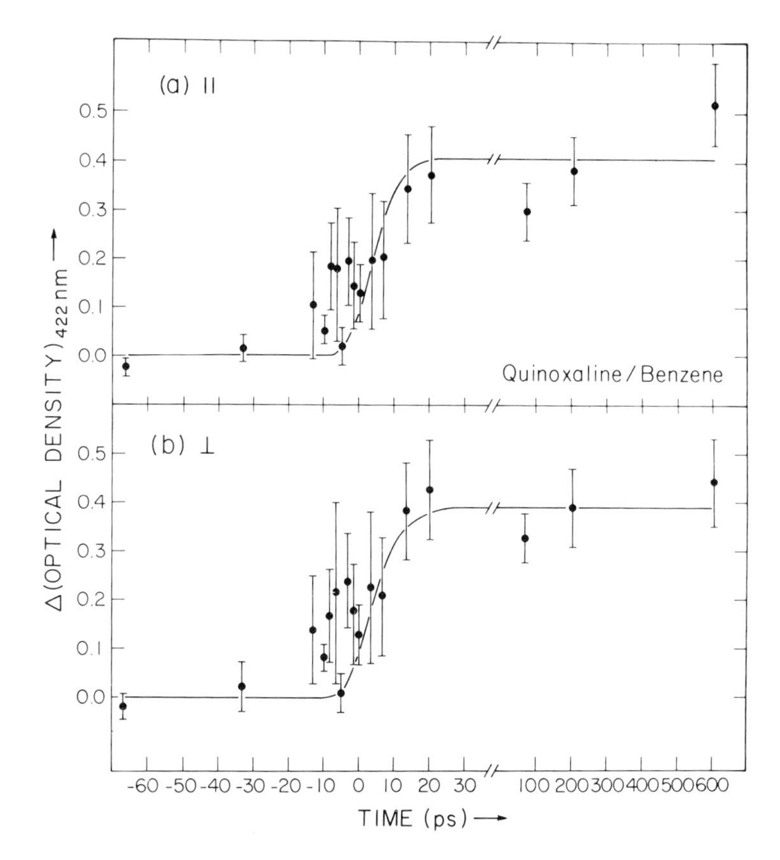

Fig.2 Optical densities at 422 nm (parallel and perpendicular polarizations) due to excited state absorption of quinoxaline in benzene as a function of delay subsequent to excitation at 355 nm

into account finite excitation and probe pulse durations [5]. (The 5 ps buildup time was actually determined in a more precise parallel polarization kinetics experiment.) These data shown here demonstrate that the buildup kinetics were unaffected by relative pump and probe polarizations. Preliminary kinetics experiments using a probe continuum pulse and monitoring at 425 nm yielded a somewhat longer buildup time. (Actually, the errors in the kinetics determination at 422 nm allow a buildup as long as ~12 ps. However, the probable value and error are less.) The shorter buildup time seems more reasonable as it is difficult to rationalize an experimental artifact with a risetime faster than an actual absorption buildup. Preliminary experiments in a triplet quencher solvent (cis-1,3-pentadiene [5]) produced quite noisy data due to the weakness of the absorption. These data indicated a rapid buildup and decay of 422 nm absorption, but the signal-to-noise ratio was too low to determine definitively the quenching time. These results would distinguish between short time contributions to the absorption due to T-T and $S_n \leftarrow S_1$ absorption if a better signal-to-noise ratio were obtained. Experiments along these lines are continuing.

Previous measurements [1] of the fluorescence decay time of quinoxaline in benzene gave τ_{fl} = 40 ps. For this measurement to be consistent with a more rapid buildup of T-T absorption would require, for example, that the intersystem crossing rate constant (k_{isc}) depends on the excitation wavelength and that this process be competitive with vibrational relaxation. To check the possibility of a vibronic level dependent k_{isc}, the phosphorescence excitation spectrum (10^{-5} M quinoxaline in 3-methylpentane at 77 K) was compared with a room temperature absorption spectrum. The two spectra are only slightly different, with the changes possibly due to temperature differences. Hence the relative phosphorescence quantum yield must be approximately independent of excitation wavelength from the 0-0 of S_1 through S_2 of quinoxaline. These measurements do not directly measure the energy dependence of k_{isc}, especially if the intersystem crossing quantum yield is nearly 1 [7]. However, a very recent report of the room temperature fluorescence of quinoxaline in isooctane (λ_{max} = 410 nm) indicates an excitation wavelength independent, fluorescence quantum yield of 7.5 x 10^{-5} [8]. Thus, it would seem that k_{isc} is independent of excitation wavelength.

A consistent interpretation of these experimental results suggests that these absorption kinetics do not directly measure an excited state buildup time. For example, a weak, decreasing $S_n \leftarrow S_1$ 422 nm absorption superimposed on a stronger, increasing $T_n \leftarrow T_1$ absorption would yield the observed kinetics. This interpretation is supported by the rapid initial excited state absorption at 422 nm (Fig.2) and also by results presented by others at this conference [9]. Of course, one cannot rule out the possibilities that there were interfering impurities contributing to the measured fluorescence and/or excited state absorption kinetics or that k_{isc} depends on the excitation wavelength in benzene solution, but not isooctane. However, these latter possibilities seem less likely.

References

[1] R. W. Anderson, Jr. and R. M. Hochstrasser, unpublished results.

[2] R. W. Anderson, Jr. and R. M. Hochstrasser, J. Phys. Chem. 80, 2155 (1976).

[3] A. J. Cox, G. W. Scott, and L. D. Talley, Appl. Phys. Letts. 31, 389 (1977).

[4] G. W. Scott and A. J. Cox, Advances in Laser Spectroscopy I, Proc. Soc. Photo-Opt. Instru. Eng. 113, 25 (1977).

[5] R. W. Anderson, Jr., R. M. Hochstrasser, H. Lutz, and G. W. Scott, J. Chem. Phys. 61, 2500 (1974).

[6] R. Astier and Y. Meyer, J. Chim. Phys. 64, 919 (1967).

[7] R. Li and E. C. Lim, J. Chem. Phys. 57, 605 (1972).

[8] K. Yamamoto, T. Takemura, and H. Baba, Bull. Chem. Soc. Japan 51, 729 (1978).

[9] L. A. Halliday, H. B. Lin, and M. R. Topp, "Measurement of Singlet Lifetimes by Upper State Fluorescence", Post-deadline paper presented at this conference.

A Novel Technique for Monitoring Transient Optical Phenomena on a Picosecond Time Scale

S. Schneider, E. Lill, and F. Dörr

Institut für Physikalische und Theoretische Chemie
Technische Universität München
D-8046 Garching, Fed. Rep. of Germany

Recently [1], we reported on the possibility of "active" mode-locking of a flashlamp-pumped dye laser by bleaching the saturable absorber contained in its resonator with picosecond light pulses from a master oscillator (LI in Fig. 1).Thereby independently tunable and highly synchronous picosecond pulses with energies between 15 μJ and 80 μJ , resp., are produced in the wavelength regime 650 nm $< \lambda <$ 550 nm. The maximum difference achievable in the output wavelength of both lasers depends both on the gain profile of the active medium and the absorption spectra of the mode-locking dye used in the slave oscillator L II. For the combination of rhodamine 6G and DODCI, the tuning ranges are found to be shifted towards shorter wavelength compared to the case of passive mode-locking (see Fig. 2).

Measurements using a picosecond streak camera have shown [2]

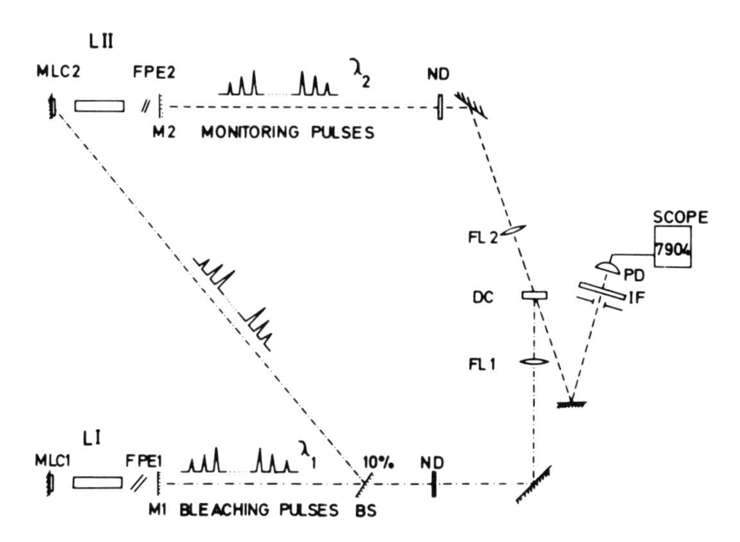

Fig.1 Experimental set-up (L I,L II: dye lasers, MLC: mode-locking cells, FPE: Fabry-Perot etalons, BS: beam splitter, ND: neutral density filters, DC: dye cell, IF: interference filter)

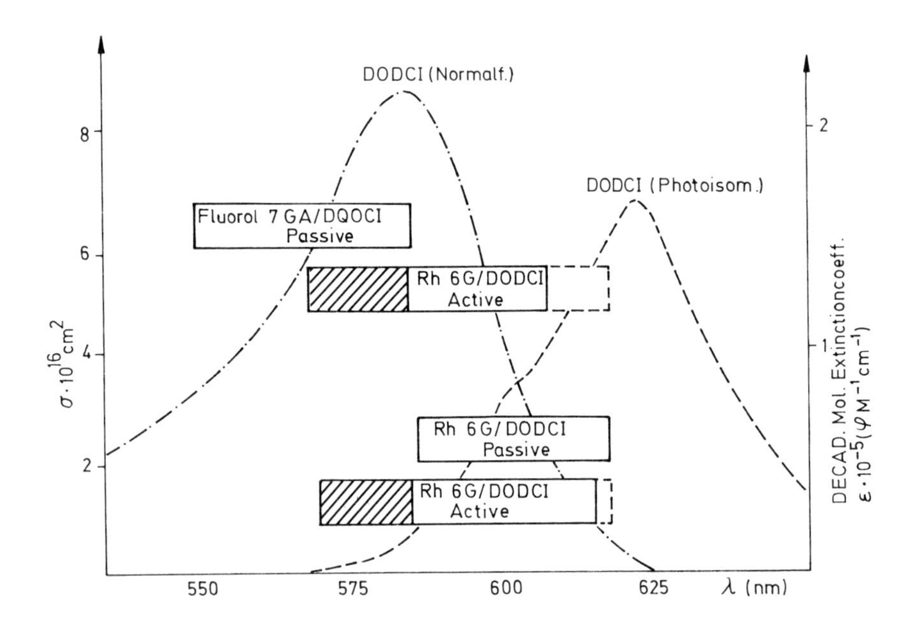

Fig. 2 Absorption spectra of DODCI ("normal" form and photo-isomer, resp.). The tuning range of the "actively mode-locked laser is displayed when operated together with a Fluorol 7GA and rhodamine 6G master oscillator, respectively

that pulse duration is typically between 2 and 6 ps in the middle and later stages of the output pulse train of the "actively" mode-locked laser L II, even then, if there is a slight mismatch in cavity length of both oscillators ($\Delta L < 40$ mm). As a consequence of this one can use the two pulse trains in order to monitor a transient optical phenomena like the short-lived transparency of a saturable absorber dye by only one shot of the two lasers [3].

To this end (see Fig. 3) the pulses of laser L I (roundtrip-time t_1) are used to excite the sample in the dye cell DC repetitively and to induce thereby the transient phenomena with relaxation time τ (Fig. 3b, 3c). The pulses from laser L II (roundtrip-time $t_2 = t_1 + \Delta t$) are used to probe the transient optical density or any other sensitive property (Fig. 3a, 3c). Since the delay between the corresponding exciting and probing pulses, resp., is encreased by Δt from pulse to pulse, the transient phenomena is sampled automatically during one shot, provided that the difference in roundtrip-time Δt is chosen properly (Fig. 3).

In Fig. 4, the relative intensities of the consecutive probing pulses transmitted by an ethanolic solution of DODCI are dis-

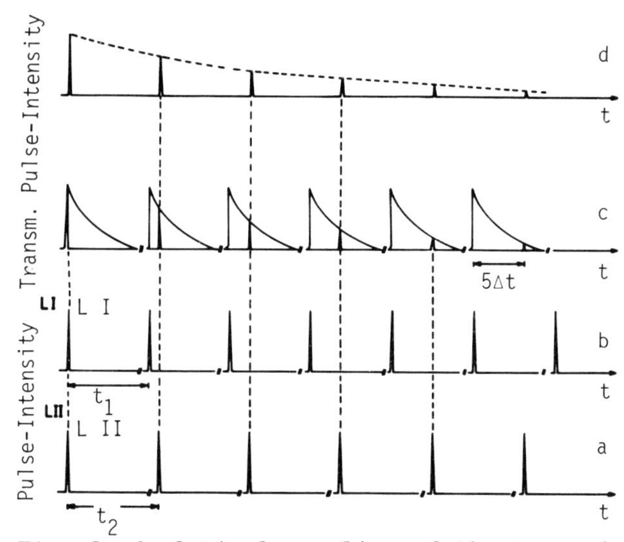

Fig. 3a-d Optical sampling of the transmission decay. (a) Series of pulses originating from L II. (b) Series of pulses originating from L I. (c) Transient phenomena (transmission) induced by each bleaching pulse and monitoring of it by the corresponding probing pulses. (d) Intensity of the transmitted probing pulses as seen on the oscolloscope

played on a logarithmic scale. The wavelength of the exciting pulses was chosen to coincide with the absorption maxima of DODCI in its "normal" form (λ_{ex} = 580 nm). The wavelength of the probing pulses, on the other hand, was tuned to 580 and 610 nm, respectively. In both cases, the biexponential nature of the decay of the induced optical transmission is apparent. Since probing at 610 nm monitors relatively more of the fast decaying transmission change whilst probing at 580 nm favors the recording of the slower decaying component, we can conclude (see Fig. 2) that the shorter aperture time is correlated to the photoisomer and the longer one to the "normal" form of DODCI. Further evidence for this interpretation can be found easily, if the number of exciting pulses (≙ photolysis pulses) which pass the sample before the measuring period begins is changed e.g. by means of a Pockels cell unit.

As in any experiment of this kind, the lifetime of the excited state of the dye under study can be deducted from the relaxation time τ of the optical density only by means of a model. Wheneverthe pulsewidth is short compared to the aperture timeand if stimulated processes can be neglected, a two state model may be sufficient. Applying it to the results of Fig.4 we find 1.3± 0.2 ns as the excited state lifetime of DODCI in its "normal" form and 430± 40 ps for its isomer. The formation of the latter is actually found to be the reason for the shift in tuning range of the "actively" mode-locked laser and its modified behavior during the build-up of mode-locking.

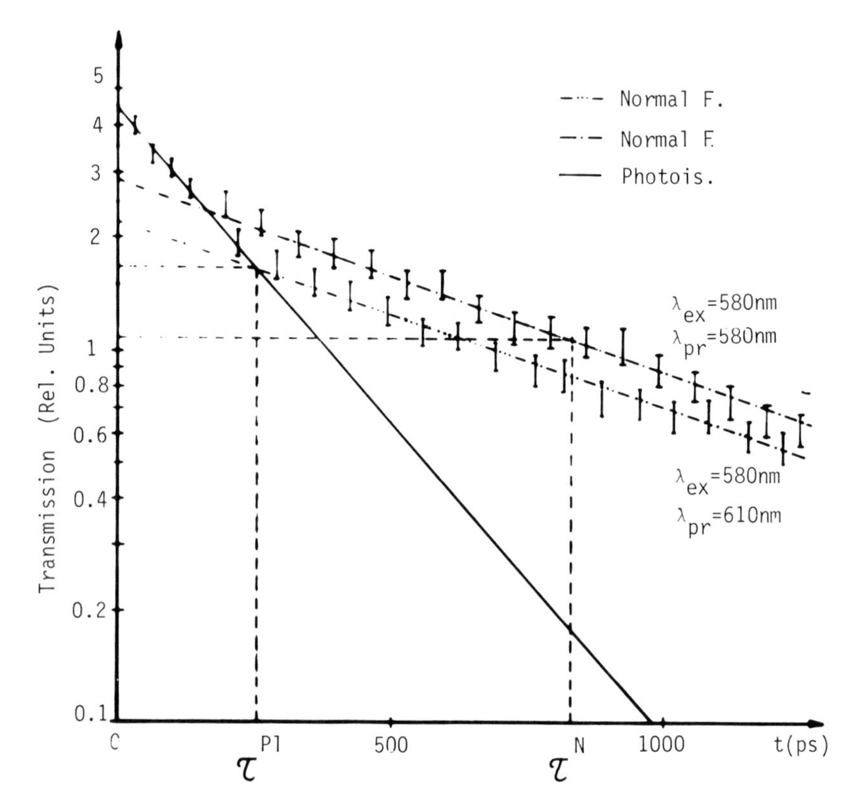

Fig. 4 Relative intensity of consecutive probing pulses. Sample is an ethanolic solution of DODCI (= 3,3'-diethyloxadicarbocyanine iodide). Excitation at λ= 580 nm, probing at λ= 580 nm (-·-·-) and λ= 610 nm (-····-), respectively

References

1. E.Lill, S.Schneider, F.Dörr: Optics Comm. 22, 107 (1977)
2. E.Lill, S.Schneider, F.Dörr, S.Bryant, J.R.Taylor: Optics Comm. 23, 318 (1977)
3. E.Lill, S.Schneider, F.Dörr: Appl. Phys. 14, 399 (1977)

Fluorescence Spectroscopy of Subpicosecond States in Liquids

Kee-Ju Choi, Horn-Bond Lin, and M. Topp

Department of Chemistry, University of Pennsylvania
Philadelphia, PA 19104, USA

1. Fluorescence from Short-lived Upper States

We have recently shown [1,2] that it is possible to make direct observations of phenomena originating from short-lived upper states of aromatic molecules in solution. The method which has proved to be the most successful in observing such upper states uses consecutive two-photon pulsed laser excitation, followed by fluorescence detection.

One of the early observations was that the shape of this type of spectrum was dependent on the excitation conditions, indicating that emission was being observed from unrelaxed vibrational states. This resulted in gross broadening of the fluorescence spectra on irradiation with high energy photons. Subsequent experiments have used two lasers such that the level of excitation could be controlled more exactly. This allowed fluorescence analysis of the levels actually excited rather than of the results of radiationless relaxation. It was shown in the cases of 3,4,9,10-dibenzpyrene [3] and 3,4-benzpyrene [4] that this approach yielded sharp vibronic progressions reminiscent of Shpol'skii-type spectra [5,6], except that the actual linewidths reflected the short lifetimes of the emitting states.

Controlled experiments in this area required careful consideration of several aspects:

(a) location of the exact states of excitation,

(b) resolution of fine-structured progressions,

(c) the effects of molecular symmetry,

(d) quantum-yield measurements.

2. States of Excitation

The consecutive two-photon approach relies for sensitivity on the presence of two allowed transitions: one to excite the first singlet state, preferably with the minimum excess energy to avoid overlap of the excitation frequency with the upper state spectrum; and the other, having well-defined frequency, for precise excitation of an upper level. Selection of upper vibronic states is governed by the absorption spectrum of the first excited singlet state.

In the absence of accurate literature data, we have adapted the principles of nanosecond laser flash photolysis to measure excited singlet state absorption spectra. We used photoelectric detection for high sensitivity and, in addition, a double-beam arrangement to allow accurate internal referencing. The apparatus is shown schematically in figure 1 [7,8]. The second harmonic of a Q-switched ruby laser was attenuated to about 10 mJ per pulse and directed unfocussed into the sample cell through a 6 mm diameter aperture. Light from the laser flashlamp, leaked through a hole in the laser head, was collimated through the sample and conveyed (with the minimum possible use of lenses) to two spectrometers. A glass slide beam-splitter sent about 10% of the probe beam into a monochromator, set to monitor triplet at the absorption maximum. The remainder of the probe beam entered a second spectrometer, which was tuned over the near infrared and visible to scan the singlet-singlet (and the short time part of the triplet-triplet) absorption spectrum, using the triplet optical density after a few microseconds as an internal standard. This procedure allowed us to generate excited state absorption spectra with an uncertainty for the most part of less than 10%.

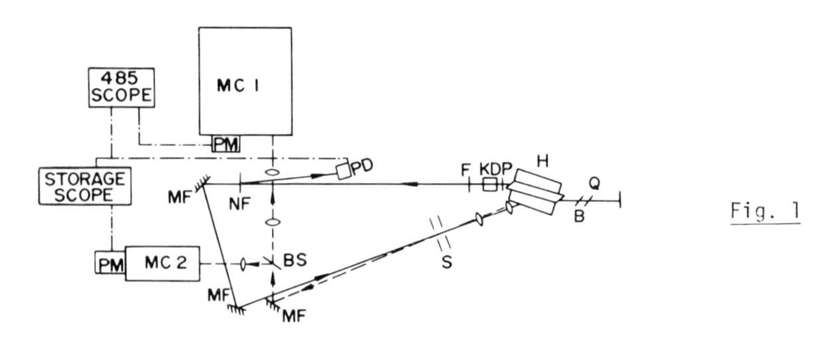

Fig. 1

A typical experimental result is presented in fig. 2. The upper frame shows a part of the ultraviolet spectrum of ground state 3,4,9,10-dibenzpyrene. The middle frame shows the absorption spectrum of S_1, recorded on the laser apparatus. The spectrum has been displaced by an energy equivalent to the 0-0 transition energy ($S_1 \leftrightarrow S_0$), so that the absolute energies of the terminal upper states in the two spectra coincide. The triplet-triplet absorption spectrum is presented together with the $S_1 \rightarrow S_n$ absorption spectrum, for reference. The extinction coefficient for the singlet-singlet absorption maximum near 18.4 kK was measured to be $1.1 \times 10^4\ M^{-1}\ cm^{-1}$.

There is for the most part a good match between the peaks in the two singlet spectra, indicating that the same levels can be reached from either lower state. The obvious exception is the intense S_1 absorption band near 16 kK, corresponding to a terminal level near 39 kK, which is not evident from the S_0 absorption spectrum, and therefore must represent a symmetry-forbidden transition. This band would be expected to play a major part in populating upper states from S_1, but it cannot contribute greatly to the emission directly to the ground state.

The shape of the singlet-singlet absorption spectra in general cannot be predicted from the ground state spectra because of residual molecular selection rules. The case of higher symmetry molecules such as dibenzpyrene(C_{2v}), or perylene (D_{2h}) is clear, but we have shown that this is also true for 3,4-benzpyrene(C_{1h}). Therefore, measurement of S_1 absorption spectra is an essential part of a study of upper state fluorescence involving selected vibronic levels.

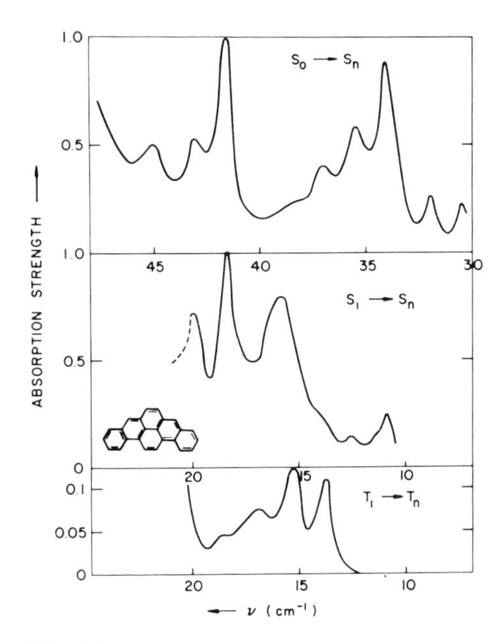

Fig. 2

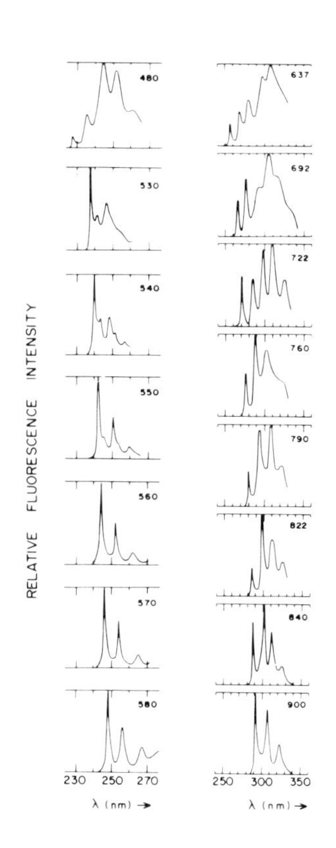

Fig. 3

3. Fluorescence from Upper States of Dibenzpyrene

In 3,4,9,10-dibenzpyrene, the state about 41 kK above the ground state is accessible by strongly allowed transitions from both S_1 and the ground state. Therefore, if we can observe emission from totally unrelaxed states, we should observe emission having a high-frequency limit at a wavelength representing the maximum energy available to the system. As fig. 3 shows, this is the case throughout the spectral range investigated, although the relative intensity of the high-frequency bands compared with the rest of the spectrum depends on the excitation conditions. Further, we observe progressions near the high-frequency limit to have narrow linewidths, showing that the emission proceeds not only from the same vibronic state reached in the excitation process but also from a very narrow energy-band of vibronic states. The widths of the fine-structured bands are of the order of 0.3 kK, which is considerably narrower than the band structure observed either in the absorption spectra or in vibrational bands downshifted in energy from the high-frequency progression. Thus, we attribute the fine-structure bands to molecular site-selection, where the narrow-band laser excited a specific energy-band of molecules from S_1 to the upper state, from which relaxation was much faster than the rate of molecular site exchange [9]. During the short lifetime of the upper state, a small amount of emitted fluorescence labelled its energy distribution. We can compare this type of result to site-selection work carried out in solid matrices at low temperature and our observations do have some similarities to absorption spectra measured in Shpol'skii matrices (so-called "quasi-line" spectra). In particular, our measured linewidths are similar to those of upper state quasi-line absorption spectra in similar molecules [6].

It can also be seen from figure 3 that the fluorescence bands broaden considerably towards lower frequencies. This is interpreted in terms of relaxation of the initial state via coupling of a discrete level to a dense manifold of levels contributed by the lower electronic states. (Because of the inhomogeneous line-broadening present in solution at 300°K, each molecule has actually a different absolute initial energy level). Emission from the essentially hot vibrational distributions generated by the relaxation process broadens out the vibrational bands.

It is informative to study the shape of the fluorescencespectrum, going from high to lower frequency, in consultation with figs. 2&3. 480 nm excitation populates the system between upper vibrational levels of the 41 kK band and an upper state. Emission from the 41 kK band has a Franck-Condon maximum near 245 nm, the exact position depending on the excitation level. The bands near the Franck-Condon maximum correspond to a variety of possible transitions, but the main feature is the small change in vibrational quantum number. Therefore, we have a range from $v'=2 \rightarrow v''=2$ in the principal progression, to a large number of possible transitions involving partially relaxed molecules, all emitting close to this position. As we move to longer wavelength excitation, particularly the range 530 nm - 580 nm, we intercept the 0-0 band of the 41 kK state and the vibronic progression (of which only the high-frequency part is shown) reduces to a set of three bands with minimum width. For these conditions, any relaxation process must involve internal conversion to the next lower electronic state and a large red shift to the vicinity of the Franck-Condon maximum of that state, i.e. $\sim$305 nm. Therefore, the spectrum excited by 550 nm - 580 nm appears uncomplicated. We note also the presence of a secondary progression with 530 nm - 550 nm excitation, indicating the selective excitation of a weak vibrational progression by laser photons having enough energy to excite molecules with a 600 cm^{-1} vibrational mode. Inhomogeneous broadening spreads out the photon energy range over which this can be observed. This progression, which was not resolved in our introductory paper on this subject [3], was responsible for the most part for the apparent continuous broadening of the emission bands with higher energy excitation.

When we excite dibenzpyrene between $\sim$590 nm - 760 nm, the shape of the upper state fluorescence spectrum depends on the relative absorption cross-sections of the 39 kK and 34 kK states at the excitation wavelength. Between 590 nm and 690 nm, the 39 kK state dominates the absorption, as can be seen from fig. 2, and the primary progression is both weak and clearly separate from the progression including the Franck-Condon maximum of the 34 kK state. This clear evidence for excitation of the molecule into a state which is weakly coupled to the ground state. Thus, most of the emission resulting from excitation in the range 590 nm - 690 nm comes from the 34 kK state. Since this is populated by radiationless relaxation, the emission band around 305 nm is diffuse. At 722 nm, we intercept the probability maximum of the $0 \rightarrow 2$ vibronic transition into the 34 kK state, and the two states appear to absorb about equally. Longer wavelength excitation causes the two progressions to coalesce as the 34 kK state is excited directly. There is also a general sharpening as less excess energy is supplied to the 34 kK state.

4. The Case of Perylene

We have seen how the absorption spectra between different states are affected by molecular symmetry. An extreme example is that of perylene, of which a partial S_1 absorption spectrum has been published by Goldschmidt and Ottolenghi [10]. As expected for a D_{2h} molecule, there is little correspondence between the $S_0 \rightarrow S_n$ and $S_1 \rightarrow S_n$ absorption spectra, and this has a profound effect on the shape of the upper state fluorescence spectra. In a molecule such as perylene therefore, the states populated directly by the laser should not be observably fluorescent to the ground state. Emission to the ground state can only occur from states populated by radiationless relaxation of the primary states. Our results indeed showed that the maximum of the fluorescence excitation spectrum did not correspond to a strong feature of the ultraviolet fluorescence spectrum. This also results in the absence of a fine-structured progression and only an indirect dependence of the high-frequency limit on the excitation level. That is, this should depend on the extent to which internal conversion can take place before energy relaxation and then, as for dibenzpyrene, on the Franck-Condon factor for lower state emission at this high-frequency limit. From the example of dibenzpyrene, it is clear that irreversible internal conversion results in a large red shift of the upper state fluorescence spectrum, even if the molecule does not actually lose energy during this process. We would therefore expect a weakly defined "turn-on" point, and it is indeed observed for perylene that the exact position of the excitation wavelength is not evident from the fluorescence spectrum, being usually down towards the noise level.

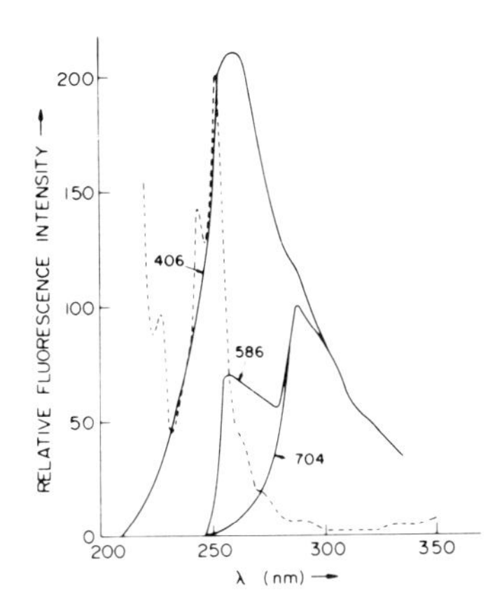

Fig. 4

Our early work on perylene [2] gave results similar to that shown in figure 4, where 406 nm irradiation is seen to yield a broad fluorescence band having a maximum near 255 nm, and a half-intensity full-width of $\sim$7000 cm^{-1}. This clearly cannot originate in a single state, since the absorption spectrum shows a half-intensity full-width of $\sim$2500 cm^{-1}. It was apparent from our preliminary work that a second state, of which a relatively long decay time compensated for the low oscillator strength to the ground state, was contributing effectively to the fluorescence spectrum. Irradiation at longer wavelengths (i.e. adding a longer-wavelength pulse to a weak 434 nm pulse) indeed showed the broad fluorescence band to become resolved into two components. The longer wavelength component is the sole contributor to the spectrum when the molecule is excited by 704 nm, as shown in fig. 4. This labels the position of the lower "u" state (B_{3u}^{-}) as about 290 nm, which corresponds to a weak absorption band.

5. Quantum-Yield Measurements

Among the most important numerical quantities to be obtained from a study of states deactivated by radiationless relaxation processes are the actual decay times of the states involved. If possible, we would like to use such lifetimes as reference parameters to study the dependence of radiationless relaxation processes on a wide range of variables, such as molecular structure, temperature, gas pressure, solvent effects, and so on. However, in order for this to be meaningful, we must be able to measure small differences in decay rates and this imposes a requirement of high accuracy on our experimental techniques. Direct lifetime measurements have not been successful in this area, the major reason being the very short lifetimes, and the limitations of laser pulse durations to > 10^{-13}s.

In order to attempt a reliable quantitative calculation of the upper state relaxation times, we have used the general principles of the consecutive two-photon method. Our approach is based on nanosecond laser flash photolysis, and allows direct calculation of upper state fluorescence quantum-yields. The essential principle of any fluorescence quantum-yield determination involves dumping a certain amount of energy into a prepared sample and measuring the fluorescence intensity with reference to some standard. Our approach employs one extra step. The solution must be prepared in the first excited singlet state before quantum-yield determinations can be made, and this is carried out by irradiating the system with an ultraviolet laser pulse of a few mJ. The sample is then irradiated by a dye laser, which is partially absorbed in the sample, and upper state fluorescence is observed. The correlation between upper state fluorescence intensity and the amount of dye laser energy dumped into the sample is established, and this is referenced against some normal fluorescence standard.

The apparatus for quantum-yield measurements is shown in figure 5. The ultraviolet pulse from a frequency-doubled ruby laser was split such that 10% - 20% was reflected off a glass beam-splitter to generate about 30% transient absorption in the sample. The remainder of the ultraviolet pulse

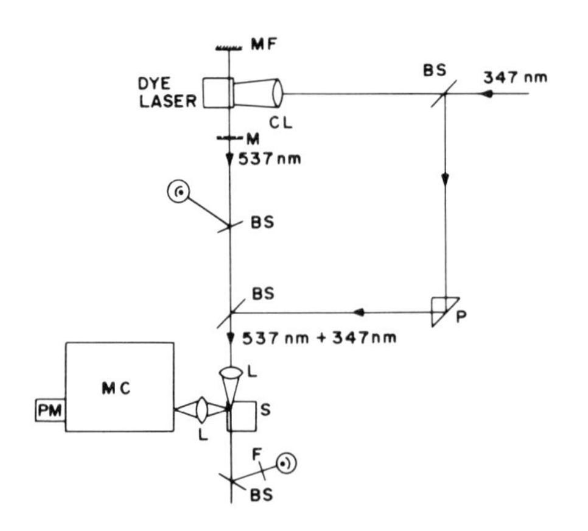

Fig. 5

was used to pump a dye laser, the absorption of which by the sample was monitored by two calibrated photodiodes. Rhodamine 6G and Rhodamine B were used as fluorescence quantum-yield standards and all measurements were ascertained to be in a linear intensity region.

Our present task is to establish a procedure for relating lifetime values calculated from the absolute quantum-yields of fluorescence to those obtained from the observed fine-structure, where the latter can be observed. It is important to determine whether a cross-check can be established, since then it would be possible to use relative quantum-yields for absolute lifetime measurements on systems where symmetry prevents fine-structure observations. Although the spectral response characteristics of the detector remain to be calibrated, we have made relative quantum-yield measurements on a number of compounds with reproducibility better than ±20%. Rough estimates based on available data place relaxation times between 0.03 and 0.1 ps.

Parallel experiments being carried out in our laboratory include measurement of relaxation times down to about 1 ps, using either fluorescence gating or time-correlated upper state fluorescence excitation experiments.

This work was supported by the National Science Foundation directly (CHE-76-10336) and through the Materials Research Programme (DMR-76-00678).

References

1. G. C. Orner and M. R. Topp, Chem. Phys. Lett., 36, 295 (1975).

2. H. B. Lin and M. R. Topp, Chem. Phys. Lett., 47, 442 (1977); 48, 251 (1977).

3. M. R. Topp and H. B. Lin, Chem. Phys. Lett., 50, 412 (1977).

4. K.J.Choi, L.A.Hallidy, H.B.Lin and M.R.Topp: *Advances in Laser Chemistry*, ed. by A.H.Zewail, Springer Series in Chemical Physics, Vol. 3 (Springer, Berlin, Heidelberg, New York 1978)

5. J. L. Richards and S. A. Rice, Chem. Phys. Lett., 9, 444 (1971).

6. J. L. Richards and S. A. Rice, J. Chem. Phys., 54, 2014 (1971).

7. C. Porter and M. R. Topp, Proc. Roy. Soc. Lond., A315, 163 (1970).

8. M. R. Topp, Chem. Phys. Lett., 32, 144 (1975).

9. R. I. Personov, E. I. Al'shits and L. A. Bykovskaya, Opt. Comm., 6, 169 (1972).

10. C. R. Goldschmidt and M. Ottolenghi, J. Phys. Chem., 75, 3894 (1971).

Picosecond Laser Studies of Electron-Hole Interactions and Double Proton Transfer

K.B. Eisenthal, K. Gnädig[1], Wm. Hetherington, M. Crawford, and R. Micheels

Department of Chemistry, Columbia University
New York, NY 10027, USA

I would like to discuss some of our preliminary results on electron-hole interactions in liquids and excited state double proton transfer.

A. Electron-Hole Interactions in Liquids

The ionization potential of a molecule in the gas phase is the energy required to remove the electron to infinity with no excess kinetic energy. For aromatic organic molecules such as pyrene and anthracene the ionization potentials are about 7.5 eV. However, if the molecule is placed in a liquid its ionization potential is changed. Some of the factors to be considered are the polarization energy of the molecular fragment produced and the ground state energy of the excess electron in the liquid, i.e. the energy at the bottom of the conduction band. Rough estimates indicate that the ionization potential can be lowered by 1-2 eV in a polar liquid [1]. Using an intense laser pulse in the near ultraviolet photoionization can be achieved by a two photon mechanism. This has been established for a number of organic molecules in a variety of liquids [1-10]. The photoejected electron is initially quasifree but quickly becomes trapped and solvated by the molecules of the liquid. The exact nature of these processes are not known. We do know the time scale for solvation in some liquids, e.g. alcohols and water [11-13]. The solvation dynamics depends on the thermalization of the electron and the dielectric and orientational properties of the liquid. One description might be the following one. The quasifree electron rapidly thermalizes and diffuses in search of preexisting solvent traps, i.e. a local solvent arrangement in which the electron energy is lowered. This trapping step is responsible for the early time absorption toward the infrared in alcohols. Following this trapping the surrounding solvent molecules reorient, the potential well deepens and the electron becomes solvated leading to the final blue shifted absorption in the visible. In this description and in previous discussions of the early time events as measured in picosecond experiments, [11-17] e.g. picosecond pulse radiolysis and picosecond laser studies, the role of the remaining fragment is justifiably neglected in the solvation dynamics. In the picosecond laser studies the photoionization was from a species which had a multiple negative charge [13]. Thus the fragment generated was still negatively charged following photoejection and did not exert an attractive interaction on the photoejected electron. In picosecond pulse radiolysis experiments [11,12, 14-17] the electrons injected into the liquid are of such high energies that even the secondary elec-

---

[1] Present address: Max-Planck-Gesellschaft zur Förderung der Wissenschaften e.V., Projektgruppe für Laserforschung, D-8046 Garching bei München, Fed. Rep. of Germany

trons have sufficient energy to avoid thermalization in the vicinity of any cations produced in the liquid.

However if we photoionize a neutral molecule then the remaining fragment is positively charged and there can be a strong attractive Coulomb interaction between the electron and the parent hole or cation. The escape of the electron and its subsequent full solvation must therefore occur against a large drag due to the attractive electrostatic field of the hole. We thus recognized that we might be able to directly probe electron-hole interactions by monitoring the early time dynamics following photoejection. A key manifestation of the electron-hole interaction would be the time dependent change in the energy levels and thus spectrum of the ejected electron as the electron-hole separation and solvation evolve in time.

In the present study of the laser induced photoionization of neutral aromatic molecules in polar solvents the effects of the electron-hole interaction is seen in the markedly longer times required for the escape and solvation processes. The systems studied were 2.5×10^{-3} M anthracene in acetonitrile and 2.5×10^{-3} M pyrene in methanol. The electrons were produced by two photon ionization of the solute

$$Ar + 2\hbar\omega \longrightarrow Ar^{+} + e^{-}$$

where $\hbar\omega$ is the second harmonic energy (λ = 347 nm) of a mode-locked ruby laser.

There are two possible excitation pathways for the two photon absorption shown above since one second harmonic photon is of sufficient energy to promote anthracene or pyrene to their excited singlet manifolds. The relative importance of simultaneous versus consecutive two photon absorption is not known in these experiments. However the relative width of the excitation pulse to vibronic relaxation time should favor the sequential route. With much longer excitation pulses, e.g. 10^{-8} sec, the sequential route can be reasonably regarded as the dominant one [8,9]. Irrespective of the excitation pathways two near ultraviolet photons are required to photoeject an electron from pyrene and anthracene in liquids so far investigated. This two photon mechanism has been well established in a number of studies on these molecules and other aromatics in a variety of liquids [8-17].

To follow the time evolution of the system the transmission of a probe pulse at 694 nm, the ruby fundamental wavelength, was measured as a function of time. The mode-locked single pulse at 694 nm, selected from the pulse train, has a full width at half maximum of 15-20 psec. as determined by two photon fluorescence. The fully solvated electron spectra are broad and have maxima at 700 nm in acetonitrile [18] and 620 nm in methanol [19].

For both pyrene in methanol and anthracene in acetonitrile the absorption levels off at about 100 psec (Fig.1). By fitting the convolution of an exponential function with Gaussian excitation and probe pulses (half-width 20 psec) to the experimental results the rise times were found to be about 40 psec. This is in marked contrast to pulse radiolysis studies on methanol in which a value of 10 psec was obtained for the electron solvation time [5,7]. Although the electron solvation time in acetonitrile is not known it is expected to be less than for methanol. This is based on the assumption that the limiting step in the solvation of an electron, not in the vicinity of a cation, is correlated with the orientational relaxation time of the solvent as has been found to be the case for alcohols [20,21].

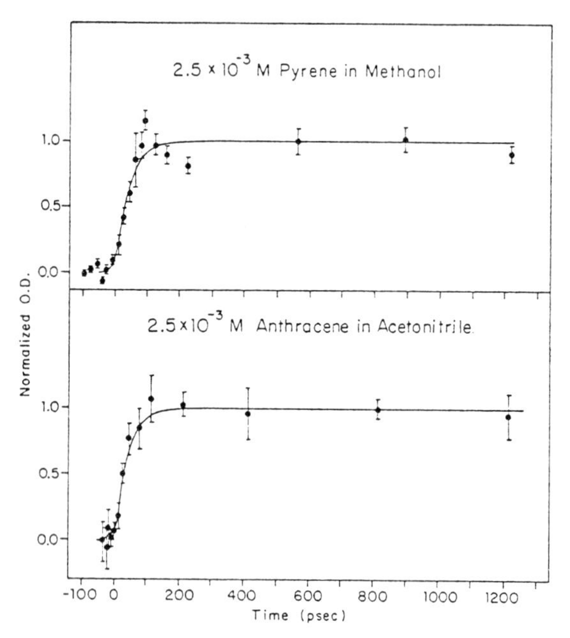

Fig.1 Experimental absorption signal (●) superimposed on theoretical curve (—)

To determine if the probe absorption was directly related to the photo-ejected electron, scavenging studies using chloroform were carried out in the pyrene-methanol system. Although chloroform scavenges electrons it does not scavenge the pyrene excited singlet or pyrene cation [17]. The decrease in absorption with 0.1 M chloroform measured at different probe delay times, or compared with the absorption signal in the absence of the chloroform scavenger, yielded scavenging constants in the range of 6-9 x 10^9 liters/mole-sec. This is consistent with the values to be expected in diffusion controlled reactions.

The possibilities of absorptions by the excited singlet or triplet states of anthracene and pyrene are discounted due to their relatively weak absorptions at the probe wavelength [22-25]. Absorption from unrelaxed excited states is deemed unimportant since vibronic relaxation is too fast, typically less than a few psec for aromatic molecules in liquids, [26] to explain the relatively slow rise time observed. Furthermore it should be noted that indirect processes such as excited singlet production by electron-hole recombination is ruled out not only by the relatively weak excited singlet absorption at 694 nm but also on consideration of the kinetics. If such indirect processes were of importance then the observed rise time indicates that they are completed in about 100 psec. It would then follow that the chloroform scavenger which does not quench excited singlet pyrene molecules would not reduce the absorption signal at longer times contrary to our findings. By the same arguments the contribution of pyrene cation solvation to the observed dynamics, including possible effects due to interactions with a nearby electron, is ruled out. The possibility of pyrene excimer formation as a factor in these experiments is rejected since the reaction times are far too long at the concentrations used to explain our data.

We therefore conclude that the dominant process we are observing is the time dependent absorption of the photoejected electrons. Further experiments are underway to more fully examine these electron-hole interactions in liquids.

B. Excited State Double Proton Transfer

I would now like to talk about another elementary particle in chemistry, the proton. The ability of a molecule to act as an acid or base, i.e. donate a proton to or accept a proton from a neighboring species, is critically dependent on the electronic structure of the molecule. It is therefore not surprising that the acid base properties of a molecule can be vastly different in an excited vs. ground state or between any states, e.g. singlet and triplet. Due to the change in acid-base properties the relaxation characteristics of excited molecules can be altered, i.e. the channels for energy degradation, lifetimes, emission yields and spectra can be vastly altered.

I would like to discuss some of our research on a very unusual excited state proton transfer discovered by TAYLOR, EL-BAYOUMI and KASHA [27]. The molecule is 7-azaindole, a heterocyclic base similar to DNA bases and possibly of importance as a model in mutagenesis.

In nonpolar hydrocarbon solvents such as hexane they found that at reasonably high concentrations ($<10^{-2}$ M) that a 1:1 complex, i.e. dimers were formed. This was dramatically seen, not in absorption changes but rather in the appearance of a new fluorescence at long wavelength [27-29]. (See Fig.2) The new emission was attributed to a tautomer resulting from a double proton-transfer (Fig.3). This unusual process was also observed at low concentrations in alcohols (Fig.4).

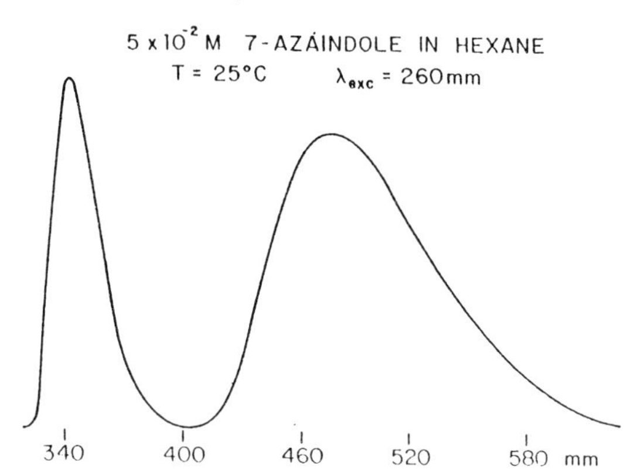

Fig.2 Fluorescence spectrum of 7-azaindole

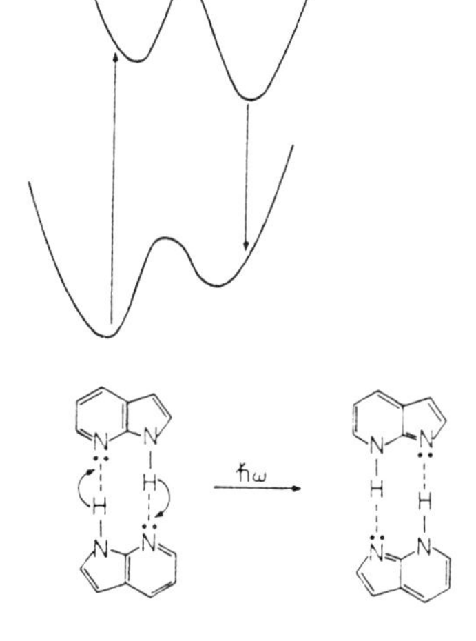

Fig.3 Double proton transfer on excitation of 7-azaindole dimer in nonpolar solvent

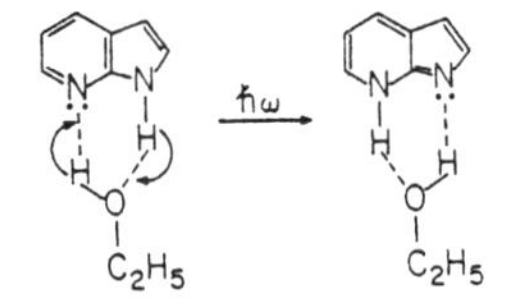

Fig.4 Double proton transfer on excitation of 7-azaindole in ethanol

The ultrafast nature of the double proton transfer requires the use of picosecond methods. Prior studies using nanosecond excitation yielded important results but had to be performed at 77^0 K in a hydrocarbon matrix [30]. For the picosecond experiments reported here a Nd:phosphate glass laser was used. A single pulse was selected from the pulse train, amplified, and then quadrupled to 266 nm. A streak camera coupled to an optical multichannel analyzer was used to monitor the fluorescence of the tautomer. Appropriate filters bandpass were used to detect only the tautomer fluorescence. The response of the streak camera at the streaking speed used in these experiments 0.03 nsec/mm was determined by monitoring the rise time of fluorescence from a 10^{-4} M Rhodamine B ethanol solution excited with a second harmonic (527 nm) pulse. In Fig. 5 is shown the excitation pulse and the fluorescence of Rhodamine B. The width of the excitation pulse is 5 psec or less. The rise time of the fluorescence corresponds to the width of the excitation pulse. This is consistent with the subpicosecond rise time observed in gain experiments using a subpicosecond dye laser for excitation of Rhodamine [26].

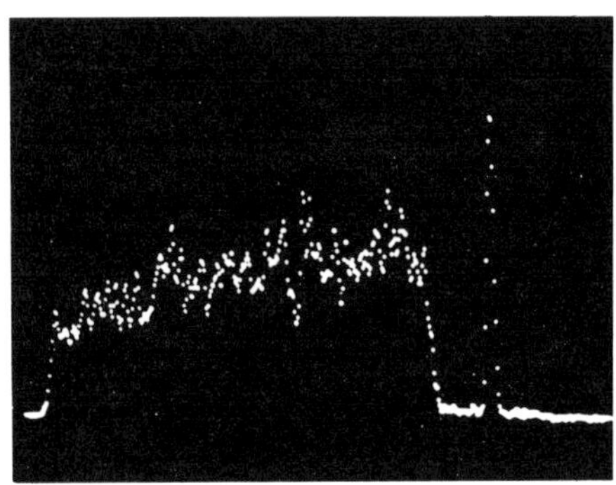

Fig.5 Streak camera-OMA output of excitation pulse and Rhodamine B fluorescence. Time scale is 1 psec/channel.

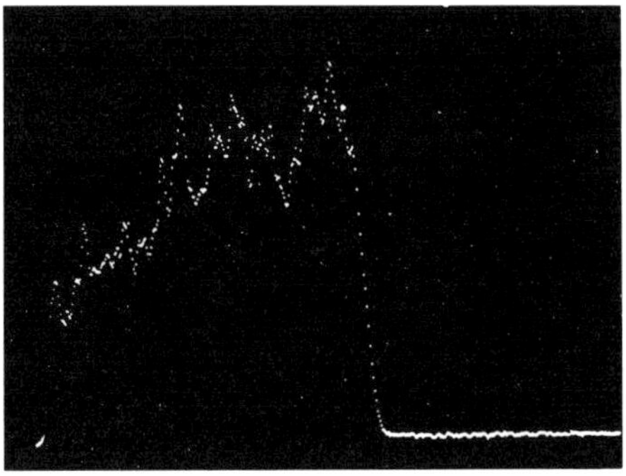

Fig.6 Streak camera-OMA output of long wavelength (tautomer) fluorescence of 7-azaindole in hexane. Time scale is 1 psec/channel

Our preliminary results on the rise time of the tautomer fluorescence, and thus the double proton jump time, is that it is faster than 5 psec at room temperature (Fig. 6). Measurements at the fastest camera streak speed and temperature dependent results will be published at a later date.

Acknowledgments

Various parts of this research were supported by the National Science Foundation, Air Force Office of Scientific Research and the Army Research Office.

References

1. P.L. Piciulo and J.K. Thomas, J. Chem. Phys. 68, 3260 (1978).
2. J.L. Metzger and H. Labhart, Chem. Phys. 7, 150 (1974).
3. G.E. Hall and G.A. Kenney-Wallace, Chem. Phys. 28, 205 (1978).
4. G.E. Johnson and A.C. Albrecht, J. Chem. Phys. 44, 3179 (1966).
5. J.T. Richards, G. West and J.K. Thomas, J. Phys. Chem. 74, 4137 (1970).
6. M. Ottolenghi, Chem. Phys. Lett. 12, 339 (1971).
7. Y. Taniguchi, Y. Yishina and N. Mataga, Bull. Chem. Soc. Japan 45, 2973 (1972).
8. M. Fisher, B. Veyret and K. Weiss, Chem. Phys. Lett. 28, 60 (1974).
9. D.K. Sharma, J. Stevenson and G.J. Hoytink, Chem. Phys. Lett. 29, 343 (1974).
10. A. Bergman, C.R. Dickson, S.D. Lidofsky and R.N. Zare, J. Chem. Phys. 65, 1186 (1976).
11. W.J. Chase and J.W. Hunt, J. Phys. Chem. 79, 2835 (1975).
12. G.A. Kenney-Wallace and C.D. Jonah, Chem. Phys. Lett. 47, 362 (1977).
13. P.M. Rentzepis, R.P. Jones and J. Jortner, J. Chem. Phys. 59, 766 (1973).
14. M.J. Bronskill, R.K. Wolff and J.W. Hunt, J. Chem. Phys. 53, 4201 (1970).
15. G. Beck and J.K. Thomas, J. Phys. Chem. 76, 3856 (1972).
16. J.H. Baxendale and P. Wardman, J. Chem. Soc., Faraday Trans. 1 69, 584 (1973).
17. D. Huppert, P.M. Rentzepis and W.S. Struve, J. Phys. Chem. 79, 2850 (1975).
18. A. Singh, H.D. Gesser and A.R. Scott, Chem. Phys. Lett. 2, 271 (1968).
19. M.C. Sauer Jr, S. Arai and L.M. Dorfman, J. Chem. Phys. 42, 708 (1965).
20. G.A. Kenney-Wallace, Chem. Phys. Lett. 43, 529 (1976).
21. L. Kevan and K. Fueki, Chem. Phys. Lett. 49, 101 (1977).
22. N. Nakashima and N. Mataga, Chem. Phys. Lett. 35, 487 (1975).
23. R. Astier, A. Bokobza and Y.H. Meyer, J. Chem. Phys. 51, 5174 (1969).
24. G. Porter and M.R. Topp, Proc. Roy. Soc. London A 315, 163 (1970).
25. M.F.M. Post, J. Langelaar and J.D.W. Van Voorst, Chem. Phys. Lett. 10, 486 (1971).
26. C.V. Shank, E.P. Ippen and O. Teschke, Chem. Phys. Lett. 45, 291 (1977).
27. C.A. Taylor, M.A. El-Bayoumi and M. Kasha, Proc. Nat'l. Acad. Sci. U.S. 63, 253 (1969).
28. K.C. Ingham, M. Abu-Elgheit and M.A. El-Bayoumi, J. Amer. Chem. Soc. 93, 5023 (1971).
29. K.C. Ingham and M.A. El-Bayoumi, J. Amer. Chem. Soc. 96, 1674 (1974).
30. M.A. El-Bayoumi, P. Avouris and W.R. Ware, J. Chem. Phys. 62, 2499 (1975).

Polydiacetylenes, a Promising Material for Nonlinear Optics with Picosecond Pulses

J.P. Hermann[1] and M. Lequime

Laboratoire d'Optique Quantique, Ecole Polytechnique
Palaiseau, France

Abstract

We set up several experiments to take advantage of the high third order susceptibility of polydiacetylene crystals; these include optical Kerr effect and parametric amplification. Problems arose from an unexpected and considerable two-photon absorption. The origin of this absorption and a possible remedy are discussed.

Efficient optical conversion and parametric oscillation have so far been obtained only in non-centrosymmetric crystals in which a laser field induces a nonlinear polarization of the second order:

$$P^{(2)} = \chi^{(2)}E^2.$$

The stimulated Raman effect is a noticeable exception, in which a good photon conversion can be obtained in a fluid, in which the lowest nonlinear susceptibility is of the third order:

$$P^{(3)} = \chi^{(3)}E^3.$$

It has been shown previously [1] that long conjugated organic molecules possess a linear and a nonlinear polarizability that increases when the length of the delocalized chain increases:

$$p = \alpha E + \gamma E^3.$$

This was due to the nonlinear behavior of the delocalized π electrons in the molecule; their influence on the polarizability α, though noticeable is small compared to that of the localized σ electrons; but the π electrons enhance considerably the hyperpolarizability γ. The main interest of this class of compounds lies in the fact that this nonlinearity is non-resonant, so that it does not require any resonant enhancement as atomic

[1] Presently at IBM Thomas J. Watson Research Center, Yorktown Heights, New York 10598.

vapors do, and also this nonlinearity should stay roughly the same for all third order nonlinear processes. Polydiacetylenes are made of parallel conjugated infinite chains, this peculiar structure results in a strong mechanical and optical anisotropy. [2] They represent an ideal packing of conjugated molecules, in which one expects a high nonlinear susceptibility $\chi^{(3)}$.

Actually, optical third harmonic generation measurements [3] showed a considerable $\chi^{(3)}$, of the order of 10^{-10} esu, comparable or superior to that of germanium. The advantage of the polydiacetylenes lies in their transparency in the red and near infrared. An elementary estimate shows that a 1 GW cm^{-2} laser pulse induces in such a crystal a cubic nonlinear polarization, $\chi^{(3)}E^3$, equal to the quadratic polarization, $\chi^{(2)}E^2$, which it would induce in $LiNbO_3$. In addition, the polydiacetylenes can withstand an intensity of 20 GW cm^{-2} for picosecond laser pulses. The polydiacetylenes are not applicable to third harmonic generation because the $\chi^{(3)}$ tensor has only one appreciable component $\chi^{(3)}_{xxxx}$ where x is the direction of the conjugated chains, so they cannot be phase-matched.

We tried several experiments on these crystals, including optical Kerr effect and parametric amplification, but found a severe limitation in the two-photon absorption, which we tried to understand.

1. Optical Kerr Effect

Due to the strong anisotropy of the tensor $\chi^{(3)}$, only the refractive index for a light polarization parallel to the x axis is affected by the laser intensity

$$n_x = n_x^o + \frac{6\pi^2}{(n_x^o)^2 c}\chi^{(3)}_{xxxx}(\omega,-\omega,\omega)I_x$$

$$n_y = n_y^o.$$

Thus a slab which is cut as a $(2k+1)\lambda/2$ plate can become a $(2k+2)\lambda/2$, $(2k+3)\lambda/2$ ... for increasing laser intensities. A polarizer should therefore select the high intensity part of the pulse; one could also expect a time modulation of the beam. Since the nonlinearity has an electronic origin, there is no response time to deal with.

Making the experiment with a 1.06 μm picosecond pulse, we observed a considerable and very anisotropic two-photon absorption which affected the amplitude of the wave as well as its phase. We actually observed a modulation of the pulse, but at the expense of the loss of half the laser energy, which makes the system far less attractive.

2. Parametric Amplification

The theory of four-photon parametric gain shows that with a nonlinearity of 10^{-10} esu, it should be possible to amplify a signal and create an idler in appreciable quantities with a pump intensity of 1 GW cm^{-2}.

The phase matching can be obtained if the signal and the idler are propagating at given angles from the pump wave vector. These angles are not too critical, due to the fast exponential growth of the signal. Thus we tried to amplify the continuum created in D_2O by a single pulse at 1.06 μm, the pulse itself being the pump; by adjusting the angles, we expected to selectively amplify a narrow band of the continuum and generate a tunable infrared idler pulse.

There again, we found a severe limitation in nonlinear losses induced by the high intensity pump pulse: it was not possible to propagate a high intensity pump pulse through all the crystals because it was absorbed by two-photon absorption.

Thus it was essential to determine where this two-photon absorption came from.

3. Two-Photon Absorption in Polydiacetylenes

Again using a 1.06 μm pulse and the continuum generated in D_2O, we plotted the two-photon absorption spectrum. We found it to be very different from theoretical predictions: it looked very similar to the absorption tail present on the long wavelength side of the one-photon absorption band. We interpreted this as the creation of defects (= localized orbitals) by the 1.06 μm pulse which is slightly absorbed even at low intensity. We also studied the dynamics of this induced absorption and found that the lifetime of the defects was less than 4 ps.

4. Conclusion

We believe that it is possible to find a solution to the light absorption induced by a strong laser in polydiacetylene crystals; an improvement would be to grow and polymerize better crystals with no residual absorption at 1.06 μm. This absorption itself seems to be due to permanent defects in the crystals, which suggests that a better crystal growth technique would help a lot.

We want to acknowledge Prof. J. Ducuing for stimulating discussions and Drs. R. Baughman and D. Bloor for providing the polydiacetylene crystals.

References

1. J. P. Hermann and J. Ducuing, J. Appl. Phys. 45, 5100 (1974).
2. R. H. Baughman, J. Polym. Sci., Polym. Phys. Ed. 12, 1511 (1974).
3. C. Sauteret, J. P. Hermann, R. Frey, F. Pradere, J. Ducuing, R. H. Baughman and R. R. Chance, Phys. Rev. Lett. 36,956 (1976).

Short Pulse Absorption Spectroscopy of Nitromethane Photolysis

W.L. Faust, L.S. Goldberg, T.R. Royt, J.N. Bradford, R.T. Williams, and J.M. Schnur

Naval Research Laboratory, Washington, DC 20375, USA

P.G. Stone and R.G. Weiss[1]

Department of Chemistry, Georgetown University
Washington, DC 20057, USA

We are applying short pulse techniques to studies of small molecular fragments, stable or reactive radical species emergent as the primary products of unimolecular decomposition of simple photolabile materials. Nitromethane (CH_3NO_2) photolysis has been the subject of extensive prior investigations under varying conditions of irradiation wavelength, phase, temperature, and with active additives. The identity of the primary products and the kinetics of the reaction remain unresolved, although several mechanisms have been advanced to explain the observed products. From nitromethane vapor the list of major final products which have been reported [1] includes CH_3ONO, CH_2O, CH_3NO, NO, and N_2O; minor final products include H_2, CH_4, N_2, and CO; and a number of reaction intermediates have been reported, notably NO_2 and HNO. As an inferred early product, NO_2 was estimated to have a quantum yield as large as 0.6, and this fragment has been our principal object of study to date.

The present experiment employs a short pulse of the Nd 4th harmonic to excite condensed-phase nitromethane on its first electronic transition $n \rightarrow \pi^*$ (λ_{max} = 270 nm). We have recorded an extensive band of absorption which we attribute to the nitro radical NO_2 (Fig. 1). The onset of the absorption is not immediate; it becomes detectable a few nanoseconds after the ultraviolet pulse and is still rising after 13 nanoseconds (Fig. 2). The intensity of the absorption is consistent with a substantial yield.

The detection of NO_2 presents a significant challenge in that its absorption cross section σ is small. Measurements on low pressure NO_2 gas, which absorbs in a very broad band extending roughly from 300 to 550 nm, give $\sigma \sim 3\times10^{-18}$ cm^2. Most previous short-pulse continuum studies, in contrast, have dealt with highly-conjugated systems such as polyenes, with cross sections two orders of magnitude larger. We have also sought HNO, which has accessible absorption in the red and which has been considered potentially an important primary product; but no clear indication of this species has been found under the current conditions.

The picosecond photolysis apparatus was constructed with the capability to observe spectra of weakly absorbing transient species. The laser system comprises a passively modelocked Nd:YAG oscillator, a single pulse selector, and a double-pass Nd:YAG amplifier to provide a 1.064 μm pulse of up to 25 mJ energy and 30 ps duration. A frequency doubler and a redoubler, both of KD*P, generate a 266 nm photolysing pulse of 3 to 4 mJ energy. The

[1]Research supported by the Office of Naval Research.

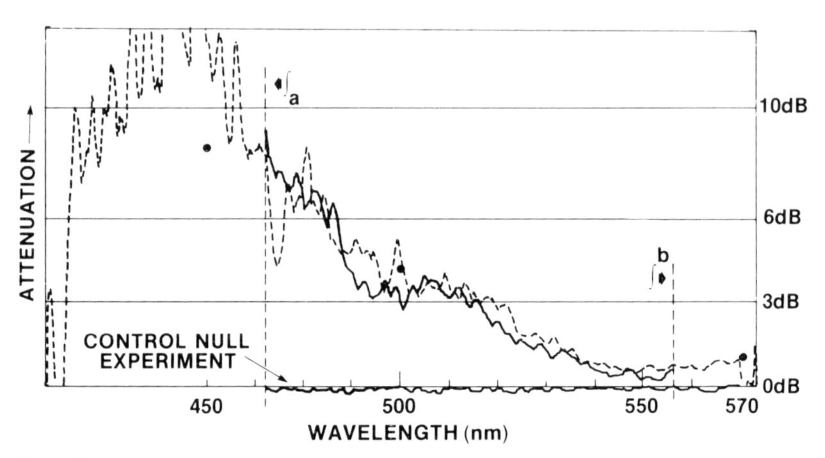

Fig. 1. Spectrum of absorption at 13.9 nsec delay. The relative NO_2 concentration at a given interrogation time was taken by the integration of the attenuation between the limits displayed in the figure, where the detected probe light signal level was adequate. Two shots are displayed.

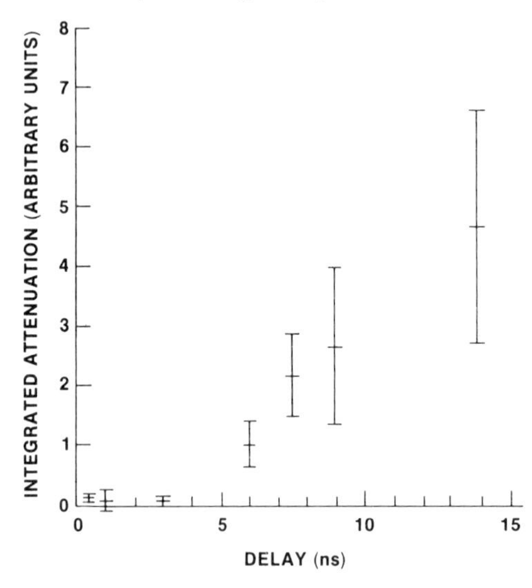

Fig. 2. Temporal development of integrated attenuation attributed to NO_2 absorption, following a photolysis pulse of about 2.5 mj energy.

ultraviolet pulse is brought to a submillimeter diameter focus (~500 μm) in the sample cell (2 mm thick) in order to provide sufficiently high energy density to produce adequate NO_2 for observation. A white light continuum pulse is used to probe this small photolysed region. The residual 1.064 μm light after frequency doubling passes through a variable optical delay and is focused into a 15 cm cell of D_2O. The emergent continuum beam is then steered through the center of a 600 μm pinhole just beyond the D_2O cell and is imaged by a lens to a 350 μm diameter spot in the sample cell. The lens is apertured to pass only a small portion of the central, relatively homogeneous core of the continuum light. A beam splitter is used to divert a replica of the focused probe beam to impinge on a second spot in the sample cell. This provides a reference beam identical to the signal beam with regard to any lateral variation of the spectral distribution. The ultraviolet and main probe beams are made collinear and are focused independently. LiF windows are used in the sample cell to avoid two-photon absorption and damage from the intense photolysing pulse. The spectrum of induced absorption is measured on individual laser shots by dual-beam digital ratiometry using a grating spectrograph, intensified vidicon multichannel recording, and computer data processing. The vidicon dark current is subtracted and instrumental variations in the beams are stripped out. The energy of the individual ultraviolet pulses is measured, and a discriminator threshold is employed to reject weak pulses.

As we noted above, there is still no firmly-established picture of nitromethane photolysis despite many years of investigation with several experimental approaches and despite various improvements in experimental technique. There is a component of confusion which arises from the varied conditions under which the reactions have been conducted. Polychromatic flash excitation of gaseous nitromethane by HIRSCHLAFF and NORRISH [2] at 101°C led them to conclude that the primary photochemical process involved the formation of the aci from I, followed by thermal conversion to formaldehyde and nitroxyl:

$$CH_3NO_2 \xrightarrow{h\nu} \underset{\text{I}}{CH_2 = \overset{\oplus}{N}}\begin{matrix} \diagup O^{\ominus} \\ \diagdown OH \end{matrix} \rightarrow CH_2O + NOH\ .$$

Later NAPIER and NORRISH [3] reexamined the flash photolysis of gaseous nitromethane in the presence of an excess of nitrogen. The results of these experiments indicated that homolytic scission of the carbon-nitrogen bond to form methyl and nitro radicals is the first step:

$$CH_3NO_2 \xrightarrow{h\nu} CH_3 + NO_2\ .$$

The latter mechanism was favored also by BROWN and PIMENTEL [4] on the basis of low-temperature photolysis work, and later by McGARVEY and McGRATH [5] who again employed flash techniques with gaseous nitromethane. Classical photochemical techniques were applied by HONDA, MIKUNI, and TAKAHASI [1], who concluded that gaseous nitromethane at 55°C under irradiation at 313 nm yields formaldehyde and nitroxyl from the excited singlet state, methyl and nitro from the triplet.

Considering the import of our own measurements, we do not suppose that NO_2 is formed directly from the nitromethane singlet in a process protracted over nanoseconds and with substantial yield, since competing processes of internal conversion and intersystem crossing should dominate this. The long delay which we have found for NO_2 formation is consistent with either a triplet or an isomer such as the aci form I of nitromethane as the precursor. We plan additional tests addressed specifically toward each of these possibilities for the intermediate.

We have found some unexpected but interesting effects associated with relatively high intensities in the probe beam. This beam itself can give rise to new absorption in the red, beyond the NO_2 region. These effects abate for adequately reduced probe intensity levels which still are readily measured with the intensified vidicon. Also, we find no such effects at small time delays such that the principal NO_2 absorption has not yet developed. It is clear that such investigations must be executed with careful regard for stepwise and nonlinear effects associated with high levels of probing flux. This concern must be heightened for work with still shorter pulses.

Since there is a diversity of potential products in nitromethane photolysis, it was essential to acquire spectral information across as broad a band as possible for the distinct identification of the species present. On the other hand, the KCl F center, for which we give a brief discussion of new results, is a case for which the gross features of the spectroscopy at least were understood at the outset. In this circumstance discrete-frequency synchronous probe beams remain very useful for studies of temporal development and for actinometric measurements. The F center yield following two-quantum excitation possesses an interesting temperature dependence (Fig. 3). It approaches unity near the melting point (1049°C), and absorption persists even into the molten state. It should be appreciated that the stable population of these centers apparent to conventional means of measurement is reduced by two orders of magnitude or more because of very fast annealing of the complementary F and H centers. The temperature dependence then also acquires additional structure.

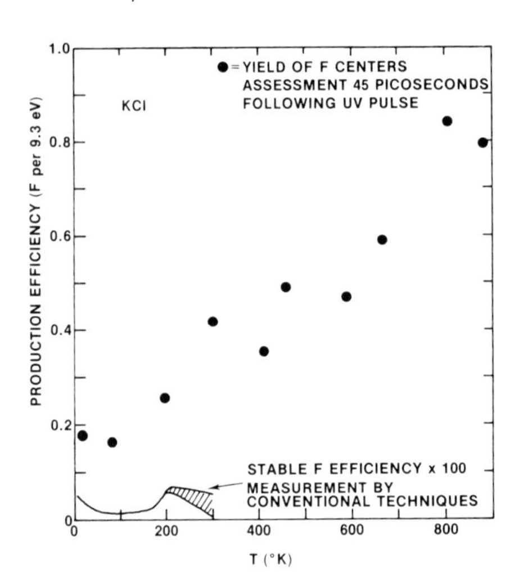

Fig. 3. The yield of F centers per two-photon absorption event as assessed with a 532 nm probe pulse. The production efficiency has been corrected for temperature dependence of the F-band spectrum.

References

1. K. Honda, H. Mikuni, and M. Takahasi, Bull. Chem. Soc. Japan 45, 3534 (1972).
2. E. Hirschlaff and R. G. W. Norrish, J. Chem. Soc. 1580 (1936).
3. I. M. Napier and R. G. W. Norrish, Proc. Roy. Soc. A 299, 317 (1967).
4. H. W. Brown and G. C. Pimentel, J. Chem. Phys. 29, 883 (1958).
5. J. J. McGarvey and W. D. McGrath, Trans. Faraday Soc. 60, 2196 (1964).

II. Poster Session

Singlet Exciton Interactions in Crystalline Naphthalene

F. Heisel, J.A. Miehe, and B. Sipp

Laboratoire de Physique des Rayonnements et d'Electronique Nucléaire
Centre de Recherches Nucléaires et Université Louis Pasteur
F-67037 Strasbourg Cédex, France

M. Schott

Groupe de Physique des Solides de l'ENS, Université Paris VII
F-75221 Paris Cédex 05, France

Intense, short light pulses provided a suitable means of excitation to study the singlet-singlet interaction in aromatic substances. Here, we report the results of an analysis of the temporal decay of naphthalene crystal fluorescence, excited by 25 ps pulses at 266 nm. A good fit is obtained with the prediction of the usual kinetic equations, provided that both the exponential attenuation with depth of the penetrating beam and its radial intensity profile are taken into account.

Naphthalene crystals were grown by the Bridgman method from zone-refined naphthalene powder and cleaved along the ab plane. The excitation light source consisted of a passively mode-locked Nd : YAG laser associated with a single pulse switch and two second-harmonic generators (CD*A and ADP crystals) delivering one single 266 nm pulse of 25 ps duration. The UV pulse energy was measured by means of a Laser Precision RkP 335 detector followed by a Rk 3230 energy meter. The transverse intensity profile of the laser beam at 266 nm was determined with a Reticon RL 512C photodiode array [1], after conversion of the UV into visible light by a polycrystalline layer of sodium salicylate. It was found to be gaussian with a spot size r_o near 0.14 cm on the sample surface. The temporal variation of the fluorescence was observed, behing the crystal, with a broad band real-time detection system consisting of a RTC-UVHC20 vacuum photocell (FWHM = 0.2 ns) fed into a 5 GHz oscilloscope [2].

During the time range actually investigated (up to about 5 ns) the triplet build-up can be ignored. The rate equation governing the singlet concentration n_S at a given depth z and a distance r from the beam axis is then given, provided exciton diffusion can be neglected, by :

$$\dot{n}_S(t) = -k_S n_S - \gamma_{SS} n_S^2 + \frac{2N_o \varepsilon}{\pi r_o^2} \exp(-\varepsilon z)\exp(-2r^2/r_o^2)U(r)U(z)\,\delta(t) \qquad (1)$$

where the last term represents the source term, U(r) and U(z) being the Heaviside function. N_o is the total number of photons per excitation pulse, ε the absorption coefficient of the sample and γ_{SS} the rate constant of the S-S interaction. Reabsorption of fluorescence is not important in naphthalene and is certainly negligible at times much shorter than the fluorescence lifetime. The monomolecular lifetime $1/k_S$ of the S_1 excitons, measured under low intensity excitation was found to be 115 ns, in agreement with previous measurements [3]. By solving (1) and integrating over the total volume, we obtain for the fluorescence decay law

$$i(t) = \frac{2\pi k_F \exp(-k_S t)}{\frac{\gamma_{SS}}{k_S}\epsilon[1-\exp(-k_S t)]} \int_0^\infty \text{Log}\{1+\frac{\gamma_{SS}}{k_S}[1-\exp(-k_S t)]\frac{2N_o \epsilon}{\pi r_o^2} \exp(-2r^2/r_o^2)\} r dr \qquad (2)$$

where k_F is the radiative singlet decay rate.

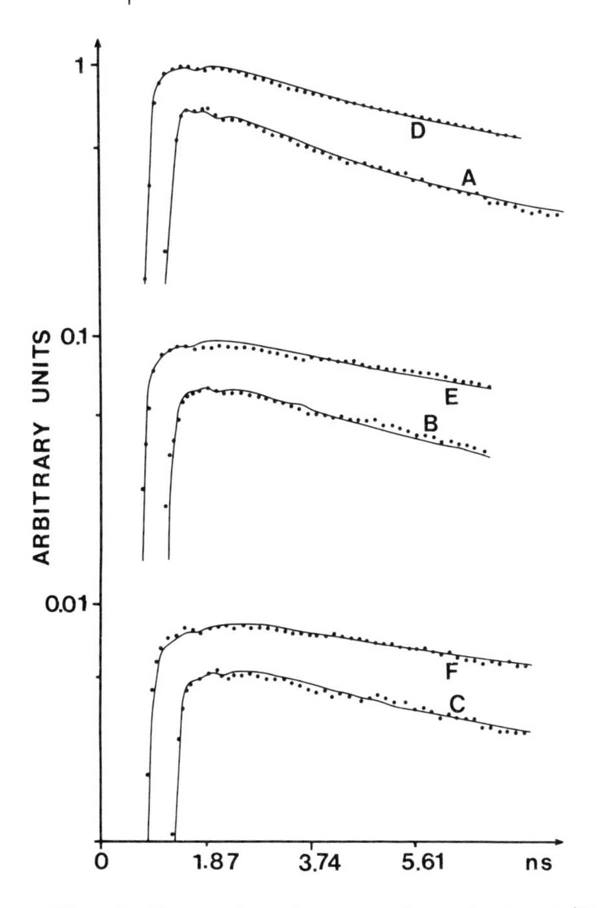

Fig. 1 Comparison between the calculated (lines) and experimental (dots) fluorescence decay curves. (A, B, C) incident radiation polarized parallel to b-axis and energies respectively equal to 2.8, 1.3 and 1 μJ. (D, E, F) incident radiation polarized parallel to a-axis and energies respectively equal to 3.3, 1.3 and 1 μJ.

In Fig. 1 (A through F) are reported the experimental decay curves (dots) obtained at room temperature with thick crystals for three energies of the excitation pulse polarized respectively parallel to the a and b axis

($\epsilon_a = 4 \times 10^4 cm^{-1}$ and $\epsilon_b = 10^5 cm^{-1}$ [3,4]). The plots are single shot oscillograms digitized by a TV scanning system [5] : the time interval between two points is equal to 75 ps. The solid lines in Fig. 1 show the calculated time evolution of the fluorescence emission, representing the convolution of the instrumental response function with the theoretical quenching curves (Eq. 2). These fits have been obtained with $\gamma_{SS} = 4 \times 10^{-10} cm^3 s^{-1}$ and $r_o = 0.14$ cm. It is not possible to adjust the experimental data if either γ_{SS} and/or r_o are varied by more than 25 % from these best-fit values.

The maximum rate constant γ_{SS} assuming immediate reaction of two singlet excitons encountering in a spherical reaction volume and assuming isotropic three-dimensional singlet diffusion is [6]

$$\gamma_{SS} = 8\pi D_S R [1 + R/\Lambda] \qquad (3)$$

R is the spherical reaction radius, D_S the diffusion constant and $\Lambda (= \sqrt{D_S/k_S})$ the diffusion length. Since R is larger than the nearest neighbor distance of 5 Å and probably smaller than 20 Å, $R/\Lambda \ll 1$ and from (3) we found $10^{-4} \leq D_S \leq 3 \times 10^{-4} cm^2 s^{-1}$ in agreement with the values inferred from energy transfer experiments in doped crystals $D_S = 1.5 \times 10^{-4}$ [7,8] and $D_S = 5 \times 10^{-5} cm^2 s^{-1}$ [9].

References

1. W. Seka, J. Zimmermann : Rev. Sci. Inst. 45, 1175 (1974)
2. B. Cunin, J.A. Miehé, B. Sipp, J. Thébault : Rev. Sci. Inst. 47 , 1435 (1976)
3. H.C. Wolf : Sol. Stat. Phys. 9 , 1 (1959)
4. A. Bree, T. Thirunamachandran : Mol. Phys. 5 , 397 (1962)
5. M. Kienlen, G. Knispel, J.A. Miehé, B. Sipp : Nucl. Instr. Methods 137 , 257 (1976)
6. U. Gösele : Chem. Phys. Lett. 43 , 61 (1976)
7. C.L. Braun, G.M. Dobbs : J. Chem. Phys. 53 , 2718 (1970)
8. V.L. Zima, A.N. Faidysh : Opt. Spectrosc. 20 , 566 (1966)
9. K. Uchida, M. Tomura : J. Phys. Soc. Japan 36 , 1358 (1974)

Intramolecular Lineshape Narrowing in the Overtone Spectrum of Benzene

D.F. Heller

Allied Chemical Corporation, Morristown, NJ 07960, USA

S. Mukamel

Department of Chemistry, University of California
Berkeley, CA 94720,USA

1. Introduction

Recently, BRAY and BERRY [1] have systematically explored the overtone spectra of gas phase benzene and certain deuterated benzene derivations (H5, H2, H1). They found that above 14000 cm^{-1} only absorptions assignable to CH local mode stretching overtones were evident and that these absorption features exhibited particularly striking lineshapes. Generally, all observed overtones were relatively structureless bands having nearly Lorentzian profiles. Even the very slight asymmetries of some of the lower overtones disappeared with increasing overtone excitation (v). The apparent linewidths were remarkably broad (~100 cm^{-1}, corresponding to dynamical processes having sub-picosecond timescales) and varied only slowly with overtone excitation or deuterium substitution. The overtone dependence of the linewidth was rather curious: the H1 linewidth increases with v but for all other species (H2, H5, H6) the linewidth narrows with v.

As benzene can be expected to be representative of other aromatic systems (and to some extent of other large polyatomic molecules in general), it provides an important test of our ideas about the nature and dynamics of highly excited vibrational states. We set ourselves the task of developing a theoretical framework for the evaluation of polyatomic overtone lineshapes [2]. Furthermore, we attempt to provide a model which can be used to interpret and predict overtone lineshape behavior. We address specifically the origin of the observed lineshape narrowing which bares some resemblence to the well known motional narrowing encountered in magnetic resonance.

2. Development of the Lineshape Formalism

Taking the absorbing overtone to be well represented as a single local (CH stretching) mode [2,3,4] |s> whose dynamics are governed by the molecular hamiltonian

$$H_M = H^{Local}(\underset{\sim}{R}) + H^{Normal}(\underset{\sim}{Q}) + V(\underset{\sim}{R},\underset{\sim}{Q}) \qquad (1)$$

where

$$H^{Local}(\underset{\sim}{R}) = \sum_i h_i^L(R_i)$$

$$H^{Normal}(\underset{\sim}{Q}) = \sum_k h_k^N(Q_k)$$

and

$$V(\underset{\sim}{R},\underset{\sim}{Q}) = \sum_{ij} C^{LL}_{ij}R_iR_j + \sum_{ik} C^{LN}_{ik}R_iQ_k + \text{h.o.t. in displacements} \tag{2}$$

specify the normal and local modes and their interactions, the optical theorem gives the absorption lineshape function

$$L(\omega) = \text{Im}\langle g|\mu G(\omega)\mu|g\rangle \tag{3a}$$

$$= |\mu_{gs}|^2 \text{Im}[G_{ss}(\omega)]. \tag{3b}$$

Using straightforward operator techniques, (3b) reduces to

$$L(\omega) = |\mu_{sg}|^2\ \Gamma_s(\omega)/\{[\omega - E_s + \Delta_s(\omega)]^2 - \Gamma_s^2(\omega)/4\} \tag{4}$$

where Δ_s and Γ_s are to be determined. Eq. (4) is effectively Lorentzian when Δ_s and Γ_s vary only very slowly with ω over the absorption feature $\omega \sim E_s \pm \Gamma_s(E_s)$. Typically, $\Delta_s \ll \Gamma_s$ and can be ignored and the line is centered at $\omega = E_s$. Thus, to evaluate the overtone lineshape we require only an explicit expression for $\Gamma_s(\omega)$.

3. Evaluation of $\Gamma_s(\omega)$ for Benzene

We wish to describe the interactions between a prepared local mode $|s\rangle = |v00000;\ \alpha\rangle$ and its energetic neighbor $|m\rangle = |v-1,10000;\ \beta\rangle$ where the explicit labels refer to local mode states and $\alpha(\beta)$ refer collectively to the remaining (normal) "bath" modes. These states are coupled by the leading term in V^{LN} in (2). The relevant coupling scheme is depicted in Fig. 1. The resultant (second order) expression for the linewidth is

$$\Gamma_s(\omega) = \sum_m |V_{sm}|^2 \Gamma_m/\{\ \omega - (E_m + \Delta_m)^2 + \Gamma_m^2/4\} \tag{5}$$

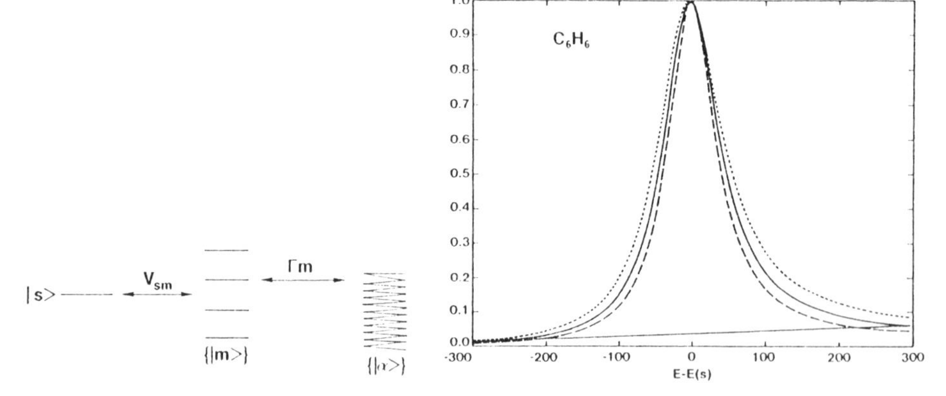

Fig. 1 Coupling schematic for the decay of local mode overtones.

Fig. 2 Calculated lineshape (6) for H6 overtones, ··· v=5, — v=6, ---v=7.

where $\Gamma_m = \sum_{m'} \Gamma_{mm'}$ depends only on the coupling between $|m\rangle$ and the bath states $|\alpha\rangle$ and is not expected to vary substantially with overtone quantum number v. For C_6H_6, one expects $\Gamma_m \gtrsim 5\ \Gamma_s(v-1) \sim 500\ cm^{-1}$. ($\Gamma_m$ may contain both relaxation T_1 and dephasing T_2 contributions.) When C-D stretches are involved, Γ_m should be even larger. Since there are only 5 (degenerate) modes of type $|m\rangle$ which can couple to $|s\rangle$ in lowest order (5) becomes

$$\Gamma_s(\omega) = 5|V_{sm}|^2\ \Gamma_m/[(\varepsilon + \Delta E_{sm})^2 + \Gamma_m^2/4] \tag{6}$$

where we have referenced our energies to line center ($\varepsilon = \omega - E_s$). To first order for a weakly anharmonic oscillator, the coupling matrix elements (2) are approximated by the harmonic result

$$|V_{sm}|^2 \approx |V_{sm}|^2_{v=1}\ \frac{v}{2}\ . \tag{7}$$

The local mode energy differences can be evaluated exactly

$$\Delta E_{sm} = 2(v-1)\omega eXe \text{ for H6} \tag{8a}$$

$$= 984-2v\omega eXe \text{ for H1} \tag{8b}$$

Thus, when $\Gamma_m/2 < \varepsilon + \Delta E_{sm}$ as is the case for H6 we see that $\Gamma_s(\omega) \sim v^{-1}$ while if $\Gamma_m/2 > \varepsilon + \Delta E_{sm}$ as may be the H1 case $\Gamma_s(\omega) \sim v$. The actual H6 lineshapes from (6) are given in Fig. 2 for the parameters $|V_{sm}|_{v=1} = 50$ cm^{-1}, $\Gamma_m = 500\ cm^{-1}$. The essentially quantitative agreement with experiment for these reasonable parametric values is encouraging. Furthermore, only one additional parameter $\Gamma_m(H1)$ is required for H1 since to first order $|V_{sm}|_{v=1}(H1) = \frac{1}{2}|V_{sm}|_{v=1}(H6)$. We find that $\Gamma_m(H1) \approx 1100\ cm^{-1}$ suffices. No further parameters are required for the other isotopic species since (5) implies

$$\Gamma_s^{C_6H_nD_{6-n}} = \frac{1}{5}[(n-1)\ \Gamma_s^{(H6)} + (6-n)\Gamma_s^{(H1)}] \tag{9}$$

A comparison of the observed and calculated widths is given in Table 1. (Calculated widths are given in parentheses).

Table 1 Benzene Overtone Linewidths

Overtone (v)	H1		H2		H5		H6	
4		(78)		(88)		(118)		(128)
5	86	(90)	111	(94)	105	(104)	109	(108)
6	107	(102)	91	(101)	97	(97)	94	(95)
7		(114)		(108)		(91)	87	(86)

4. Summary

We have shown that the observed benzene overtone lineshapes can be understood within a reasonably simple framework. The Lorentzian profile and narrowing (or broadening) behavior are natural consequences of the relatively weak coupling between the prepared local mode state and the remaining molecular vibrations. Excitation "flows" from the excited local mode state into other nearby local modes and finally on a possibly much longer timescale into the "bath" of other molecular vibrations. The lineshape includes both T_1 and T_2 (relaxation and dephasing) contributions which cannot be deconvoluted without additional information. Thus, it is entirely possible that the prepared overtone lives much longer than the inverse linewidth would suggest. The gas phase widths do permit some insight into the liquid phase results of SWOFFORD ET AL. [5]. Apparently most (~200 cm^{-1}) of the observed liquid linewidth must be due to intermolecular interactions on the sub-picosecond timescale.

References

1. R. G. Bray and M. J. Berry (submitted to J. Chem. Phys.).

2. A more comprehensive discussion is given by the authors in D. F. Heller and S. Mukamel (submitted to J. Chem. Phys.).

3. B. Henry and W. Siebrand, J. Chem. Phys., 49, 5369 (1968).

4. D. F. Heller, "On the Preparation of Local Modes" to be published.

5. R. L. Swofford, M. E. Long, and A. C. Albrecht, J. Chem. Phys., 65, 179 (1976).

Kinetics of the Sensitized Fluorescence of Anthracene-Doped Naphthalene Following Two-Photon Picosecond Excitation

H. Auweter, U. Mayer, and D. Schmid

Physikalisches Institut, Teil 3, Universität Stuttgart
D-7000 Stuttgart, Fed. Rep. of Germany

1. Introduction

Energy transfer processes in organic molecular crystals have been investigated quite extensively by studying the sensitized fluorescence of doped crystals [1]. It is evident, for instance, that the time dependence of the fluorescence of the guest molecules following the excitation of the host crystal with a short light pulse yields direct information about the rate of energy transfer from the host to the guests. The question if the energy transfer rate is time dependent or not is not yet answered unambiguously.

In order to study the characteristics of the energy transfer rate we investigated the dynamics of the sensitized fluorescence of anthracene-doped naphthalene crystals. The crystals were excited by a single picosecond pulse from a modelocked dye laser, inducing a two-photon transition to the first singlet exciton band of the host crystal. The advantage of this excitation is a truely homogeneous bulk excitation. The resulting exciton density is below 10^{15} cm^{-3} and thus bimolecular exciton-exciton annihilation processes are clearly ruled out.

2. Experimental Results

The time-resolved sensitized fluorescence of the anthracene-doped naphthalene crystals was measured in the concentration range from 4.9×10^{15} cm^{-3} to 2.1×10^{17} cm^{-3} and in the temperature range from 3.8 K to 300 K.

The decay of the host fluorescence as well as the rise and the decay of the guest fluorescence depend strongly on the doping concentration. At high concentrations ($N_A > 5\times10^{16}$ cm^{-3}) the decay of the host fluorescence is purely exponential, (Fig.1b). At low concentrations ($N_A < 6\times10^{15}$ cm^{-3}) pronounced deviations from the purely exponential decay were observed at short times after the excitation, (Fig.1d). The rise of the guest fluorescence was exponential at high concentrations (Fig.1a), whereas at low concentrations the rise was much faster than exponential, (Fig.1c).

From these results it is evident that the energy transfer rate is time independent in the high concentration range, whereas it is time dependent at low concentrations [2].

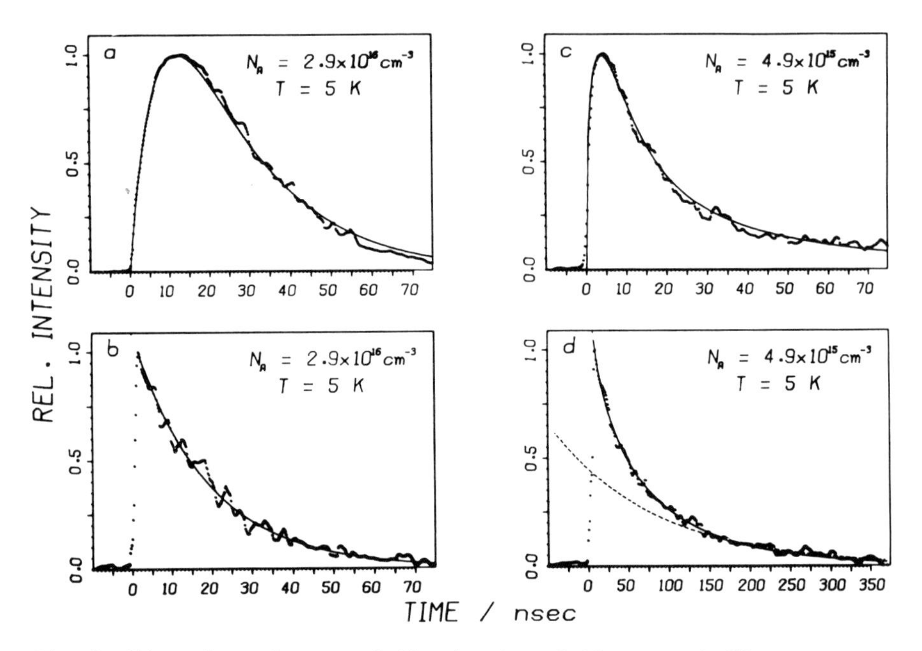

Fig.1 Time dependence of the host and the guest fluorescence for a highly doped sample (curves a and b) and a low concentration sample (curves c and d). The full lines are calculated using (1) and (2). Note the different time scale for figure d. In addition in Fig.1d an exponential curve (dashed curve) is plotted in order to show the nonexponential decay for the sample with low guest concentration.

The time dependence of the host fluorescence was strongly temperature dependent in the highly doped samples: The energy-transfer rate increased by a factor of 50 if the temperature was lowered from room temperature to 3.8 K. In the lightly doped samples on the other hand the transfer rate was almost independent on the temperature [3].

In order to evaluate the dynamics of both, the host and the guest fluorescence quantitatively, we used the following rate equations to describe the time evolution of the excited state-population numbers of the host and the guest molecules, n_H and n_G respectively following a δ-shaped excitation pulse:

$$\dot{n}_H = - k_H^O n_H - k_{HG} n_H \tag{1}$$

$$\dot{n}_G = - k_G n_G + k_{HG} n_H \tag{2}$$

Using a diffusion model for the exciton migration the energy transfer rate k_{HG} can be calculated as

$$k_{HG} = 4\pi DRN_A(1 + R(\pi Dt)^{-1/2}) = a + bt^{-1/2} \tag{3}$$

(D = diffusion constant, R = capture radius of the guest molecule, k_H^O, k_G = reciprocal lifetimes of host and guest, respectively, N_A = anthracene concentration.)

Eqs.(1) and (2) have been solved [4] and the solutions can be fitted to the experimentally observed time dependencies of host and guest fluorescence. This fitting procedure yields the parameters a and b defined by (3), which in turn give the diffusion constant D and the capture radius R independently. Some typical results are compiled in Table 1.

Table 1 Fitting parameters a and b and secondary results thereof at T = 30 K ($L = \sqrt{2D\tau}$ = diffusion length, τ = exciton lifetime)

N_A $[cm^{-3}]$	a $[s^{-1}]$	b $[sec^{-1/2}]$	D $[cm^2 s^{-1}]$	L [Å]	R [Å]	$d=N_A^{-1/3}$ [Å]	comments
2.1×10^{17}	7.5×10^{7}	$<1.8\times10^{1}$	$>1.7\times10^{-3}$	>627	<2	169	hopping motion
2.9×10^{16}	2.4×10^{7}	7.4×10^{1}	5.3×10^{-4}	565	13	327	Diffusion bottleneck
6.2×10^{15}	1.8×10^{6}	1.6×10^{2}	2.7×10^{-5}	214	83	544	
4.9×10^{15}	2.0×10^{6}	1.8×10^{3}	7.3×10^{-6}	127	439	588	Long range transfer

3. Discussion

The essential result of this work is the observation that the energy transfer rate k_{HG} depends strongly on the guest concentration. At medium concentrations both, a and b, are relevant, indicating the existence of a diffusion bottleneck. This is reasonable since the diffusion length L is less than or in the order of the average distance d between the guest molecules.

At high concentrations, we observe b = 0 and thus the energy transfer rate is time independent. We conclude that in this case the energy transfer is not diffusion limited, but that it is determined by the trapping mechanism at the guest site.

At low concentrations the energy transfer rate is governed by the coefficient b. The estimated diffusion length is significantly smaller than the observed trapping radius R. Thus it seems that the energy transfer in this case is controlled by a long range capture mechanism.

The temperature dependence can be described in terms of phonon relaxation of the excitons into shallow traps in the neighbourhood of the guest molecules and phonon assisted detrapping[3].

In summarizing our results we want to point out that a clear understanding of the mechanism of exciton migration and the mechanism of exciton trapping can only be achieved in studying the time-resolved fluorescence in a large range of concentration and temperature. In view of our results, it seems possible to settle the controversy whether the energy transfer rate is time dependent [1] or time independent [5].

This work was supported by the Deutsche Forschungsgemeinschaft, SFB 67.

References

1 R.C.Powell and Z.G.Soos, J.Luminescence 11, 1-45 (1975)
2 H.Auweter, U.Mayer and D.Schmid, Z.Naturforsch. in print
3 H.Auweter, Dissertation, Universität Stuttgart 1978
4 J.Heber, phys.stat.sol. b48, 319-326 (1971)
5 A.J.Campillo, S.L.Shapiro and C.E.Swenberg, Chem.Phys.Lett. 52, 11-15 (1977)

Picosecond Gain Kinetics of 1,4-Bis (2-Methylstyril) Benzene[1]

D.E. Damschen, J.R. Richards, G.W. Scott, and L.D. Talley

Department of Chemistry, University of California, Riverside
Riverside, CA 92521, USA

A.J. Cox

Department of Physics, University of Redlands
Redlands, CA 92373, USA

1. Introduction

The production of picosecond pulses in the blue region of the spectrum [1,2] has stimulated our interest in methods of amplifying such pulses. The photophysics and photochemistry of stilbene laser dyes have not previously been studied by picosecond spectroscopy. We report the gain kinetics and spectra of all trans-1,4-Bis(2-methylstyryl)benzene (Bis-MSB). This laser dye apparently undergoes very little excited-state conversion to a cis-form. Indeed, in related molecules the all trans-form is favored by photoisomerization [3]. Recently, picosecond kinetics of photoisomerization in cis- and trans-stilbene were reported [4]. The gain kinetics reported in this paper are discussed in terms of the excited state conformation, vibrational level, orientation, and population.

2. Spectra

Absorption and fluorescence spectra of Bis-MSB in p-dioxane at room temperature are shown in Fig. 1. The vibronic structure, present in the fluorescence, is absent in the room temperature absorption. This is also true for Bis-MSB in cyclohexane solution for which a fluorescence quantum yield of 0.94 has been reported [5]. At 77 K in methylcyclohexane, we observe vibronic structure in the absorption spectrum. The lack of mirror image symmetry in the room temperature absorption and fluorescence spectra indicates the possibility of a conformational change in the excited state which is rapid compared to the fluorescence lifetime. The presence of a forbidden state below the long axis polarized, lowest energy allowed state is considered an unlikely explanation of these observations in view of the short fluorescence decay time and high quantum yield. (See below.)

3. Gain Kinetics

We have measured gain kinetics at 422 nm for Bis-MSB (2.3×10^{-4}M dioxane solution) following excitation at 355 nm using the apparatus shown in Fig.2. A delayed probe pulse at 422 nm

[1]This research was supported in part by the National Science Foundation, the Research Corporation, and the Committee on Research of the University of California, Riverside.

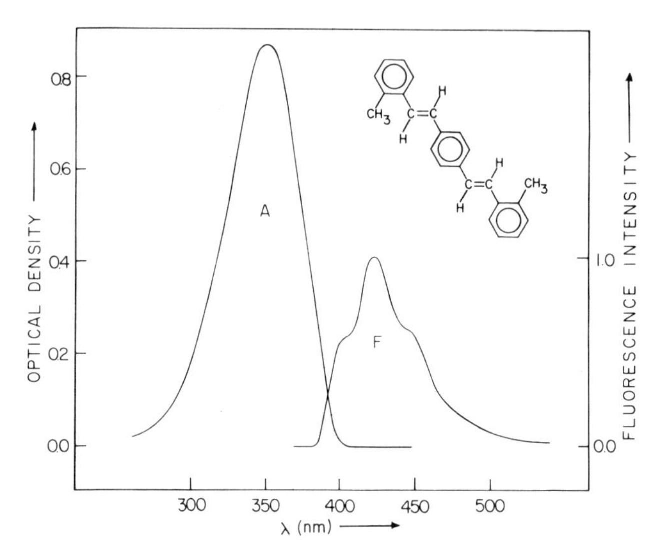

Fig.1 Absorption (A) and fluorescence (F) (uncorrected for instrument response) of Bis-MSB (1.5×10^{-4}M in dioxane) in a 1 mm cell

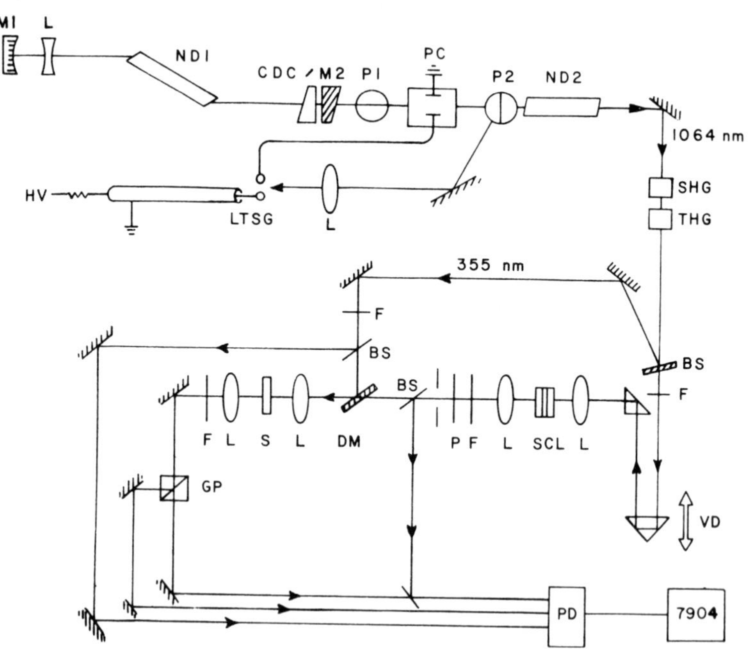

Fig.2 The experimental apparatus

($\Delta\lambda$ ~ 8 nm, Δt = 3 ps) was generated in a short cavity dye laser (SCL) [1,2]. The Bis-MSB sample (S), in a 5 mm cell, was excited with a 355 nm pulse (Δt = 8 ps), suitably attenuated to avoid amplified spontaneous emission. The probe pulse was polarized at $\pi/4$ to the excitation pulse, while $\parallel$ and $\perp$ polarized pulses were separated after the sample by a glan polarizer (GP) for independent detection.

The gain results for both polarizations are shown in Fig. 3. The experimentally determined gains, shown in Fig.3, were calculated from

$$G_{\parallel}(t) = \log \frac{I_{\parallel}(t)}{I_o} \; ; \; G_{\perp}(t) = \log \frac{I_{\perp}(t)}{I_o} \, , \tag{1}$$

where I_o is the incident probe intensity and $I_{\parallel}(t)$ and $I_{\perp}(t)$ are pulse intensities after the sample at delay, t.

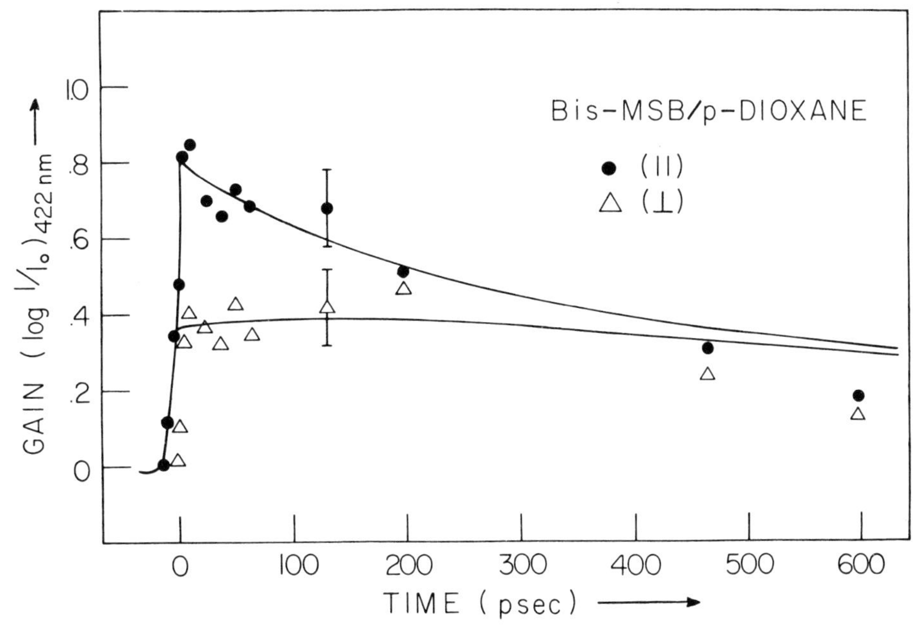

Fig.3 Picosecond gain kinetics at 422 nm of Bis-MSB in dioxane following excitation at 355 nm (The smooth curves are the calculated $G_{\parallel}(t)$ and $G_{\perp}(t)$ using τ = 1.14 ns and k_{or}^{-1} = 215 ps as discussed in the text.)

4. Discussion

The experimental results were analyzed by calculating the following functions:

$$K(t) = G_{\parallel}(t) + 2G_{\perp}(t); \quad r(t) = \frac{G_{\parallel}(t) - G_{\perp}(t)}{G_{\parallel}(t) + 2G_{\perp}(t)} \, , \tag{2}$$

where r(t) is the orientation decay function and K(t) is the excited state decay function, analogous to functions used for fluorescence decay analysis [6]. Assuming r(t) and K(t) are exponentials,

$$r(t) = r_o e^{-k_{or} t} \text{ and } K(t) = K_o e^{-t/\tau}. \quad (3)$$

Decay times were obtained by a least squares analysis. The short delay time data shown in Fig.3 yields k_{or}^{-1} = 215 ± 210 ps. Data in the 0 - 3 ns range gives τ = 1.14 ± 0.46 ns. Measurements of the fluorescence decay, using a biplanar photodiode and Tektronix 7904 oscilloscope (500 MHz bandwidth), yield τ_{fl} = 1.37 ± 0.23 ns. A literature value is τ_{fl} = 1.35 ns [5]. Thus K(t) probably measures the excited state lifetime, although the experimental decay may be nonexponential.

These measurements on Bis-MSB indicate a fast buildup at 422 nm of ≲5 ps. Possible molecular explanations for a finite buildup time of the gain include (1) "slow" vibrational relaxation in S_1, (2) an excited state conformational change such as slight rotation about the double bonds leading to a less planar excited state, or (3) an increase in planarity in the excited state due to relief of steric hindrance of the methyls with the ethylenic hydrogens after lengthening of the double bonds. Thus, although a rapid conformational change may be occuring in Bis-MSB, the data provide no definite evidence for cis-trans photoisomerization. The observed orientation rate presumably measures rotation about an axis to the long molecular axis. (Loss of gain anisotropy by energy transfer should not be important at the low concentration used.) This rate is comparable to that reported in similar viscosity solutions for ionic, oblate top molecules of a similar size but different shape [6]. The neutral aromatic hydrocarbon, Bis-MSB, is approximately a prolate top.

References

[1]A. J. Cox, G. W. Scott, and L. D. Talley, Appl. Phys. Letts. 31, 389 (1977).

[2]G. W. Scott and A. J. Cox, Advances in Laser Spectroscopy I, A. H. Zewail, ed., Proc. Soc. Photo-Opt. Instr. Eng. 113, 25 (1977).

[3]S. Misumi, M. Kuwana, and N. Nagakura, Bull. Chem. Soc. Japan 35, 143 (1962).

[4]O. Teshke, E. P. Ippen, and G. R. Holtom, Chem. Phys. Letts. 52, 233 (1977).

[5]I. B. Berlman, Handbook of Fluorescence Spectra of Aromatic Molecules, 2nd ed. (Academic Press, N.Y., 1971), p. 327.

[6]G. R. Fleming, J. M. Morris, and G. W. Robinson, Chem. Phys. 17, 91 (1976).

Evidence for Nearly Transform-Limited Pulses from a Short Cavity Blue Dye Laser[1]

A.J. Cox

Department of Physics, University of Redlands
Redlands, CA 92373, USA

D.E. Damschen, C.D. Merritt, G.W. Scott, and L.D. Talley

Department of Chemistry, University of California, Riverside
Riverside, CA 92521, USA

1. Introduction

We report a short cavity dye laser which generates picosecond pulses shorter than the uv pumping pulse. An extremely short laser cavity length (0-500 μm) and an average mirror reflectivity of 0.6 result in a short photon cavity lifetime, τ_c. Pulse shortening is predicted by the laser rate equations for photon cavity lifetimes shorter than the pumping pulse [1], [2].

2. The Dye Laser

The laser consists of two mirrors ($M_1{:}R_1 \simeq .95$; $M_2{:}R_2 \simeq .40$ at 410 to 460 nm). This optical cavity is filled with dye solution in contact with the mirrors. At a typical cavity length of 100 μm, τ_c = 1.2 ps. The short cavity dye laser is axially pumped through M_1 with a focused third harmonic (355nm) single pulse from a modelocked Nd^{+3}:glass laser. The coating of M_1 was designed to transmit >85% of the uv pump pulse.

3. Spectral Output

The short cavity dye laser output consists of a series of evenly spaced axial modes which extend over a range of ~10 nm (depending on the dye) near the center of the dye gain curve. Figure 1 shows a densitometer trace of a spectrographic record of typical multimode laser output. The film record was obtained using a 1 m spectrograph with 0.020 nm resolution. The dye laser cavity length was 79 μm, and it contained a 3.2×10^{-3}M Bis-MSB in dioxane solution. For this recording the free spectral range (FSR) was 0.79 nm, the representative linewidth ($\Delta\lambda$) was 0.09 nm, and the bandwidth was 7.1 nm, centered at 421.9 nm. Other dyes which we have investigated include α-NPO (408 nm center), Coumarin 120 (440 nm), Coumarin 1 (455 nm) and Coumarin 102 (462 nm), all in an ethanol solvent.

---

[1]This research was supported by the National Science Foundation, the Research Corporation, and the Committee on Research, University of California, Riverside.

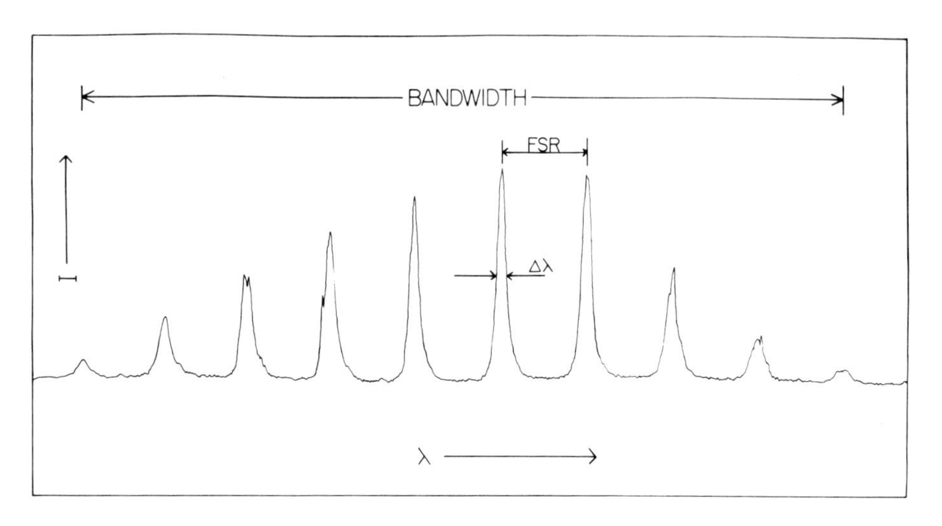

Fig. 1 Spectral output of the short cavity dye laser

4. Pulse Duration

Determination of the dye laser pulse duration was accomplished by utilizing the buildup of an excited state absorption ($S_n \leftarrow S_1$) of coronene (toluene solution) subsequent to excitation of the S_2 state of coronene at 355 nm. The experimental procedure was as follows: A single third harmonic pulse from a modelocked Nd^{+3}:glass laser (λ_o = 355 nm, E ~ 0.5 mJ, Δt = 8 ps FWHM) was divided into two parts with a beam splitter. The major fraction (~75%) was used to excite the $S_2 \leftarrow S_0$ absorption of coronene in solution, contained in a 2 mm cell. The minor fraction was delayed and focused into the short cavity dye laser containing Bis-MSB. The 422 nm output ($\Delta\lambda$ = 8 nm) of the dye laser was used to probe the $S_n \leftarrow S_1$ excited state absorption of coronene. The excited state buildup kinetics were measured in a multishot experiment by varying the delay of the blue dye laser probe pulses. Figure 2 shows this buildup which depends on the shape and duration of both the uv pump pulse and the blue probe pulse as well as the $S_2 \rightsquigarrow S_1$ relaxation time of coronene. This latter time is known to be 2 ps [3]. The experimental points, ♦, represent an average of several shots with error bars giving the standard deviation in the average.

The uv pump pulse duration was Δt = 8 ps (FHWM) as determined in other experiments. For computational purposes, the pulses were assumed to be Gaussian, and computer calculations were done to determine the dye laser probe pulse length which gave a best fit to the observed overall absorption buildup.

The resulting short cavity dye laser blue pulse durations were Δt = 2 ps for a 79 μm cavity (these data are shown in Fig. 2),

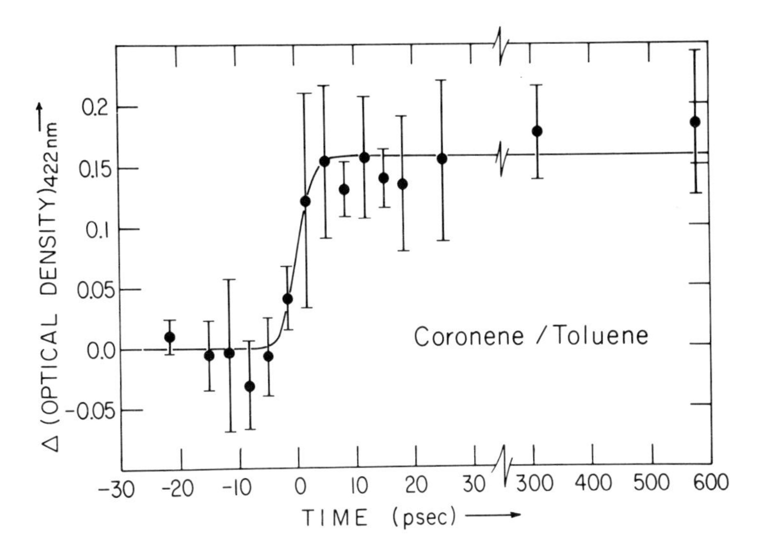

Fig. 2 Optical densities at 422 nm due to excited state absorption of coronene in toluene resulting from singlet state excitation at 355 nm

and Δt = 4 ps for a 305 μm cavity. (Experiments using probes polarized both parallel and perpendicular to the polarization of the uv pump pulse gave identical kinetics.) These measured dye laser pulse lengths may be broadened due to jitter in the time of the dye laser output pulse relative to the arrival of the dye laser pump pulse. But the measured dye laser pulse durations agree with the rate equation predictions for this laser [2]. Preliminary single shot streak camera measurements of pulse duration are consistent with the multishot results.[1] The streak camera pulse showed no background or tail.

5. Time-Modewidth Products

The measured modewidths, $\Delta\lambda$, of individual modes were 0.09 nm to 0.12 nm (79 μm cavity) and 0.07 nm to 0.12 nm (305 μm cavity). The dye laser is not itself modelocked, so each mode may be associated with a different temporal pulse. However, all pulses overlap in time and are simultaneously measured by the excited state absorption buildup. Using 0.1 nm as a representative modewidth, $\Delta\lambda$, for the 79 μm cavity yields a frequency width, $\Delta\nu$, of 1.7×10^{11} Hz at 422 nm, and a time-modewidth product, $\Delta\nu\Delta t = 0.34$. For the 305 μm cavity, and a representative $\Delta\lambda = 0.09$ nm, $\Delta\nu\Delta t = 0.61$. A Fourier transform limited Gaussian pulse would give $\Delta\nu\Delta t = 0.44$. Other workers have investigated a rhodamine B short cavity dye laser, lasing between

---

1. We wish to thank Prof. G. W. Robinson for the use of his streak camera facilities for these measurements.

600 nm and 630 nm [4]. They reported pulse durations similar to those reported here, but wider modewidths resulted in a larger time-modewidth product of 1.25.

References

[1] D. Roess, J. Appl. Phys. 37, 2004 (1966).

[2] G. W. Scott and A. J. Cox, Adv. in Laser Spectroscopy I, A. H. Zewail, ed., Proc. Soc. Photo-Opt. Instr. Eng. 113, 25 (1977).

[3] C. V. Shank, E. P. Ippen, and O. Teshke, Chem. Phys. Letts. 45, 291 (1977).

[4] B. Fan and T. K. Gustafson, App. Phys. Letts. 28, 202 (1976).

Applications of the Mode Locked Near UV Output from a Krypton Ion Laser in Chemical Analysis [1]

L.L. Steinmetz, J.H. Richardson, B.W. Wallin, and W.A. Bookless

University of California, Lawrence Livermore Laboratory
Livermore, CA 94550, USA

The ever increasing demand for spectroscopic information on a subnanosecond time scale has motivated the development of picosecond sources. The two principle requirements for any excitation source are tunability and repetition rate. Many applications in chemistry and biology require tunability in order to optimize sample excitation and distinguish mixtures. The use of a picosecond excitation source greatly augments the number of compounds which can be studied by time resolved fluorescence and hence increases the ability to use the powerful techniques developed for fluorescence in mixture analysis.

A high repetition rate is desirable from the standpoint of signal-to-noise (signal averaging) and sample turn-around time. A high repetition rate implies low energy per pulse, hence the avoidance of both sample decomposition and many non-linear and saturation effects which perturb the system and distort the phenomena being measured.

Two types of laser systems satisfy the above requirements for chemical analysis: a passively mode-locked dye laser [2,3] and a synchronously pumped dye laser. [4] These systems generally use the 514 nm Ar^+ line, and hence use rhodamines in the dye laser. Consequently, the output consists of picosecond pulses (less than 100 ps), tunable over the orange-red region of the spectrum, and repetition rates variable up to the megahertz range with acoustooptic output coupling. Frequency doubling accesses a limited portion of the ultra-violet (∿270-350 nm), but there is still no good picosecond source for the near uv and blue-green region of the spectrum. The lack of such a source prevents the undertaking of many interesting spectroscopic investigations.

As a first step in providing a tunable source for this region we have mode-locked the near uv lines (406.7, 413.1 nm) of the krypton ion laser. The maximum cw power for the 406.7 nm line is 400 mW. The average power when mode-locked is 120 mW, corresponding to a 70% reduction. The mode-locked train consists of pulses with 12 ns separation whose width (FWHM) was electronically measured to be 80 ps (detection limited). The linewidth of the 406.7 nm line was measured as 0.082 Å (15 GHz) using a 1 m monochromator. This linewidth has a Fourier transform of 30 ps, for a gaussian pulse. A high speed streak camera trace is shown in Fig. 1 where the pulse width is 55 psec thus $\Delta\nu\Delta t$ = 0.8. Similarly, the 413.1 nm line has been mode-locked yielding 400 mW average power. Output cw power is 1.3 watts. In the mode-locked condition the linewidth at 413.1 nm is 0.058 Å (10 GHz). We estimate the 413.1 nm pulsewidth to be about 80 psec using $\Delta\nu\Delta t$ = .8 and the widened oscilloscope display of 95 psec (detection limited). We presume the broad lasing linewidth is due to isotopic broadening [5] from several naturally abundant krypton isotopes.

One significant application of the mode-locked krypton laser is as a source for synchronously pumping a tunable dye laser. While the technology for synchronous pumping is not new, the new wavelength of 406.7 and 413.1 nm for

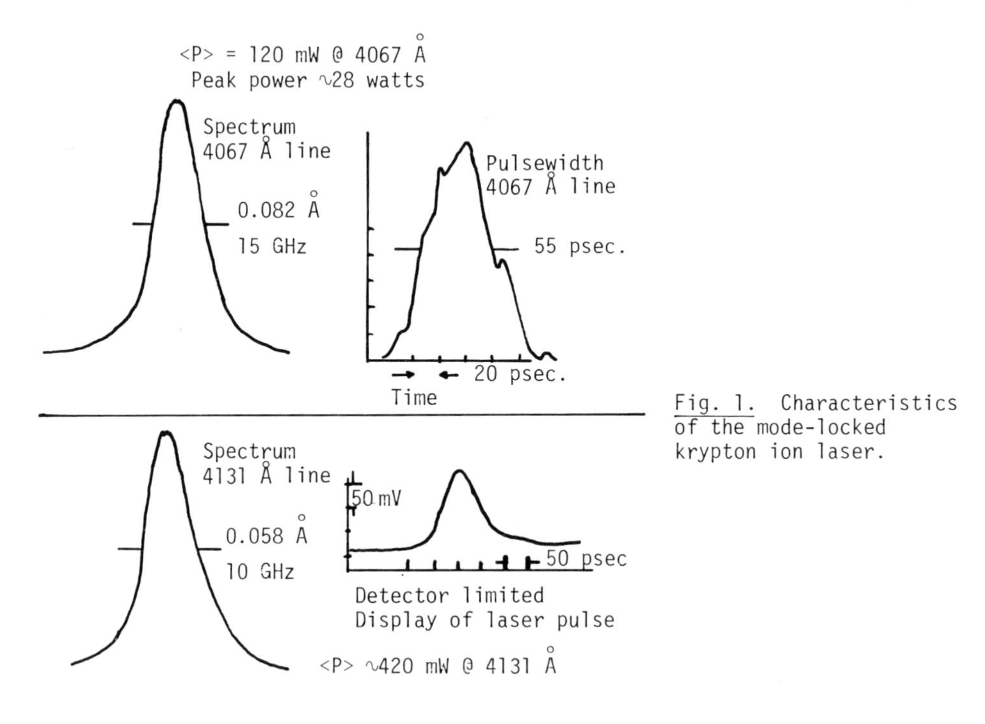

Fig. 1. Characteristics of the mode-locked krypton ion laser.

pumping reported here extend the tuning range of picosecond, continuous pulse trains into the blue and green region of the spectrum. This accesses a large number of absorbing species for analytic studies. With this interest we report the synchronous pumping and mode locking of 3 coumarin dyes (C-30, C-102, C-7) previously used for cw operation. Their tuning curves are presented as examples of the tuning capabilities, although no attempt was made to optimize mirror transmission spectrum. Not only is the violet pump wavelength a key property but the pump pulsewidth will allow generation of subpicosecond dye laser pulses [6] provided the tuning element has sufficient bandwidth (4 Å in our laser). Useful mode-locked output power can be achieved over a tuning range of 800 Å using these 3 dyes. A striking feature is the broad tuning capability of C-30 (over 600 Å). We have insufficient operating time on the dyes to determine whether the reduced average pump power in the mode-locked condition will extend the dye operating lifetime. All of the tuning curves were generated using mode-locked 4131 Å light as the pump. With the 4 Å bandwidth tuning element in place, lasing threshold requires 120 mW of pump power except for C-102 which required 380 mW. Dye laser peak efficiencies for C-102 is 2%, C-30 is 8% and 7%, and C-7 is 10% and 9% with the second figure representing the efficiency using green optics.

We report also the use of the mode-locked Kr^+ laser as an excitation source for measuring fluorescence decay times of the fluorescent cytochemical probe, chromomycin. Chromomycin has been shown [7] to be specific to DNA in cellular staining and the staining procedure is quite simple. Fluorescence yield and lifetime show sensitivity to local environment. The 406.7 nm Kr^+ laser line is a good match to the absorption peak of both free (410 nm) and DNA bound (425 nm) chromomycin. Fluorescence decay time for the stain in TMS solution (0.1 M NaCl, 1.5 mM $MgCl_2$, and 0.05 M TRIS @ ph 7.5) is 13 ± 1 nsec., 19 ± 1 nsec. in water and 3.2 ± 1 nsec. when bound to naked

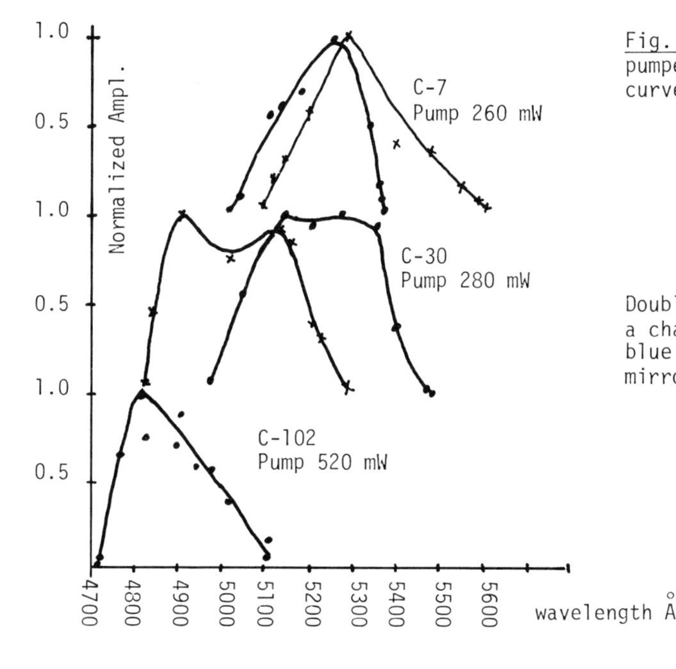

Fig. 2. Synchronously pumped dye laser tuning curves.

Double curves result from a change in optics from blue mirrors to green mirrors.

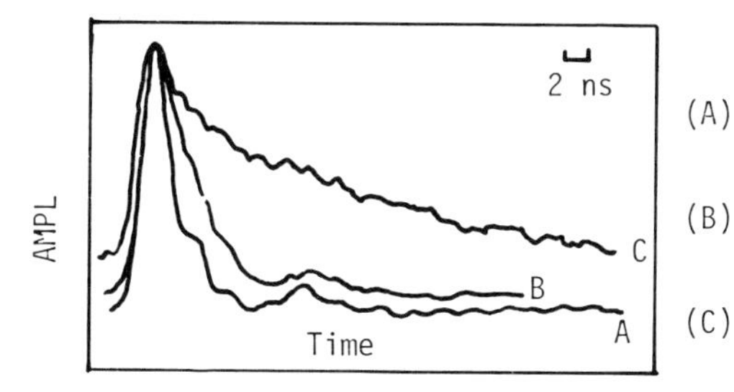

(A) Rayleigh scatter from Ludox solution.

(B) Fluorescence decay from DNA bound chromomycin A_3 in TMS.

(C) Fluorescence decay from 10^{-5} M chromomycin A_3 in TMS.

Fig. 3. Time resolved fluorescence signal using mode-locked krypton laser as an excitation source.

DNA. The reduction in decay time for DNA bound stain is at present unexplained. However, the demonstrated change in decay time represents a significant parameter for determining the ratio of bound to free stain in cell or chromosome suspensions for flow cytometry.

The krypton laser has also been used to measure the fluorescence lifetime of several coumarin laser dyes. A notable feature of the measurements is a constancy of fluorescence lifetime in water with concentration changes from 10^{-3} M to 10^{-6} M. This result suggests the absence of significant molecular aggregation. [8]

Finally, the mode-locked violet lines of the krypton laser are inherently useful or can be used to generate other wavelengths by synchronous pumping. The short temporal width of the violet lines makes this a source well suited to producing tunable picosecond pulses using coumarin dyes as the lasing medium. Time resolved spectroscopic investigations of previously unstudied blue-green absorbing compounds can now be undertaken.

References

1. Work performed under the auspices of the U.S. Department of Energy by the Lawrence Livermore Laboratory under contract number W-7405-ENG-48.

2. E.P. Ippen, C.V. Shank and A. Dienes, Appl. Phys. Lett. 21, 348 (1972).

3. Z.A. Yasa and O. Teschke, Appl. Phys. Lett. 27, 446 (1975).

4. J.M. Harris, L.M. Gray, M.J. Pelletier and F.E. Lytle, Mol. Photochem. 8, 161 (1977).

5. Private communication with D. Wright of Spectra Physics.

6. J.P. Heritage and R.K. Jain, Appl. Phys. Lett. 32, 101 (1978).

7. R.H. Jensen, J. Histochemistry and Cytochemistry 25, 573 (1977).

8. J. Knof, F.J. Theiss and J. Weber, Z. Naturforsch, 33A, 98 (1978).

Amplified Spontaneous Emission in Rhodamine Dyes

A. Penzkofer

Fachbereich Physik der Universität Regensburg
D-8400 Regensburg, Fed. Rep. of Germany

W. Falkenstein

Physik-Department der Technischen Universiät München
D-8000 München, Fed. Rep. of Germany

Introduction

The excitation of dye solutions with intense short laser pulses results in an amplification of the fluorescence light by stimulated emission [1,2,3]. Here, the amplification of spontaneous emission (ASE) in rhodamine dyes is investigated with intense picosecond light pulses [4,5]. The purpose of the investigations is i) the generation of intense picosecond fluorescence emission, ii) the analysis of the effects of the various pump laser and dye parameters on the amplified spontaneous emission, and iii) the determination of molecular parameters, especially excited state absorption cross-sections and relaxation rates from highly excited singlet states.

Experiments

In our experiments, we worked with single picosecond light pulses at $\lambda = 530$ nm (second harmonic of a mode-locked Nd-glass laser, duration $\Delta t_L \simeq 4$ps; width $\Delta\tilde{\nu}_L \simeq 5\text{cm}^{-1}$). Rhodamine 6G and Rhodamine B dissolved in ethanol at various concentrations were investigated:

i) The increase of fluorescence energy within a fixed solid angle $\Delta\Omega = 1.4\times10^{-3}$ sr in forward direction is depicted in Fig. 1a for 10^{-4} and 2×10^{-4} M rhodamine 6G. An increase by a factor of 300 is observed.

ii) The amplification is accompanied by a fluorescence shortening as shown in Fig.1b. The fluorescence durations (FWHW) reduce from $\Delta t_F \simeq 3$ ns to $\Delta t_F \simeq 20$ ps (streak camera measurements).

iii) The divergence of the amplified fluorescence light was studied with apertures of varying size. The emission angle depends on the penetration depth of the pump pulse (dye concentration) and reduces with increasing input intensity. At high intensities $\Delta\Omega \simeq 2.5\times10^{-3}$ sr was observed for 10^{-4} molar solutions.

iv) The amplification is wavelength dependent according to the spectral distribution of the spontaneous emission. The width of the emission is reduced from 35 nm to 10 nm at high intensities (10^{-4} molar solutions).

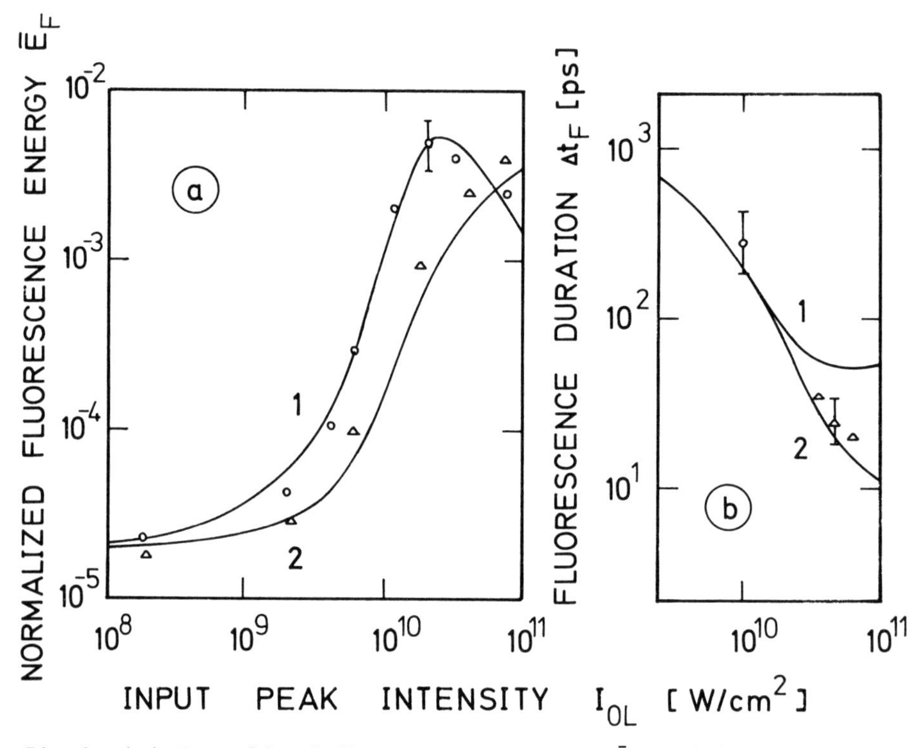

Fig.1 (a) Normalized fluorescence energy $\bar{E}_F$. (b) Fluorescence duration. Substance: Rhodamine 6G; ℓ = 1.13 cm. Curves 1 (open and closed circles), c = 10^{-4} M; Curves 2 (triangles), 2×10^{-4} M.

v) The absorption and emission bands of the dyes overlap and the spontaneous emission is shifted to longer wave lengths at high dye concentrations. In the excitation process the ground state absorption is bleached and the fluorescence emission shifts towards the wavelength of maximal stimulated emission-cross-section.

vi) The intensity dependent energy transmission indicates the drastic bleaching of the dyes (see Fig.3).

Discussion

The effects of the various laser and dye parameters were studied extensively [5]. The molecules were described by a multilevel system (see Fig.2). In brief, the involved states are: 1,4, levels in S_0; 2,3,5, levels in S_1; 6, S_n-state; 8, triplet or decomposition state. The solid arrows indicate absorptions and emission (σ), the dashed arrows illustrate relaxations (k,τ).

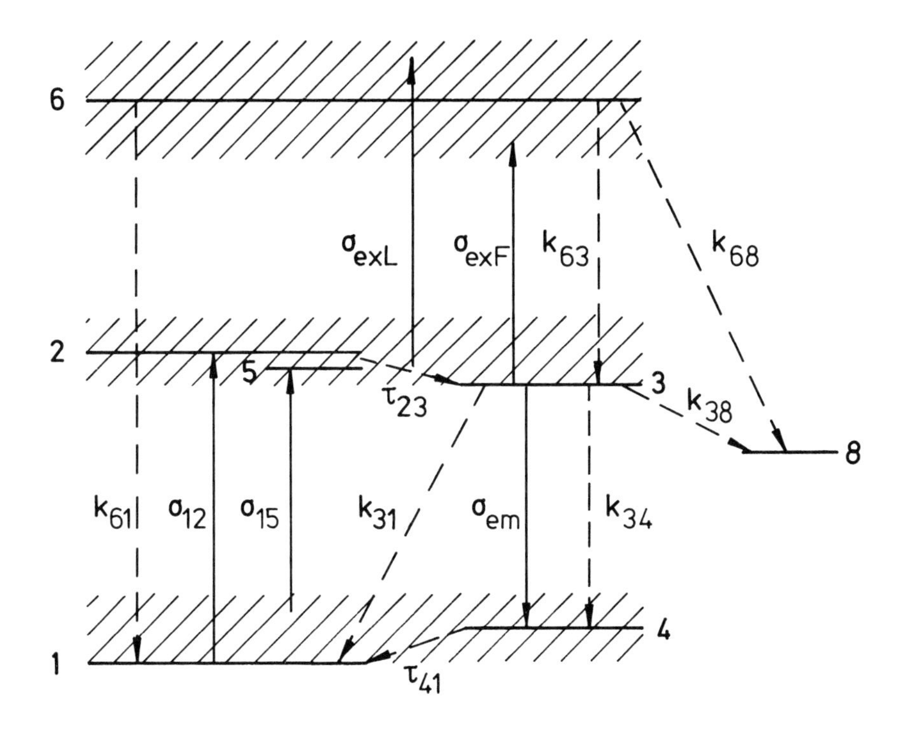

Fig.2 Multi-level system for rhodamine molecules

Roughly speaking, the ASE is proportional to $\exp\{\int_0^{\ell}[(N_3-N_4)\sigma_{em}-(N_3-N_6)\sigma_{ex}]dx\}$. This expression demonstrates the dependence on excitation (N_3-N_4; effected by dye concentration and absorbed pump energy), on penetration depths of pump laser $\ell_{eff} \leq \ell$, on emission cross-section σ_{em}, excited state absorption cross-section σ_{ex} and on the S_n-population N_6.

The dependence of the ASE and the energy transmission T_E on the excited state absorption cross-sections and relaxation rates was applied to determine these quantities (curves in Fig.1 and 3 are calculated).

i) The energy transmission depends on σ_{exL} and k_{61}, k_{63}, k_{68} (sum k_6). The ratios k_{63}/k_6 and k_{68}/k_6 are determined below. The remaining dependence on σ_{exL} and k_6 allows to determine these values by T_E-measurements for two different pulse durations (see Fig.3).

ii) The ratio k_{63}/k_6 is obtained by S_1-S_0 fluorescence quantum yield measurements after excitation at λ = 530 nm and 265 nm. (Equal yields indicate that all molecules relax to S_1 state).

iii) The fluorescence amplification at high pump intensities

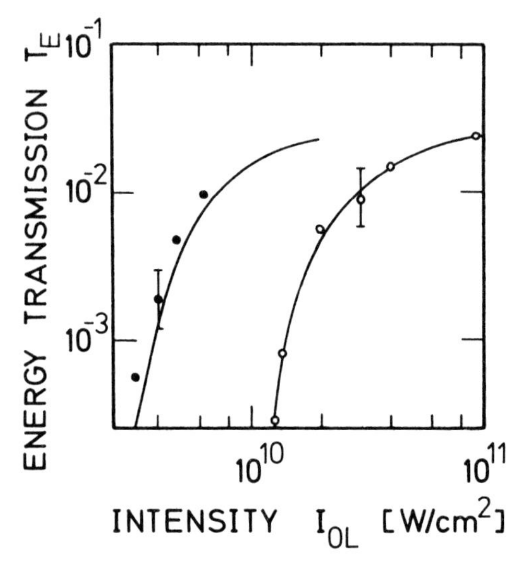

Fig.3 Energy transmission T_E of 10^{-4} M rhodamine 6G. Curve 1, Δt_L = 20 ps; 2, Δt_L = 4 ps.

is strongly dependent on k_{68}/k_6. Molecules transfered to level 8 are lost for fluorescence. The decrease of $\bar{E}_F$ at $I_{OL} > 3\times10^{10}$ W/cm^2 (see Fig.1a) was used to determine k_{68}/k_6.

The excited state absorption cross-section σ_{exF} is different from σ_{exL} due to the spectral shift. The efficiency of amplification $\bar{E}_F$ allowed the measurement of σ_{exF}.

The obtained results are listed in Table 1.

Table 1

	Rhodamine 6G	Rhodamine B
σ_{exL}	$(5\pm1)\times10^{-17}$ cm^2	$(5\pm1)\times10^{-17}$ cm^2
σ_{exF}	$(8\pm2)\times10^{-17}$ cm^2	$(1\pm0.2)\times10^{-16}$ cm^2
k_6	$\geq 10^{13}$ s^{-1}	$\geq 5\times10^{12}$ s^{-1}
k_{63}/k_6	0.95 ± 0.05	0.7 ± 0.05
k_{68}/k_6	0.01 ± 0.01	0.02 ± 0.01

References

1. M.E. Mack, Appl. Phys. Lett. 15, 166 (1969).
2. C. Lin, T.K. Gustafson, and A. Dienes, Opt. Comm. 8, 210.
3. G.R. Fleming, A.E.W. Knight, J.M. Morris, R.J. Robbins, and G.W. Robinson, Chem. Phys. 23, 61 (1977).
4. W. Falkenstein, A. Penzkofer, and W. Kaiser, to be published.
5. A. Penzkofer and W. Falkenstein, Opt. and Quant. Electr. 1978.

III. Sources and Techniques

Anti-Stokes Emission as a VUV and Soft X-Ray Source[1]

S.E. Harris, J. Lukasik[2], J.F. Young, and L.J. Zych

Edward L. Ginzton Laboratory, Stanford University
Stanford, CA 94305, USA

A VUV and soft x-ray light source based on spontaneous anti-Stokes scattering from atomic population stored in a metastable level is described. Unique properties of this source include: narrow linewidth, tunability, linear polarization, picosecond time scale, and quite high spectral brightness. We show how the maximum source brightness, within its narrow linewidth, is that of a blackbody at the temperature T of a metastable storage level. Experimental results showing laser induced emission at 569 Å and 637 Å from a He glow discharge are described. The use of the anti-Stokes process for direct, internal energy transfer from a storage species to a target species is discussed.

In this paper we discuss some of the properties of a new type of vacuum ultraviolet and soft x-ray light source [1,2]. The source is based on spontaneous anti-Stokes scattering from atomic population which is electrically stored in an appropriate metastable level. The source has several unique properties which include: narrow linewidth, tunability, picosecond time scale operation, linear polarization, and relatively high peak spectral brightness. We will see shortly that this peak spectral brightness corresponds to that of a blackbody at the temperature T of the storage level. A schematic of the anti-Stokes light source is shown in Fig.1.

Though anti-Stokes scattering is usually described in terms of a spontaneous scattering cross section, it is better for our purpose to describe it in terms of a spontaneous emission rate $A(\omega)$ induced by the laser pump field E_p at frequency ω_p [3]. This spontaneous emission rate at the vacuum ultraviolet frequency ω may be written

[1] Work jointly supported by the Office of Naval Research, the National Aeronautics and Space Administration under Contract No. NGL-05-020-103, the Advanced Research Projects Agency, and the Joint Services Electronics Program through the Office of Naval Research.

[2] On leave from the Laboratoire d'Optique Quantique du CNRS, Ecole Polytechnique, 91120 Palaiseau, France; recipient of the French-American CNRS-NSF Exchange Award.

$$A(\omega) = \frac{\omega^3 |\mu_{13}|^2}{3\pi\hbar e_0 c^3} \left[\sin^2 \left(\frac{1}{2} \tan^{-1} \frac{\mu_{23} E_p}{\hbar\Delta\omega} \right) \right] g(\omega - \omega_{VUV})$$

$$= \left(\frac{\omega}{\omega_{31}} \right)^3 A_{31} \left[\sin^2 \left(\frac{1}{2} \tan^{-1} \frac{\mu_{23} E_p}{\hbar\Delta\omega} \right) \right] g(\omega - \omega_{VUV}) \qquad (1)$$

The quantity A_{31} is the Einstein A coefficient for spontaneous emission from level $|3\rangle$ to level $|1\rangle$. The lineshape $g(\omega - \omega_{VUV})$ is the convolution of the Doppler- or pressure-broadened linewidth of the $|1\rangle - |2\rangle$ transition, and $\Delta\omega$ is $\omega_{31} - \omega_{VUV}$.

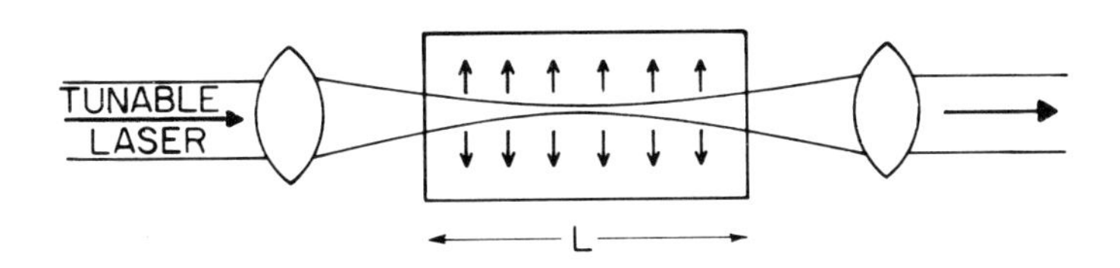

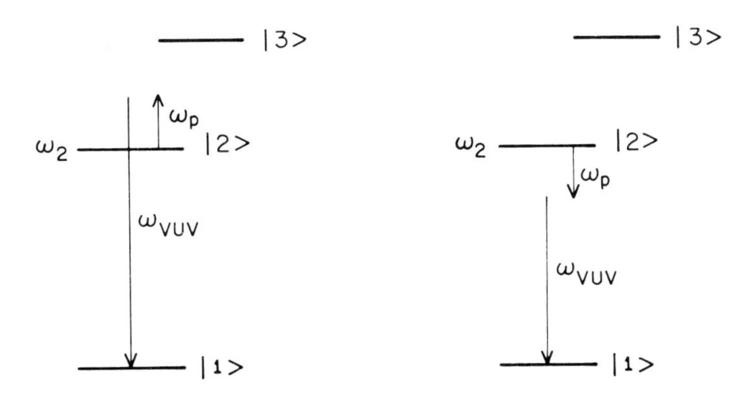

Fig.1 Schematic and energy level diagram for spontaneous anti-Stokes light source. An upper and lower sideband is obtained.

We see from (1) that as the laser pump field becomes large, the anti-Stokes emission rate approaches the Einstein coefficient A_{31} ; which at fixed oscillator strength f_{31} , increases as the square of the VUV frequency. The cross section for spontaneous scattering is related to the emission rate by $\sigma_{sp}(\omega) = \hbar\omega A(\omega)/(P/A)$, where P/A is the incident laser power density.

The key to understanding and optimizing this light source is the two-photon absorption which is created at the ultraviolet frequency ω_{VUV} in the presence of the laser pump frequency ω_p . For the range of laser power densities of interest here, both $A(\omega)$ and the two-photon absorption cross section $\sigma(\omega)$ increase linearly with laser power density, and are related to each other in the same manner as are the emission and absorption coefficients for single-photon processes, i.e., $\sigma(\omega) = (\pi^2 c^2/\omega^2)\, A(\omega)$.

The brightness of the light source, $B(\omega)$ photons/(sec cm^2 sterradian cm^{-1}), is determined by the interplay of the emissive and absorptive processes, and for an infinitely long cylinder of outer radius r_0 is given by [1,2]:

$$B(\omega) = \frac{\hbar\omega^3}{4\pi^3 c^2}\left[\frac{1}{\exp(\hbar\omega_{21}/kT) - 1}\right]\left\{1 - \exp\left[-\sigma(\omega)(N_1 - N_2)r_0\right]\right\} \tag{2a}$$

$$\sigma(\omega) = \frac{\pi\omega}{6c^2\epsilon_0^2\hbar^3}\left[\sum_i\left(\frac{\mu_{2i}\mu_{i1}}{\omega_i - \omega_{VUV}} + \frac{\mu_{2i}\mu_{i1}}{\omega_i + \omega_{VUV}}\right)\right]^2 \frac{P_p}{A}\, g(\omega - \omega_{VUV}) \tag{2b}$$

(mks units). T is the temperature of the metastable level, i.e., $N_2/N_1 = \exp - (\hbar\omega_{21}/kT)$; μ_{ij} are matrix elements; ω_i are the frequencies of the intermediate states; and P_p/A is the power density of the pump laser.

In the (two-photon) optically thin case, i.e., $\sigma(\omega)(N_1 - N_2)r_0 \ll 1$, $B(\omega)$ increases linearly with the laser power density and is the same as obtained from the usual spontaneous scattering cross section point of view. As the laser power density is increased and the medium becomes nominally two-photon opaque, i.e., $\sigma(\omega)(N_1 - N_2)r_0 = 1$, the brightness approaches a constant value equal to that of a blackbody radiator at the temperature T of the metastable level. Once the two-photon opaque or blackbody regime is attained on line center, the primary effect of a further increase in laser power density, cylinder radius r_0 , or ground state density N_1 is to increase the emission linewidth. The total number of emitted photons continues to increase slowly, and the brightness remains constant.

Before proceeding further, we note that anti-Stokes scattering in the VUV has been observed by BRÄUNLICH and LAMBROPOULOS [4], and has been discussed by ZERNIK [5] and VINDOGRADOV and YUKOV [6].

Experimental Results

In our first experiments [2] on this type of light source a glow discharge was used to store population in the $2s^1S$ level of He at 601 Å ≅ 166,272 cm^{-1} (Fig. 2). The cw He glow discharge was produced in a 40 cm long quartz tube with a cylindrical hollow cathode and pin anode at opposite ends. Typically, the discharge current was 120 mA and the pressure was about 1 torr. A 0.9 cm long slit was cut through the side wall of the 4 mm ID capillary and served as an input slit for the VUV spectrometer. An actively mode-locked Nd:YAG oscillator-amplifier system produced a train of mode-locked pulses, each with a pulse length of ~ 100 psec. Approximately 10 pulses occurred within the half-power points of the train envelope. The laser was propagated down the discharge capillary tube and focused to an area of about 3×10^{-4} cm^2 and a confocal parameter $b \cong 5$ cm parallel to the input slit.

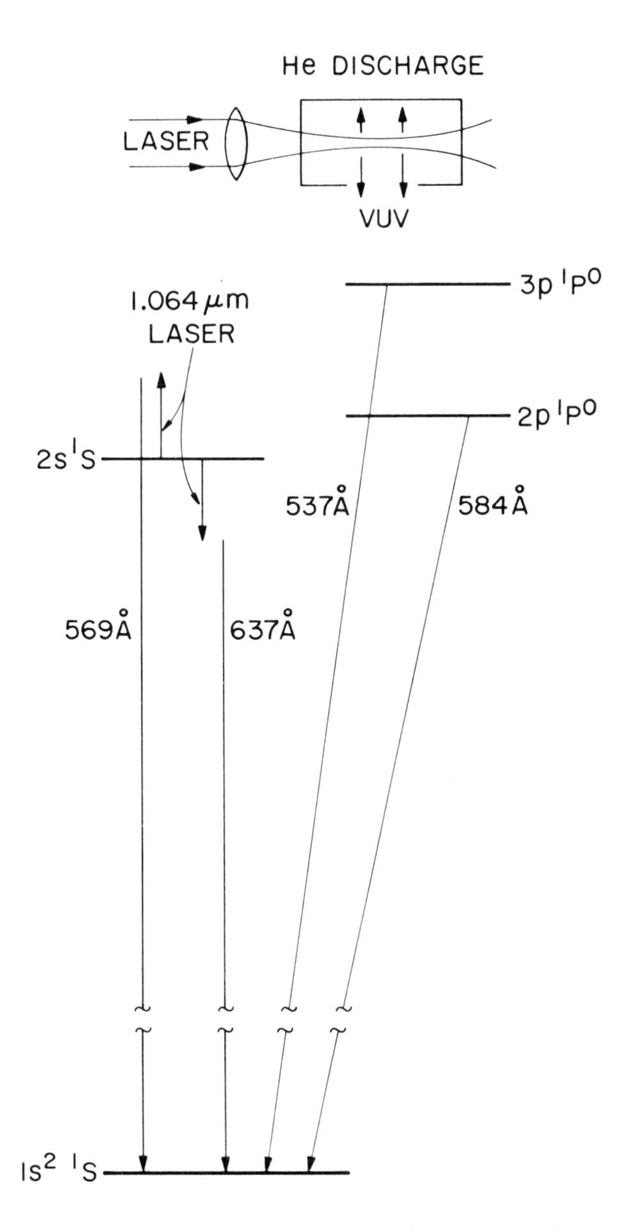

Fig.2 Energy level diagram for laser induced emission in He.

The detection system consisted of a spiraltron, a fast preamplifier, a pulse height discriminator, a coincidence gate, and a counter. The coincidence gate was set to a width of 90 nsec overlapping the laser pulse train. Typically, 10^2 counts were registered per minute with a signal-to-noise ratio of about 100. The relatively low count rate was a result of the 3×10^{-9} laser duty cycle and the 2.5×10^{-8} ratio of detected photons to total photons generated. Since the source linewidth, even far into the blackbody regime, was well below the resolution of the spectrometer, the observed count rate was proportional to the integrated brightness, $\tilde{B} \equiv \int B(\omega)\, d\omega$.

Using this system we observed laser induced emission at 569 Å, 637 Å, and 591 Å, as well as He resonance line emission at 584 Å, 537 Å, 522 Å, and 516 Å. The emission at 591 Å resulted from anti-Stokes scattering from the He $2s^3s$ level, and had an intensity of about 1/50 of the 569 Å radiation. Figure 3 shows the relative integrated brightness of the 569 Å radiation as a function of laser peak power density for three He pressures. The points represent the average of 5 one-minute counting intervals. The solid curves are theoretical calculations of $\tilde{B}$. As described in [2], the curves were drawn using a value of the two-photon absorption coefficient of $\sigma(\omega_{VUV}) = 1.8 \times 10^{-26}\ (P_p/A)\ W/cm^2$. The magnitude of each theoretical curve was determined by a least squares fit to the experimental points. Our results indicate that the ratio of metastable population to ground state population, $N_2/N_1 = 2.6 \times 10^{-5}$, which corresponds to a temperature $T = 22{,}700°K$. These numbers were independent of pressure in this range to within ± 5%. At the highest power density and pressure of Fig. 3, the two-photon source had a laser induced optical depth of $r_0 N_1 \sigma(\omega_{VUV}) \sim 7$, well into the blackbody regime.

The relative intensities of the laser induced emission and the He resonance lines are compared in Table I for a pressure of 1.6 torr and a laser power density of 600 GW/cm^2. The instantaneous count rate was calculated from the accumulated count using the laser repetition rate and either the coincidence gate aperture time (for the resonance lines) or the effective 1 nsec total laser on-time. In order to estimate the brightness we calculated the linewidth of the 537 Å and 584 Å resonance lines for our geometry and pressure. Based on a Voight profile, these are 3.2 cm^{-1} and 5.6 cm^{-1} respectively. The linewidth of the laser induced emission for these operating conditions was calculated as 1.3 cm^{-1} at 569 Å and 1 cm^{-1} at 637 Å. The second row of Table I also includes a geometrical factor of 2 to account for the larger effective radiating area of the resonance line source. Thus, we estimate that the peak induced emission at 569 Å is 140 times brighter than the strongest He resonance line. As a result of its greater detuning from the intermediate $2p^1P^0$ level, the brightness of the 637 Å emission is about 7 times smaller than that of the 569 Å emission.

One of the key properties of a laser induced two-photon radiator is that its geometry is dominantly controlled by the pumping laser beam, instead of by the geometry of the discharge. This allows a two-photon radiator to have a temperature characteristic of the <u>interior</u> of a plasma or discharge. In a glow discharge similar to ours, in the interior, electron collisions cause the population of the $2p^1P$ level to be within a factor of three of the $2s^1S$ population. However, it is the <u>exterior</u> $2p^1P$ level atoms which to a large extent determine the temperature of the single-photon 584 Å radiator. As a result of the fact that these atoms are continuously radiating, as well as due to the lower electron density and temperature near the walls, their temperature may be significantly lower; thereby probably accounting for the factor of 140 in relative brightness which we have observed. The attenuation and self-reversal of single-photon radiators which results from cold atoms, is also avoided in the two-photon radiator.

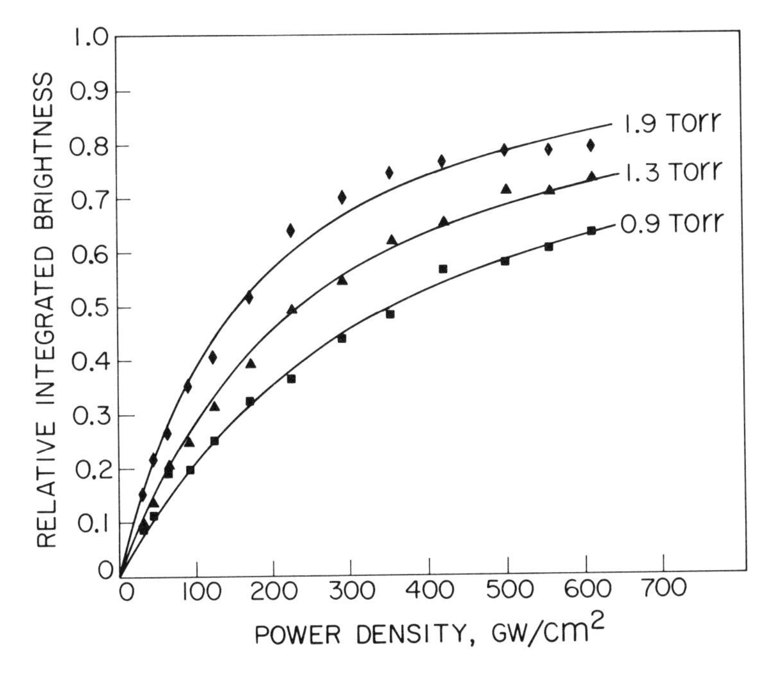

Fig.3 Relative integrated brightness at 569 Å as a function of 1.06 μm laser peak power density. The theoretical curve at each pressure was determined by numerically integrating (2); the magnitude was determined by a least squares fit to the experimental points at that pressure.

Flashlamp Applications

One of the uses of this type of light source may be as a flashlamp for short wavelength lasers. To avoid the inefficiencies of short wavelength optics, it may be best to mix the target or lasing species directly with the lamp species. For example, neutral potassium at a density of perhaps 10^{14} atoms/cm^3 might be mixed with He at a density of about 10^{19} atoms/cm^3. The mixture would then he heated either electrically or by a CO_2 laser beam. At an appropriate time, an incident tunable laser pulse would cause the generation of spontaneous anti-Stokes radiation. This radiation would be absorbed by the neutral potassium, and cause the production of excited-state K^+. A simplified energy level diagram for this type of interaction is shown in Fig. 4. As shown here, the anti-Stokes source would be tuned to an energy of 172,732 cm^{-1} so as to cause an inner shell transition from the $3p^64s$ level to the $3p^54s4d$ level. A second laser beam of energy greater than 28,739 cm^{-1} would carry this excited electron into the continuum causing the formation of the excited $3p^54s$ K^+ ion. By tuning the anti-Stokes source to a discrete

Table 1 Comparison of resonance line radiation and laser induced emission at 1.6 torr and 600 GW/cm^2.

	Resonance Lines		Laser Induced Emission	
	537 Å	584 Å	569 Å	637 Å
Instantaneous Count Rate 10^6 Counts/Sec	0.8	32.0	544	57
Estimated Peak Brightness* 10^{15} $\frac{\text{Photons}}{\text{sec cm}^2 \text{ sr cm}^{-1}}$	0.014	0.33	46	6.3

*The time averaged value is obtained by multiplying by the laser duty cycle of 3×10^{-9}.

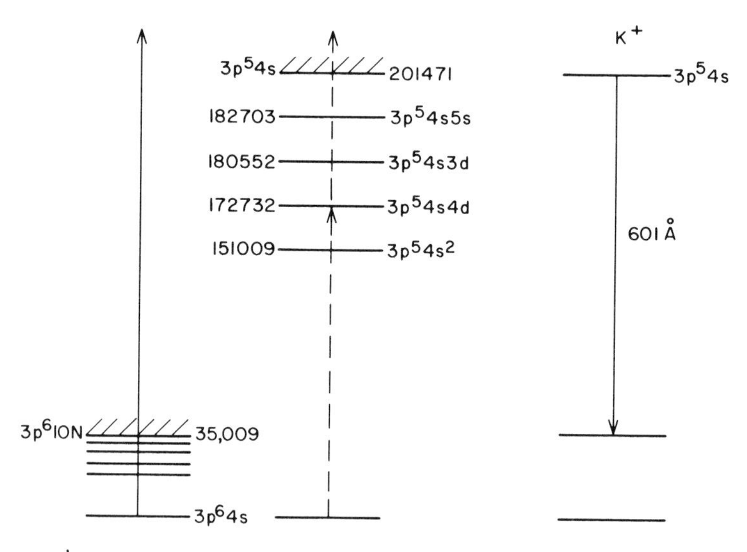

Fig. 4 Simplified energy level diagram for anti-Stokes pumping of K. The (left) solid arrow shows direct pumping to the continuum of an inner shell electron. The dashed arrows show two-photon pumping via an intermediate state.

intermediate state such as shown here, instead of tuning it directly into the continuum as shown by the solid arrow on the left side of Fig. 4, it should be possible to increase the cross section for absorption of the anti-Stokes radiation by about two orders of magnitude. This, in turn, allows operation at a K density much lower than would otherwise be possible, and mitigates, at least somewhat, the problem of the formation of ground state K^+ formed by collision with free electrons.

We should note that the energy required to cause lasing on the 601 Å line of K^+ is quite small. The calculated gain coefficient is (2.1×10^{-13}) N cm^{-1}, where N is the density of the potassium ions. A gain of e^{10} in a path length of 1 m requires an ion density of 5×10^{11} atoms/cm^3. Assuming a confocal volume of about 1 cm^3, this requires an energy of about 1 μJ. This energy must be deposited in a time short compared to the 0.6 nsec spontaneous decay time of the K^+ ion.

Before going further we should note that there are two problems associated with the K-He combination of Fig. 4. The first of these is the formation of ground state K ions by autoionization from the $3p^5 4s 4d$ level; the second and possibly more severe problem is the formation of ground state K ions by hot electrons in the discharge. During the afterglow, the density of these ions may rapidly reduce by formation of HeK^+ molecules.

There is an important advantage to the internal or mixed configuration which, in a sense, allows the blackbody limitation to be overcome: to the extent that the product of the single-photon absorption cross section and density of the target species is greater than the two-photon absorption cross section and (ground state) density of the storage species, the anti-Stokes photons will be absorbed by the target instead of reabsorbed by the generating species. For appropriate conditions, the effective anti-Stokes emission rate may then approach that of (1).

We should also briefly address the question of efficiency. At a detuning of 100 cm^{-1} from the 2p resonance line of He the cross section for spontaneous anti-Stokes scattering is 6×10^{-20} cm^2, and varies as the inverse square of the detuning from the 2p level. Assuming an excited state He density of 10^{14} atoms/cm^3, a 1 m path length, and allowing for the energy conversion gain of a factor of 34, the ratio of anti-Stokes power generated to laser power incident is about 2%. This assumes that the media is kept (two-photon) optically thin, or that equivalently, as discussed above, all of the energy is absorbed by the target species; and that the excited state population is not depleted.

In conclusion, the anti-Stokes light source has potential for producing radiation in the VUV and soft x-ray spectral regions with many laser-like properties. These include narrow linewidth, tunability, picosecond operation, and controllable polarization. The source may provide a valuable tool for studying the spectroscopy, fluorescent yield, and autoionizing rates of inner shell transitions. Its use as a pump for VUV and soft x-ray lasers is promising.

The authors acknowledge helpful discussions with Roger Falcone and John Willison.

References

1. S. E. Harris, Appl. Phys. Lett. **31**, 498 (October 1977).

2. L. J. Zych, J. Lukasik, J. F. Young, and S. E. Harris, "Laser Induced Two-Photon Blackbody Radiation in the VUV," Phys. Rev. Lett. (to be published).

3. E. Courtens and A. Szoke, Phys. Rev. A **15**, 1588 (1977).

4. P. Braunlich and P. Lambropoulos, Phys. Rev. Lett. **25**, 135 (1970); Phys. Rev. Lett. **25**, 986 (1970); and P. Braunlich, R. Hall, and P. Lambropoulos, Phys. Rev. A **5**, 1013 (1972).

5. Wolfgang Zernik, Phys. Rev. **132**, 320 (October 1963); Phys. Rev. A **133**, 119 (January 1964).

6. A. V. Vinogradov and E. A. Yukov, Sov. J. Quant. Elect. **3**, 163 (September-October 1973).

Picosecond Phase Shift Measurements at 358 MHz Using Synchrotron Radiation

A.P. Sabersky

Stanford Linear Accelerator Center, Stanford University
Stanford, CA 94305, USA

I.H. Munro

Department of Structural Biology, Sherman Fairchild Center
Stanford University School of Medicine
Stanford, CA 94305, USA

For our purposes the most important property of synchrotron radiation is the time modulation of the emitted intensity, although it possesses many other qualities, such as broad spectral range, high brightness and 100% polarisation, which are being extensively utilised in many research fields [1,2]. The radiation from the S.R.* is modulated in exactly the same manner in which the electron beam is modulated. The electron pulse or bunch shape is nominally Gaussian [3] and is determined by S.R. parameters. The bunch length increases with stored current, this effect being different from one S.R. to another [4]. The S.R. radio frequency is an integer multiple of the revolution frequency, this being a constant since the electrons are highly relativistic ($v/c \simeq .9999998$). It is possible to have as many electron or positron bunches around the ring as the RF harmonic number, which is 280 in SPEAR. In our experiments SPEAR ran with only one circulating bunch, and we discuss this case.

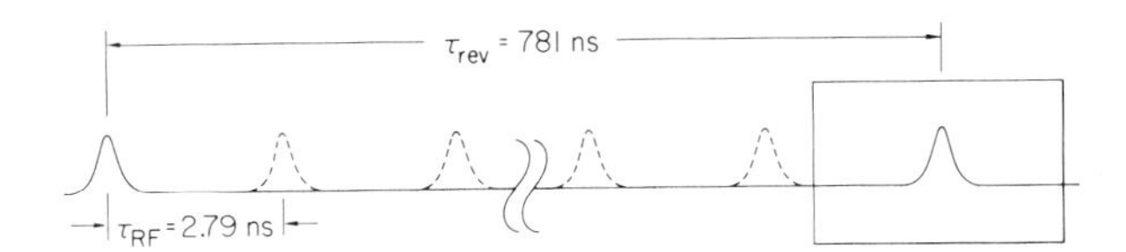

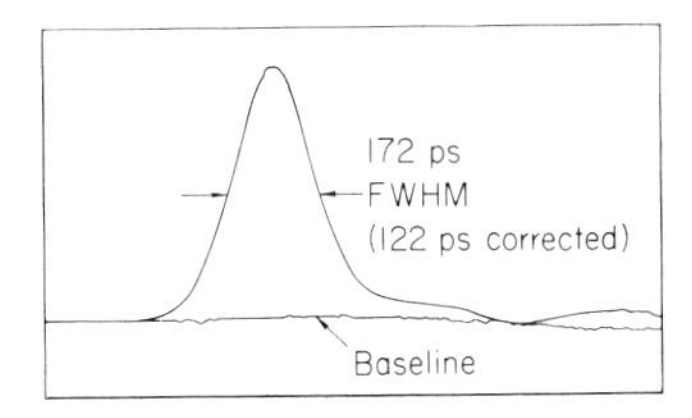

Fig. 1 Time structure of light pulses: T_{rev} is the going-around frequency for a single bunch, T_{RF} is the period of the S.R. radiofrequency. The lower bunch picture is experimental data taken for the paper of [4].

* Storage Ring

For the case of a stable bunch and repetition period, the spectrum of the *modulation* of the radiation consists of a train of delta functions spaced at the revolution frequency, (1.28 MHz in SPEAR) their amplitude envelope being the Fourier transform of the pulse shape. For SPEAR, a typical bunch length is about 200 ps FWHM (see Fig. 1). A Gaussian pulse with this width has Gaussian spectrum of 3.5 Ghz FWHM, i.e. amplitude down to half of the central peak value at 1.8 Ghz. It is possible to operate SPEAR with pulses as short as 100 ps [4]. Real S.R. pulses are generally not Gaussian or symmetric [5] thus the phases of the different components have no simple relationship with each other. Multiple-frequency phase-shift measurements must have the reference phase set up individually for each frequency.

The spectral purity of the revolution frequency harmonics is set by the stability of the S.R. RF system and the stability of the electron bunch. The bunches are prone to shape instabilities [6], their intensity varying with operating conditions and machine parameters. The instabilities appear as sidebands about the revolution-frequency harmonics. For sufficiently strong instabilities and high harmonics, the sidebands can be larger than the central line. Under most operating conditions, this is not a serious problem with SPEAR.

The spectral shape of the output from the detector phototube is, naturally, not identical with that of the source (see Fig. 2). A phototube is not a linear causal filter, since there is random time variation in electron transit times. In spite of this effect, the spectral purity of the harmonic lines is unaffected even at low intensities when the tube is in the single-photon counting mode. The combined effect of timing dispersion and tube noise is to degrade the signal-to-noise ratio of harmonic lines at the output of the detector.

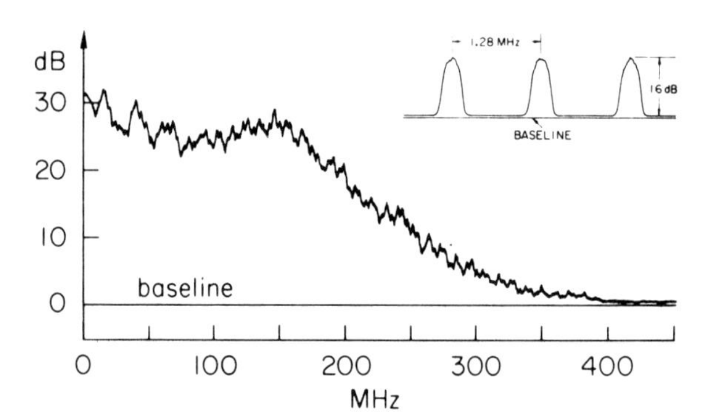

Fig. 2 The power spectrum of the modulation of synchrotron radiation from SPEAR measured using an RCA 8850 phototube. The amplitude profile represents primarily the frequency response of the phototube. Inset shows the response of the phototube in the vicinity of 358 MHz.

The power-frequency spectrum of the source and photomultiplier tube is measured with a spectrum analyzer (Tektronix 7L13). The spectrum has a S:N ratio at 358 MHz of 20dB and shows that it should be possible to make phase-shift measurements at frequencies up to about 400 MHz with the RCA 8850 tube. This power spectrum (Fig. 3) is in effect the product of the frequency spectrum of the source (which can be measured directly) and of the response of the phototube, although all relative phase information is lost. The frequency response of any phototube with resolution time > S.R. bunch length therefore can be simply and rapidly characterised and this has been done for several phototubes and contrasted with their "single electron" time response [7].

The decay time (τ) of a single fluorescent species can be deduced by measuring the phase delay (θ) between excitation and fluorescence radiation modulated at a single suitable frequency (ω). The relation between ω and τ is: $\tan\theta = \omega\tau$ [8]. The technique is restricted usually to measurement on samples having only a single exponential decay, and that within a range defined by the limited frequency range of conventional modulated sources.

However, it is possible to derive the impulse response of a causal filter from only the imaginary part of its Fourier transform [9] via the Hilbert transform. Thus it is possible to completely characterise the time (impulse) response or the complex frequency response of a linear system with only phase measurements.

Results

A number of test experiments were made using apparatus designed for time resolved fluorescence emission anisotropy studies of trytophan in proteins [11].

The operating conditions of the apparatus made direct observation of fluorescence emission difficult and all tests were made using a beam reflected directly onto the detecting photomultiplier tube. Phase shifts were produced by insertion of 1 cm water filled quartz cells into the beam and by the introduction of a thick glass slab. The total optical path length between the source and detector also could be altered by linear motion of the phototube along the axis of the beam. The aperture at the photocathode of the phototube was stopped down until rotation of the tube through 360° in its mount produced less than $\pm 0.5^{o}$ change in the phase of the direct beam in light of wavelength 320 nm. The photomultiplier was moved along the optical axis by a measured distance and the corresponding phase reading observed with respect to the RF reference. The results yielded $4.7 \pm .5$ degrees per cm of motion.

Signal-to-noise at the VVM (Vector Voltmeter H-P mod. 8405A) input limited the accuracy of the measurements we report. For conditions of poor S/N, the VVM would not yield a reading at all. Our reference signal comes directly from the S.R. master oscillator. It is possible to derive reference signals at any revolution frequency harmonic by filtering an electromagnetic pulse coupled from the beam by an antenna in the accelerator vacuum vessel. The spectral shape of this pulsed signal is very similar to that of the pulsed radiation, having the same combination of harmonics. The filter used for selecting a reference should be identical with the filter following the phototube (Fig.3) and the two should be simultaneously temperature stabilized to eliminate thermal drifts. Other techniques may be used to measure phase shifts, some of which are perhaps better able to deal with noise problems, i.e. lock-in techniques, heterodyning.

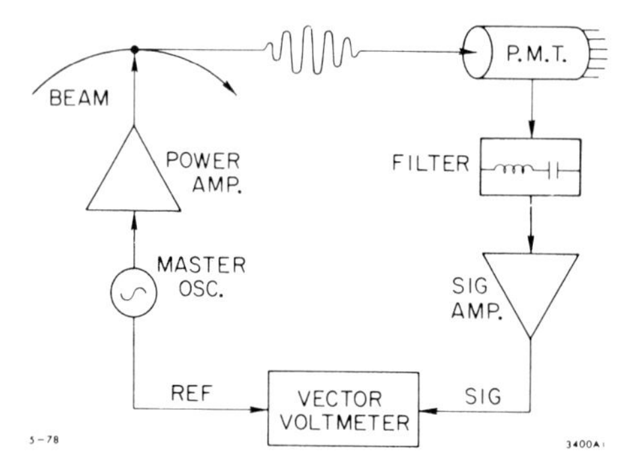

Fig. 3 Experimental apparatus: The filter attenuates harmonics 1.28 MHz away from 358 MHz by 16 dB. The signal amplifier has a gain of 36 dB.

Conclusions

Using synchrotron radiation from SPEAR and a simple arrangement of apparatus it has been shown possible to make phase shift measurements at 358 MHz. Synchrotron radiation source properties are such that it will be simple to apply this procedure to study short atomic and molecular fluorescence lifetimes using any optical system at any wavelength or to rapidly scan the excitation spectrum of a material in the time domain with sufficiently good time resolution to observe reaction kinetics in the picosecond range. Experiments could be carried out at any frequency in the range from 1.28 MHz to about 1 GHz limited only by the high frequency response of the detectors.

References

1. National Academy of Sciences, Washington, D.C., 1976, "An assessment of the national need for facilities dedicated to the production of synchrotron radiation."

2. I. Lindau & H. Winick (Ed.), Synchrotron Radiation Research, SSRP Report 77/01, Jan 1977.

3. M. Sands, Proc. Int'l. School of Physics "Enrico Fermi", Course 46, Varenna 1969, p. 257-411. Also SLAC-121, Nov. 1970.

4. P. B. Wilson, R. Servranckx, A. P. Sabersky, J. Gareyte, G. E. Fischer, A. W. Chao, M. H. R. Donald, IEEE Trans. Nucl. Science Vol. NS-24, #3, June 1977.

5. D. Germain and H. G. Hereward, CERN internal report CERN/MPS/DL 75-5.

6. F. J. Sacherer, IEEE Trans. on Nucl. Science Vol. NS-20, #3, June 1973.

7. I. H. Munro and A. P. Sabersky, to be published.

8. R. D. Spencer and G. Weber, J. Chem. Phys. 52, (1654) 1970.

9. "The Fourier transform and its applications" by R. Bracewell, McGraw-Hill 1965, p. 267.

10. I. H. Munro, I. P. Pecht, and L. Stryer, Proc. Nat. Acad. Sciences, 1978 (in press).

High Power, Nearly Bandwidth Limited, Tunable Picosecond Pulses in the Visible and Infrared

A. Laubereau[1], A. Fendt, A. Seilmeier, and W. Kaiser

Physik Department der Technischen Universität München
D-8000 München, Fed. Rep. of Germany

For the investigation of ultrafast processes, spectroscopists desire light pulses with picosecond pulse duration, wide tuning range and large peak power. Existing laser devices fulfill these requirements only in part. As an example, we point to studies of the vibrational dynamics of polyatomic molecules in condensed phases where ultrashort pulses in the infrared are required. Liquid dye lasers are not applicable in this region; mode-locked solid-state lasers, on the other hand, do not possess tunability.

In this paper, parametric three-photon amplification is considered as a pulse generation scheme. Two different experimental systems will be discussed: (i) A single path, high gain, parametric generator consisting of two $LiNbO_3$ crystals is investigated. Intense pulses of $\sim$ 3.5 ps are produced in the visible and IR; (ii) a novel parametric generator-amplifier system is presented which allows to adjust the pulse duration. Subpicosecond pulses with an energy conversion of several percent are generated. Our experimental results demonstrate that parametric pulse generation meets the experimentalist's need in picosecond investigations.

(i) The parametric generator is depicted schematically in Fig. 1. An intense single pulse at a fixed frequency serves as a pump pulse and enters the system from the left. Passing a first nonlinear crystal (e.g. $LiNbO_3$) parametric emission at the signal and idler frequency positions builds up from quantum noise in a cone around the forward direction. The signal and idler frequency (ω_s,ω_i) are determined by energy and k-vector conservation of the three-photon process; they may be tuned changing the crystal orientation. Part of the parametric emission which occurs close to forward direction is amplified by the parametric process in the carefully oriented second crystal. Theoretical estimates for $LiNbO_3$ as nonlinear material show that a pump intensity of $\sim$ 5GW/cm^2 and an interaction length of several cm is sufficient for the generation of intense signal and idler pulses. The frequency width of the parametric pulses is influenced by the optical dispersion of the nonlinear material and by several geometrical parameters of the generator system. The effect of the finite divergence of the parametric emission is illustrated

[1] Present address: University of Bayreuth, 8580 Bayreuth, Fed. Rep. of Germany

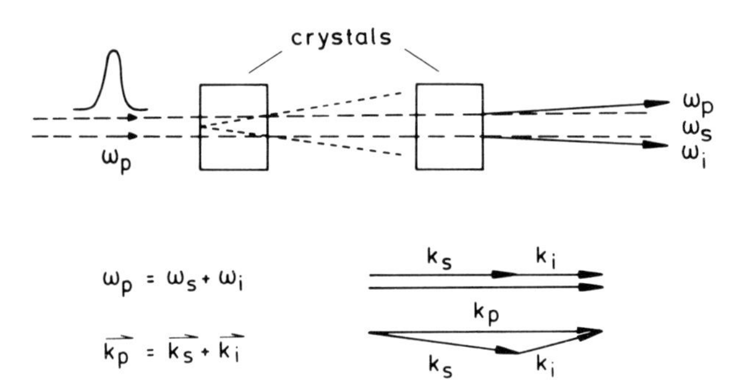

Fig. 1

Schematic of a travelling wave, high gain generator consisting of two nonlinear crystals

in Fig. 2. The k-vector diagram for the off-axis geometry is depicted in the inset. The angle between the signal and the pump pulse is denoted by α. Calculations of the phase-matching condition $\Delta\vec{k} = 0$ show that the idler and signal frequencies depend on the emission angle α. Numerical results for type I phase-matching in $LiNbO_3$ are shown in Fig. 2. The frequency difference $\Delta\omega = \omega_i(\alpha) - \omega_i(0)$ is plotted versus α for several crystal orientations represented by the setting of the idler frequency $\omega_i/2\pi c$.

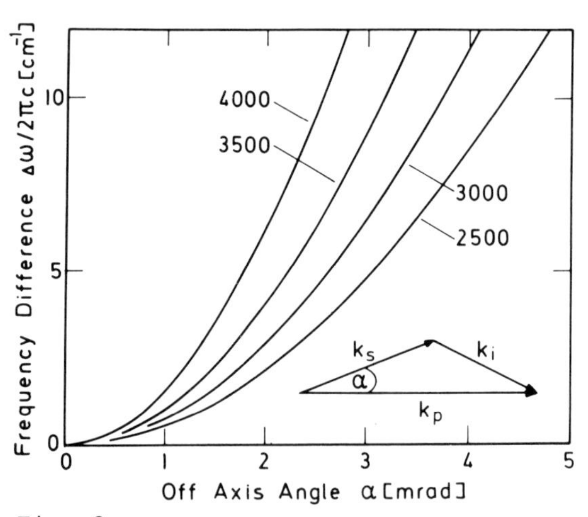

Fig. 2

Angular dispersion of the phase-matched idler (and signal) waves in off-axis direction versus emission angle α for several crystal orientations ($LiNbO_3$) denoted by the idler frequency.
Inset: Wave vector diagram

We see from Fig. 2 the fast rise of the frequency width for off-axis angles of the order of 10^{-3} rad. If we desire a frequency spread of less than 10 cm^{-1} we have to work with or have to select (by an appropriate aperture) a very narrow beam of a few milliradians.

Our experimental systems for the generation and for the detailed analysis of the parametric pulses are depicted schematically in Fig. 3. A passively mode-locked Nd:glass laser is followed by an electro-optic switch which selects one pulse from the early part of the pulse train. In this way we obtain bandwidth limited pulses of t_p = 7 ps duration and $\Delta\nu$ = 2.5 cm^{-1} bandwidth ($t_p \times \Delta\nu$ = 0.5).

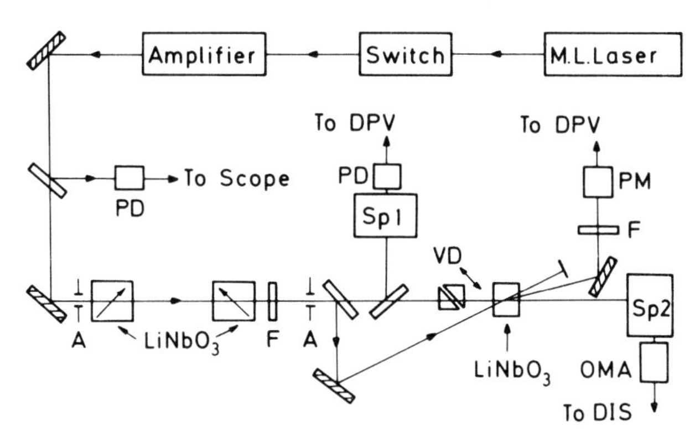

Fig. 3

Schematic of the single path parametric generator and of the experimental systems used to study the pulse properties

The single laser pulse is boosted by a factor of 200 in an optical amplifier rod before passing through two $LiNbO_3$ crystals (length 3 cm, distance 60 cm) which represent the parametric generator. Frequency tuning over a large frequency range is possible by simultaneous rotation of the crystals. It should be stressed that careful adjustment of the crystals is necessary for proper frequency matching.

The properties of the infrared pulses were analyzed by several experimental systems (see Fig. 3). (1) The bandwidth of the idler pulse was measured directly with an infrared spectrometer (SP1) and an InSb-detector. (2) The signal pulse passes through a third $LiNbO_3$ crystal for second harmonic generation. The signal wave is up-converted to a frequency range where a spectrometer (SP2) in conjunction with an optical multichannel analyser (OMA) is able to display the total spectrum of each signal pulse. (3) The duration of the signal pulse is determined by an autocorrelation experiment using again the third $LiNbO_3$ crystal.

The results are listed in Fig. 4 (left hand side). Using a

Pulse Properties

	IR	VIS
Tuning Range	2500 to 7000 cm^{-1}	12500 to 17000 cm^{-1}
Frequency Width	6.5 cm^{-1}	~6 cm^{-1}
Pulse Duration	3.5 psec	~4 psec
$\Delta\nu \times t_p$	0.7	~0.7
Divergence	3 mrad	3 mrad
Intensity	10^9 W/cm^2	10^9 W/cm^2

Fig. 4

Table of pulse properties of the parametric generator using pump pulses at 9455 cm^{-1} (left column) and at 18910 cm^{-1} (right column), respectively

pump pulse at 9455 cm^{-1} a tuning range in the infrared of more than 4000 cm^{-1} was achieved. For idler frequencies around 3000 cm^{-1} picosecond pulses with a small frequency width of 6.5 cm^{-1} were generated close to the Fourier transform limit. The energy conversion of pump pulse to parametric emission was measured to be several percent. The high intensity of the pulses allows to populate well selected modes far above the thermal equilibrium value and to determine their population lifetime with subpicosecond time resolution. The system is very useful for investigations of ultrafast dynamical processes. In fact, during the past three years we have successfully applied these systems for the excitation of vibrational modes of polyatomic molecules in the liquid and gaseous phase.

The tuning range of the parametric process is determined by the pump frequency and the nonlinear material. We have operated a similar parametric device as shown in Fig. 3 using the second harmonic of the Nd:laser at 18910 cm^{-1} as a pump source. The properties of this system are listed in the right column of Fig.4. The frequency of the signal pulse is tunable in the red part of the visible spectrum over a range of several thousand cm^{-1}. Favorable values were obtained for the frequency width and the duration of the parametric pulses which are shown in the Figure. Of special interest is the product $\Delta\nu \times t_p \simeq 0.7$ indicating that the red pulses are nearly bandwidth limited. Due to the large tuning range, short pulse duration,and high intensity, the parametric generator in the visible is very attractive for spectroscopic applications (see paper by Kaiser et al. of this issue).

(ii) In many experimental investigations with picosecond pulses it is desirable to adjust the duration of the pulses to the specific needs of the measurement. We have devised a parametric generator-amplifier set-up which gives considerable pulse shortening and allows to vary the duration of the generated pulses. In this new technique, the pulse duration is controlled simply by adjusting the time delay between the pump pulse and the input signal pulse in a parametric amplifier. The experimental system is depicted in Fig. 5.

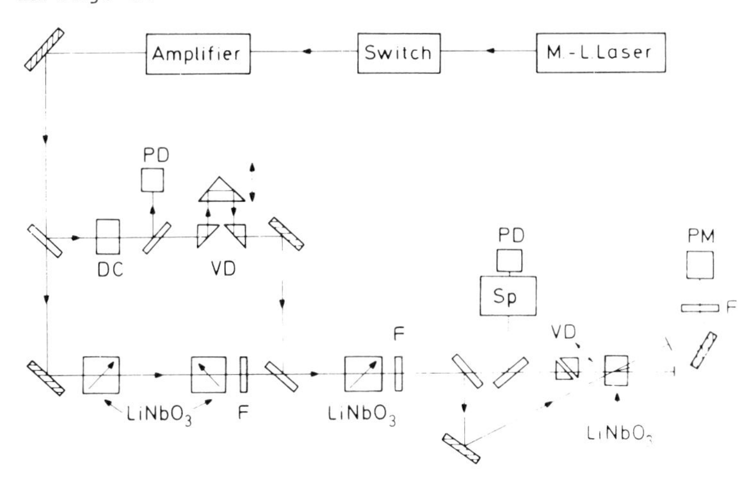

Fig. 5

Schematic of the parametric generator-amplifier system and of the autocorrelation set-up. The measured pulse duration is adjusted with the help of the variable delay unit VD.

The first part of the experiment consists of a mode-locked laser system and a parametric generator which was described above (see Fig. 3). The generated parametric signal serves as input pulse for parametric amplification in a following third $LiNbO_3$ crystal. The pump pulse for the amplifier stage is produced with the help of a beam splitter in front of the generator; it travels collinearly with the incident signal through the $LiNbO_3$ amplifier. The time delay t_D between the pump pulse and input signal is adjusted by a delay system VD. A liquid cell DC with a nonlinear absorber is used to steepen the rising part of the pump pulse. The duration of the amplified parametric pulse is measured by an autocorrelation scheme using another short $LiNbO_3$ crystal for second harmonic generation.

The pulse shortening effect achieved with the parametric amplifier is shown in Fig. 6. The measured pulse duration t_p^s of the signal pulse is plotted versus the setting of the delay unit. For $t_D = 0$ the maxima of the pump pulse and incident signal pulse coincide in the parametric amplifier. Fig. 6 vividly demonstrates that t_p^s decreases for larger values of t_D. This result is well understood by the time dependence of the parametric gain which is determined by the temporal shape of the pump pulse and the sharply rising and decaying wings of the incident signal pulse. For a time delay of several picoseconds, only the part of the signal pulse which

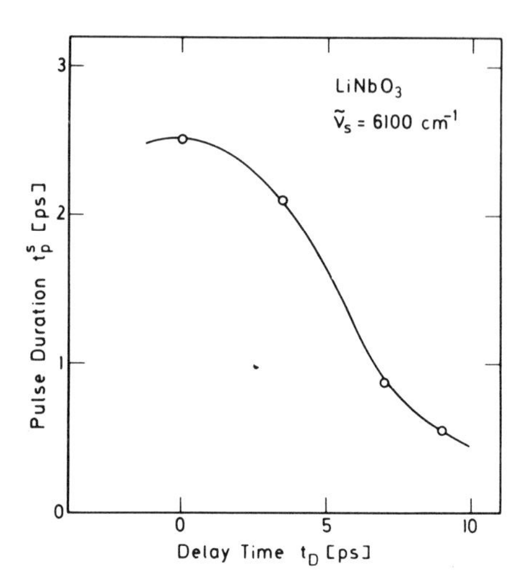

Fig. 6

Experimental values of the duration t_p^s of the amplified signal pulse versus delay time t_D between pump and input signal

overlaps with the pump pulse finds parametric gain; i.e. a shorter pulse is generated. Of particular interest are values of t_D where subpicosecond pulses are produced. An example is given by Fig. 7 showing a pulse duration of 0.54 ps for t_D = 9 ps. It should be

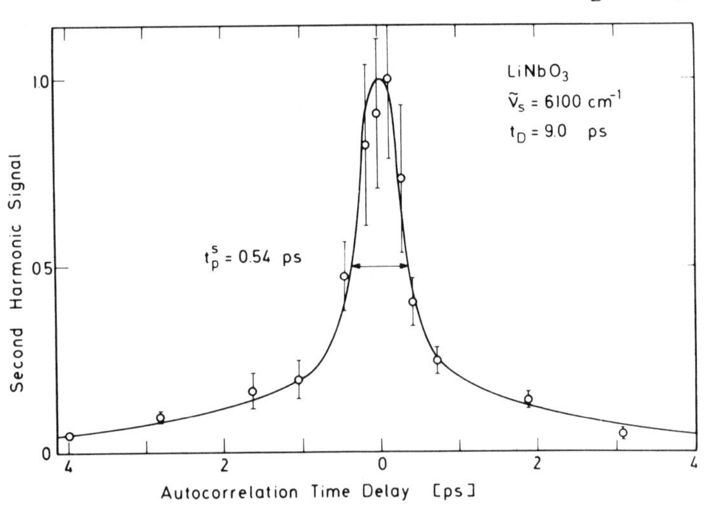

Fig. 7

Autocorrelation curve of the amplified signal pulse. A pulse duration of 0.54 ps is observed.

noted that these very short pulses are generated in the parametric device with an energy efficiency of several percent.

In conclusion, we wish to say that the parametric process has favorable properties for the generation of intense, widely tunable pulses in the visible and infrared. We demonstrated that pulses of picosecond and subpicosecond duration may be generated under properly devised experimental conditions. It is expected that the number of applications of parametric pulses to dynamical investigation will increase in the near future.

For further references of the authors:

A. Laubereau, L. Greiter, and W. Kaiser, Appl. Phys. Lett. 25,87 (1974)

A. Seilmeier, K. Spanner, and W. Kaiser, Optics Commun. 24,237 (1978)

A. Fendt, W. Kranitzki, A. Laubereau and W. Kaiser, to be published.

Sub-Picosecond Optical Gating by Optical Wavefront Conjugation

D.M. Bloom, C.V. Shank, R.L. Fork, and O. Teschke[1]
Bell Telephone Laboratories, Holmdel, NJ 07733, USA

We report the demonstration of a new technique for ultrashort optical gating. Our technique provides low background optical gating without the need for polarizers or dispersive elements. Furthermore, unlike the optical Kerr shutter of Duguay and Hansen [1] gate times shorter than the molecular relaxation times are easily obtained. The operation of this gate is based upon optical wavefront conjugation [2-8].

Optical wavefront conjugation is a form of degenerate four wave mixing in which two counter propagating pump waves are mixed with an incident object wave. In this geometry, the two counterpropagating pump waves and each plane wave component of the object wave, all at the same frequency, interact through the Kerr nonlinearity to produce a plane-wave polarization with a wavevector equal in magnitude and counterdirectional to the incident object wave. The phasor amplitude of the polarization is proportional to the product of the amplitudes of the two pump waves and the complex conjugate of the object wave. Since this interaction generates the conjugate wave for each plane-wave component of an incident wavefront, complex wavefronts can be time reversed. A physical picture of this interaction can be described in the following manner. In the Kerr medium the object wave and each of the pump waves interact through the nonlinear refractive index to produce a phase grating and the counterdirectional pump wave is diffracted by this grating to produce the backward image wave. Furthermore, while the induced hologram will remain for the molecular relaxation time of the Kerr medium, the backward wave is generated only if both pump waves are present. Exact time coincidence between the two pump pulses is not essential. It is possible to use a first pump pulse to "write" the hologram and a second delayed counterpropagating pulse to "readout" the hologram. However, the delay between the write and read pulse must be less than the molecular relaxation time. The noise background of this technique is limited only by scattering since the generated signal is produced in a direction far different from that of the incident beams. In addition, a conjugate wave will be generated only if the frequency of the object light is close to that of the pumps. Phasematching restricts the frequency bandwidth to $\Delta f = c/L$,

[1]Work done while on a leave of absence from the Universidade Estadual de Campinas, Campinas, S. P., Brazil.

where L is the interaction length. Since the interaction length is typically of order c x τ_p (where τ_p is the pump pulse length), the acceptance bandwidth is approximately equal to the pump pulse bandwidth. The wavelength discrimination provided by this technique is greater than that obtained by frequency mixing [9] and does not have an acceptance angle limitation imposed by phasematching.

In our experiment (Fig. 1) we used a passively modelocked cavity dumped dye laser pumped by an argon ion laser. The cavity dumped pulses were then amplified in a dye amplifier system pumped by a frequency doubled Nd:YAG laser. This system provides 100 μJ pulses with 0.5 x 10^{-12} sec pulse length at 10 pps. The output was split into two beams using a modified Michelson interferometer to provide displaced but parallel pump and object beams.

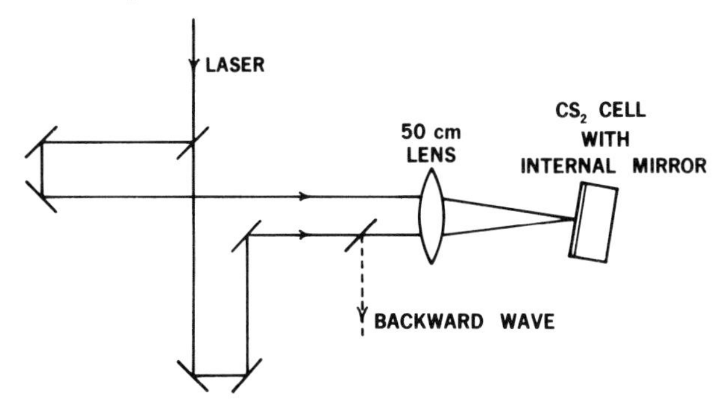

Fig. 1 Schematic of experimental arrangement for sub-picosecond optical gating.

The object beam passed through a beamsplitter and both the pump and object beams were focussed by a single 50 cm lens. A specially designed quartz cell was placed at the focus where the beams crossed. The cell containing CS_2 consisted of a quartz window and a high reflecting mirror separated by about 50 μm. This thin cell was aligned to retro-reflect the incident pump wave in order to produce the counterpropagating pump. The interferometer was adjusted to provide time coincidence between the pump and object wave. The backward generated wave which was picked off by the beamsplitter was easily detected. The energy in the detected backward wave was about equal to the incident object wave. The output was a well collimated beam demonstrating phase conjugation since the focusing of the lens was "undone" in the return pass of the backward generated wave. The autocorrelation of the generated wave was measured by second harmonic generation (Fig. 2) and was found to be slightly narrower than that of the pump which had an autocorrelation width of 1 x 10^{-12} sec FWHM. A cross-correlation measurement obtained by measuring the backward wave signal as a function of the delay between the pump and object wave verified the sub-picosecond time resolution of the gate.

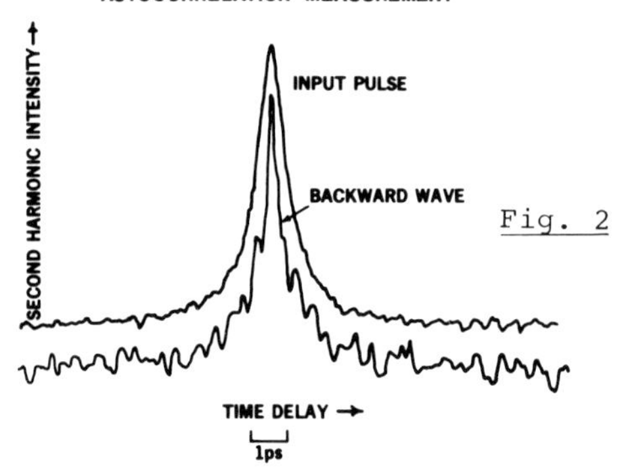

Fig. 2 Comparison of autocorrelation traces of input pump wave and generated backward wave.

This technique may prove useful for range gating applications such as medical diagnosis and nondestructive testing. The fact that rapid recovery of the nonlinear medium does not limit the time resolution of this gate greatly widens the choice of nonlinear materials. For instance, saturated absorption in ruby has been used to obtain cw wavefront conjugation at modest average powers [8]. Ruby should also be suitable for sub-picosecond gating of a wide range of cw modelocked laser sources even though its recovery time is 4 msec. The frequency selectivity of this technique makes it espccially useful in applications where inelastic scattering or fluorescence are a problem. In optical filtering applications where frequency selectivity is required but time discrimination is not critical, wavefront conjugation has been used to provide filter bandpasses varying from several hundred hertz to many gigahertz [10].

Filter bandwidths limited by either phasematching or the relaxation time of the nonlinear medium were demonstrated in that work. When the bandwidth is limited by the nonlinear material response, the relaxation time can be directly determined from the lineshape. This technique was used in Ref. 10 to measure the relaxation time in CS_2 without the need for short pulses.

References

1. M. A. Duguay and J. W. Hansen, Appl. Phys. Lett. 15, 192(1969)
2. R. W. Hellwarth, J. Opt. Soc. Am. 67, 1 (1977).
3. A. Yariv and D. M. Pepper, Opt. Lett. 1, 16 (1977).
4. D. M. Bloom and G. C. Bjorklund, Appl. Phys. Lett. 31, 592 (1977).
5. S. M. Jensen and R. W. Hellwarth, Appl. Phys. Lett. 32, 166 (1978).
6. D. M. Bloom, P. F. Liao, N. P. Economou, Opt. Lett. 2, 58 (1978).
7. P. F. Liao, D. M. Bloom, N. P. Economou, Appl. Phys. Lett. 32, 813 (1978).
8. P. F. Liao, D. M. Bloom, Opt. Lett. 3, 4 (1978).
9. H. Mahr, M. D. Hirsch, Opt. Commun. 13, 96 (1975).
10. G. C. Bjorklund, D. M. Bloom, P. F. Liao, unpublished

Parametric Study of Double Mode-Locking of the cw Dye Laser

E. Bourkoff and J.R. Whinnery

Electronics Research Laboratory, University of California
Berkeley, CA 94720, USA

A. Dienes

Department of Electrical Engineering, University of California
Davis, CA 95616, USA

The generation of subpicosecond pulses by the technique of double mode-locking has been numerically investigated [1] and experimentally confirmed [2]. This method utilizes a saturable absorber which is itself a laser dye and can be inverted at the trailing edge of the mode-locked pulse. The resulting stimulated emission leads to the generation of a secondary pulse and to a rapid restoration of the nonlinear losses, which enhances compression of the trailing edge of the primary pulse.

The dye laser configuration is similar to the one reported earlier [2]. The thickness of both Ethylene Glycol jet streams have been reduced to 125 [μm] and a single element Lyot filter of 1.5 [mm] thickness has been used to study the low pump-power properties of double mode-locking. The shorter path length leads to a smaller average area of the focused laser beam in the absorber dye; for the R6G medium, the change is negligible since the confocal length of the beam is about a factor of 4 larger. The duration of the primary pulse is predicted to decrease as the ratio of the average beam area in the laser dye compared to that in the absorber dye increases [3]. This is in contrast to conventional passive mode-locking. In the latter case, producing too high a value of s (the ratio of effective absorption cross-section of the absorber dye to the emission cross-section of the laser dye) can cause an excessive amount of steady-state pulse energy to go to the trailing edge, increasing the pulse duration [4]. In double mode-locking, the trailing edge is automatically compressed due to the stimulated emission of the absorber dye. Hence, a larger value of s is predicted to result in shorter pulses. We have found that the shortest pulses occurred near $\lambda_{R6G} = 600$ [nm], where the s value is nearly maximized for this dye combination.

Due to the larger beam area ratio it was found that the threshold absorber concentration and pump power for the onset of double mode-locking were reduced. One would expect that as the absorber concentration is increased, the primary pulse width will shorten at the expense of higher pump power requirements, since additional absorber molecules can contribute to the compression of the trailing edge. This has been shown recently by the generation of shorter pulses at higher absorber concentrations [2]. We now report the generation of subpicosecond pulses comparable in duration to those previously reported, but at a lower threshold pump power. The 1.5 [mm] Lyot filter is suitable for these measurements because the wavelength separation between transmission peaks is nominally 350 [Å].

In Fig.1 we indicate the experimental arrangement used for measuring and monitoring continuous trains of ultrashort pulses. The ¼-m monochromator had 4.0 [mm] slits which did *not* restrict the detected UV spectrum during an

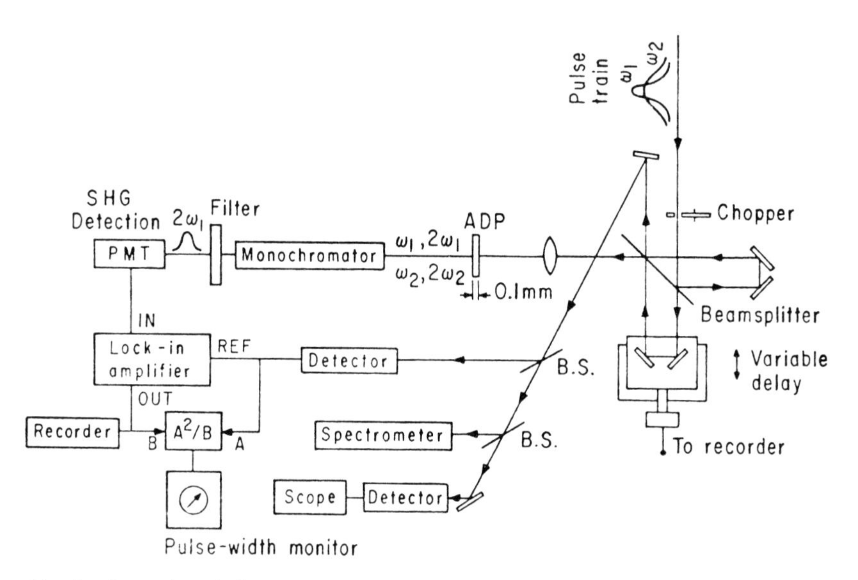

Fig.1 Experimental arrangement for measuring and monitoring continuous trains of ultrashort pulses

autocorrelation trace but still provided discrimination between the primary and secondary pulse wavelengths. No attempt was made to further shorten the pulses by temporal compression or spectral selection [5] but rather to measure the actual pulses generated by the laser. The combination of a pulse width monitor [6] and scope display was used to track the dependence of the pulse width on alignment and pump power.

The rhodamine 6G concentration was kept at 1.33×10^{-3} [M] and the cavity round-trip time was 6.5 [nsec] (the following results were similar with a cavity round-trip time of 8 [nsec]). With the Lyot filter removed from the cavity, double mode-locking was observed at a cresyl violet concentration of $C_{CV} = 6 \times 10^{-6}$ [M] and $P_p \simeq P_p^{th} = 500$ [mW]. The absorber concentration was increased up to about $C_{CV} = 2.2 \times 10^{-5}$ [M], where $\lambda_{R6G} = 594$ [nm] and $\lambda_{CV} = 629$ [nm]. A SHG autocorrelation trace repeatedly showed a noise correlation spike with the proper 3:2 contrast ratio corresponding to the large lasing bandwidths of about 50 [Å] (mirror displacement was about 20 [μm]). The shortest R6G pulse generated without a Lyot filter in the cavity (assuming a sech^2 pulse shape throughout) was 55 [psec] and for the cresyl violet pulse it was about 100 [psec]. The pump power had to be near threshold for the minimum pulse widths. For all concentrations, as the pump power increased beyond threshold, the A^2/B monitor indicated shorter pulse widths whereas the scope display indicated increased pulse widths, showing that amplitude modulation was occurring consisting of intense subpicosecond pulses. With a further increase of pump power, the A^2/B monitor correlated more closely with the scope display, indicating pulsewidths of about 2 [nsec] at the higher power levels. At high pump power, a CW background was produced. Figure 2a shows a 500 [psec] detector-limited display at threshold, whereas Fig.2b gives an indication of the pulse envelope substructure above threshold. These measurements demonstrate that amplitude substructure (in addition to possible frequency sweeps) contributes to the excess bandwidth.

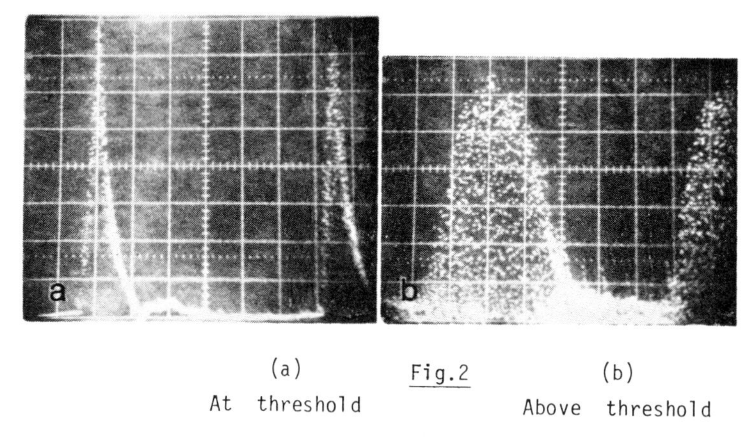

(a) At threshold Fig.2 (b) Above threshold

Time scale: 1 [nsec/div]

The Lyot filter was then placed in the cavity. With $P_p \simeq P_p^{th} = 800$ [mW] ($C_{CV} = 9.2 \times 10^{-6}$ [M]) and the filter tuned for $\lambda_{R6G} = 600$ [nm] and $\lambda_{CV} = 633$ [nm], a pulsewidth of 0.68 [psec] was measured, as shown in Fig.3. The R6G

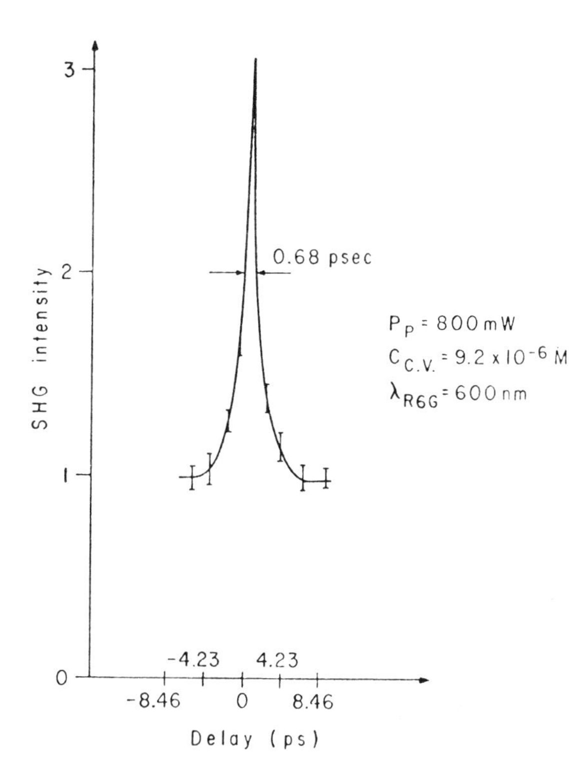

Fig.3 Autocorrelation plot for $C_{CV} = 9.2 \times 10^{-6}$ [M], $P_p = 800$ [mW], $\lambda_{R6G} = 600$ [nm]

pulse was transform-limited to within a factor of 1.7. Tuning the Lyot filter near $\lambda = 600$ [nm] consistently gave pulses between 0.7 - 1.0 [psec] at this concentration. Pump power had to be maintained near threshold and alignment was critical, in common with conventional mode-locking schemes [6]. At high absorber concentrations, pump power was not as critical. Beyond threshold, incomplete mode-locking resulted with amplitude substructure, similar to Fig.2b. At these low concentrations, the cresyl violet pulses were almost always incompletely mode-locked. They contained at most 25% of the total laser average power. At the other concentrations used, the R6G pulsewidths were typically 1 - 4 [psec]. Shortest pulses were observed at $\lambda_{R6G} \simeq 600$ [nm] but there were some 1 and 4 [psec] pulses observed at $\lambda = 590$ [nm].

A possible technique to further reduce the width of the *secondary* pulse is to incorporate a third laser dye within the cavity (triple mode-locking). A tertiary ultrashort pulse can be generated at an emission wavelength of this dye. One proposed dye combination having the proper spectral properties is shown in Fig.4. The cavity configuration can remain identical to that used

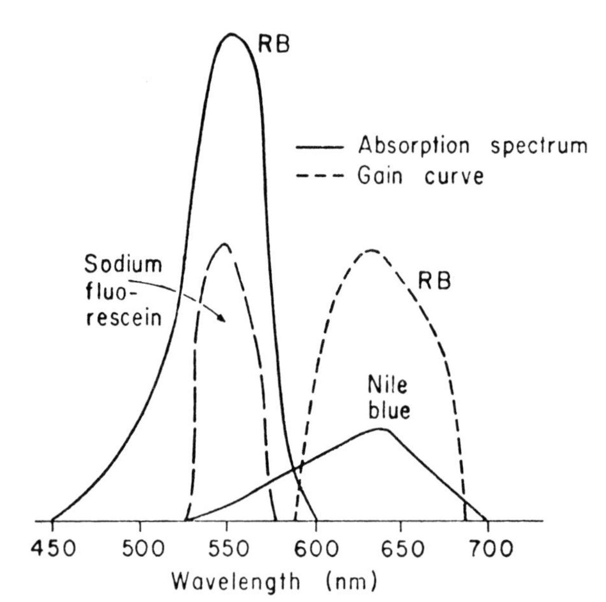

Fig.4 Spectral properties of a possible dye combination for triple mode-locking

for double mode-locking by adding the third dye to the same jet stream as the primary dye, there being minimal overlap between these dyes. Argon-pumped sodium fluorescein lases near the absorption peak of rhodamine B, which subsequently lases at the absorption peak of nile blue. The latter dye then lases at the peak of its gain curve. Other dye combinations are also possible.

REFERENCES

1. Z. A. Yasa, J. Appl. Phys. 46, 4895 (1975).
2. Z. A. Yasa, A. Dienes and J. R. Whinnery, Appl. Phys. Lett. 30, 24 (1977).
3. Z. A. Yasa, Ph.D. Dissertation, University of California, Berkeley (1976).
4. Z. A. Yasa, E. Bourkoff and A. Dienes, to be published.
5. E. P. Ippen and C. V. Shank, Appl. Phys. Lett. 27, 488 (1975).
6. C. V. Shank and E. P. Ippen, Appl. Phys. Lett. 24, 373 (1974).

Subpicosecond Spectroscopy

E.P. Ippen and C.V. Shank
Bell Telephone Laboratories, Holmdel, NJ 07733, USA

The availability of subpicosecond optical pulses at high repetition rates now greatly extends the capabilities of time-resolved spectroscopy. Powerful signal averaging techniques can be applied to picosecond studies. Measurements with 10^{-13} sec resolution or better can be made without distorting the process under investigation. In this paper we describe the further extension of these subpicosecond techniques to high intensity and to broadband wavelength tunability.

Key to our system for subpicosecond studies is a passively-modelocked cw dye laser that has been described previously [1,2]. It is pumped by 5 Watts of continuous power from an argon laser and utilizes two free-flowing ethylene glycol streams, one containing Rh6G for gain and the other containing a saturable absorber mixture of DODCI and malachite green. No active modulation or cavity-length stabilization is necessary. The system runs for days at a time without adjustment and without appreciable changes in output pulse characteristics. Single subpicosecond pulses can be acousto-optically dumped from the laser at any desired repetition rate up to about 10^5 pps. Each dumped pulse has an energy of about 5×10^{-9}J.

A schematic of an experimental arrangement for monitoring and utilizing these subpicosecond pulses is shown in Fig.1. The pulse train from the laser is divided into two beams with a relative delay that can be accurately controlled by automated positioning of a stepper-motor-driven stage. The two beams may be used for a correlation-type measurement as shown at the top of the figure or for a pump-probe type experiment as indicated in the center. The correlation technique is generally invoked for pulse monitoring by either second-harmonic generation in KDP or by two-photon fluorescence [3]. Recently, this same part of the apparatus has also been applied to the study of nonradiative relaxation of excited states [3,4]. In the particular example of azulene, two-step excitation to the fluorescing, second singlet state S_2 is induced via the short-lived first singlet state S_1. The lifetime of the intermediate (S_1) state can be deduced from the observed broadening of the pulse autocorrelation. Experiments with both azulene-d_0 and azulene-d_8 yielded an S_1 lifetime of 1.9 ± 0.2 psec independent of solvent or deuteration. To confirm that this lifetime is governed by recovery to the ground state, pump-probe measurements were also carried out [4]. For

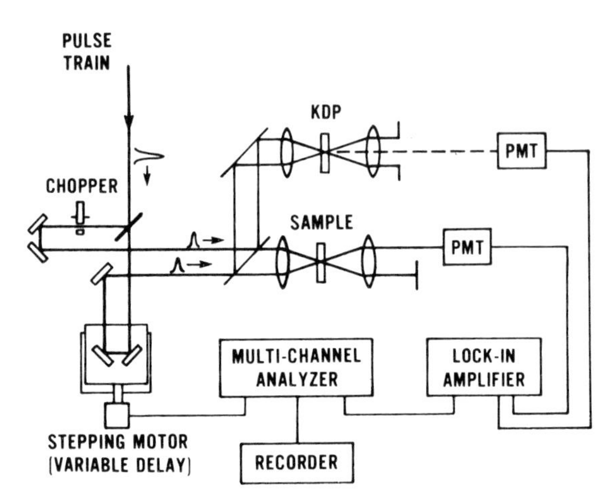

Fig.1 Schematic of cw subpicosecond pulse measurement scheme.

these latter experiments amplified pulses were used to produce a detectable ground state bleaching transient but the detection arrangement was the same as that shown in Fig.1. Complete recovery of the ground state with an exponential time constant of 1.9 ± 0.5 psec was observed.

The system described above has already been adapted and applied to a variety of different studies utilizing both visible (∿ 615 nm) and ultraviolet (∿ 308 nm) subpicosecond pulses [5,6,7]. One way to expand its range of applicability is to add a second, wavelength-tunable laser. A point we emphasize here is that this second laser need not be a picosecond source. The detection technique that makes this possible is shown in Fig.2. A cw (or long pulse) laser is used to detect the response at ω_2 of a sample to a subpicosecond pump pulse at ω_1. This response is then detected by up-conversion with a second, delayed pulse. Temporal resolution is determined in this way entirely by a single, subpicosecond source; and preliminary experiments indicate that the detection capabilities of this "read-in" - "read-out" scheme are similar to those of a simple pump probe experiment. [8] It provides new wavelength coverage in conjunction with low power, high repetition rate measurements.

Entirely new possibilities are created by the amplification of subpicosecond pulses to high powers. High power pulses facilitate the study of low density media and weak interactions. Subpicosecond durations make it easier to avoid material breakdown and allow for the use of high intensities.

High power amplification has been accomplished in the manner illustrated in Fig.3. Single subpicosecond pulses from the cw modelocked laser are selected in synchronism with the frequency doubled output of an amplified, Q-switched Nd:YAG laser. The

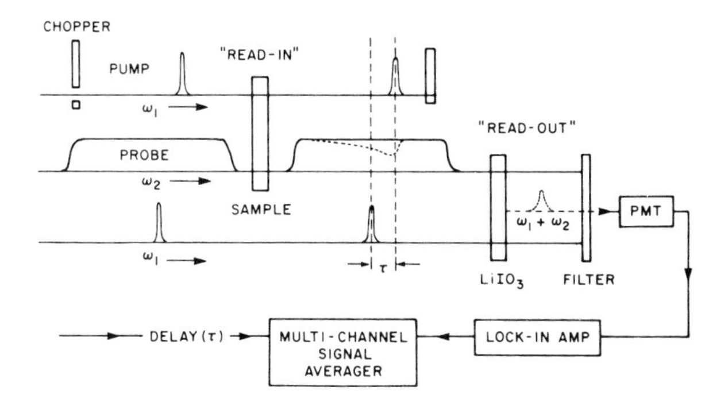

Fig.2 Schematic of "read-in" - "read-out" measurement scheme.

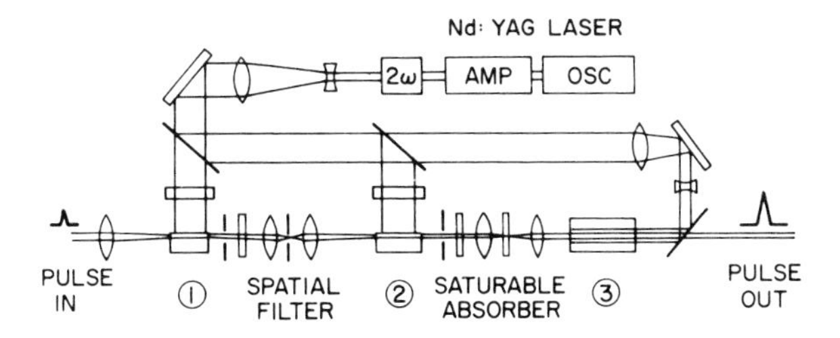

Fig.3 Subpicosecond pulse amplifier.

timing jitter is less than 2 nsec which is sufficient to ensure that each pulse arrives within the peak gain period of the amplifier. Amplification is accomplished in three stages. The beam diameter is expanded appropriately between stages. Spatial filters and a stream of the fast saturable absorber malachite green before the final stage keep amplified spontaneous emission from being a problem. In this way 0.5 psec pulses have been amplified to peak powers of more than a gigawatt. The repetition rate of the entire system is 10 ppS. Pulse measurements before and after amplification show no significant distortion or change in pulse shape in the process.

These amplified pulses are easily converted to different wavelengths by various nonlinear optical techniques. One of the most dramatic of these is continuum generation. By focussing 615 nm amplified pulses into H_2O or D_2O we are able to generate a continuous spectrum from the near infrared to the near ultraviolet. Measurements on selected portions of this continuum have shown that its duration is also subpicosecond.

An experiment made possible by amplified subpicosecond pulses is diagrammed in Fig.4. This arrangement has been used to study the dynamics of optically induced changes in the near band-gap transmission of GaAs [9]. The output of the amplifier is divided

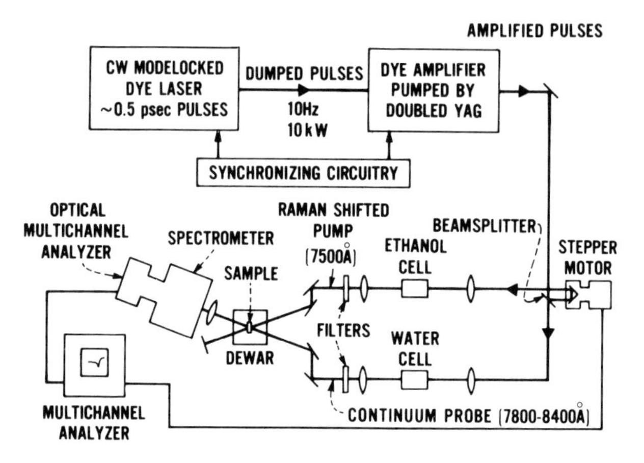

Fig.4 Measurement scheme using amplified picosecond pulses and continuum generation.

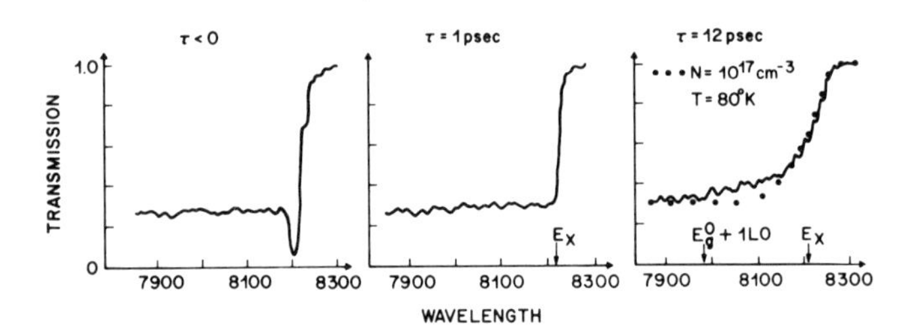

Fig.5 Excited state time resolved absorption spectra of GaAs.

into two beams with variable relative delay. One beam of pulses is converted to 750 nm by stimulated Raman scattering in ethanol. These pulses excite the thin (0.5 μ), cooled (80°K) sample of GaAs. The second beam is used to generate the continuum probe. Filters attenuate and select the wavelength range of interest (780 nm-8400 nm) to probe near the band edge of GaAs. This spectrum, after transmission through the sample, is dispersed by a spectrometer onto the face of an optical multichannel analyzer (OMA). The complete spectrum may then be registered and averaged over many pulses at each point of delay. Transmission at selected wavelengths was also measured as a continuous function of delay.

The development of the transmission spectrum with time is illustrated selectively in Fig.5. Before the 750 nm pump pulse arrives ($\tau < 0$) one sees the unperturbed band edge absorption as well as a sharp absorption peak due to free excitons. Immediately following excitation (τ = 1 psec) the exciton resonance is no longer evident. This may be attributed to screening of the exciton resonance by the excited carriers [10]. At the same time the band edge has shifted to just below the exciton peak. The spectrum taken at τ = 12 psec shows the effects of band filling after the carriers have come into equilibrium with the lattice. Intermediate measurements reveal the time constant of this equilibration to be about 4 psec.

These recent results demonstrate, we think, that subpicosecond technology based on a stable, continuous source is now in an increasingly better capability to contribute to important advances in time-resolved spectroscopy.

The authors gratefully acknowledge collaborations on various parts of this work with R. L. Ford, R. F. Leheny, A. Migus, Jagdeep Shah, O. Teschke, and J. Wiesenfeld.

References

1. E. P. Ippen and C. V. Shank, Appl. Phys. Lett. 27, 488 (1975).
2. E. P. Ippen, C. V. Shank: In *Ultrashort Light Pulses*, ed. by S. L. Shapiro, Topics in Applied Physics, Vol. 18 (Springer, Berlin, Heidelberg, New York 1977).
3. E. P. Ippen, C. V. Shank and R. L. Woerner, Chem. Phys. Lett. 46, 20, (1977).
4. C. V. Shank, E. P. Ippen and R. L. Fork, to be published.
5. C. V. Shank, E. P. Ippen and O. Teschek, Chem. Phys. Lett. 45, 291 (1977).
6. D. H. Auston, S. McAfee, C. V. Shank, E. P. Ippen and O. Teschke, Solid St. Elect. 21, 147 (1978).
7. E. P. Ippen and C. V. Shank, Physics Today, p. 41, May 1978.
8. J. Wiesenfeld, E. P. Ippen, and C. V. Shank, to be published.
9. C. V. Shank, R. L. Fork, R. F. Leheny and Jagdeep Shah, to be published.
10. J. Shah, R. F. Leheny and W. Wiegman, Phys. Rev. B15, 1577 (1977).

Synchronized Ultra-Short Pulse Generation and Streak-Camera Measurement with cw Mode-Locked Lasers

M.C. Adams, D.J. Bradley, and W. Sibbett

Optics Section, Blackett Laboratory, Imperial College
London SW7 2BZ, England

Abstract

A picosecond streak camera operating in synchronism with C.W. mode-locked lasers provides real-time measurement of repetitive luminous phenomena with a temporal resolution of ∿ 10 ps. The durations of pulses from an actively mode-locked argon ion laser (514.5 nm) have been measured to be ∿ 90 ps and the system is shown to be particularly suitable for the measurement of dye fluorescence lifetimes.

Electron-optical streak-cameras [1], which are now capable of subpicosecond time resolution [2], are being increasingly employed in experiments involving picosecond optical phenomena. Frequency tunable light pulses of picosecond and sub-picosecond durations can be conveniently generated at repetition rates of ∿ 100 MHz from mode-locked CW dye lasers [3]. While the temporal intensity profile of a single pulse (Peak power ∿100 - 1000W) can be directly measured with a streak-camera [4], for the study of fluorescence, or other luminous phenomena, generated by the pulse train, it is desirable to accumulate the streak records to improve the overall signal-to-noise ratio. This approach is particularly useful for the study of picosecond events when, in order to avoid nonlinear effects in the experimental media, it is necessary to use the lower intensity pulses from CW dye lasers rather than mode-locked flashlamp pumped dye lasers or solid state lasers [3]. The advantages [5] of combining a mode-locked CW laser with a streak-camera driven synchronously at the repetition rate of the pulse train became apparent while carrying out the work of reference [4]. Experiments with a burst of pulses from a mode-locked Nd:YAG laser [6] and with a CW dye laser train [7] demonstrated the feasibility of the concept. In particular, it was shown that it is possible to precisely superimpose successive dye laser pulse streak records on the image-tube phosphor at a rate of 140 MHz, while maintaining a time-resolution of < 25 ps [7]. With the low values of instantaneous photo-electron currents involved, no further image intensification is needed to avoid photocathode saturation and space charge effects, which occur when high-power picosecond pulses are employed [1-4]. The steady streak image is clearly visible on the streak-phosphor and can be readily photographed for microdensitometry. We wish to report further developments carried out in the past year including the achievement of a time-resolution of ∿ 10 ps and the use of an optical multichannel analyser (OMA) to record and store the superimposed streak-images. The linear intensity profiles can then be subsequently displayed on a slow oscilloscope screen or printed out on a pen-recorder. The system has been employed to study active mode-locking of an argon ion laser and, as a demonstration of its capability, to measure the fluorescence lifetime of dyes.

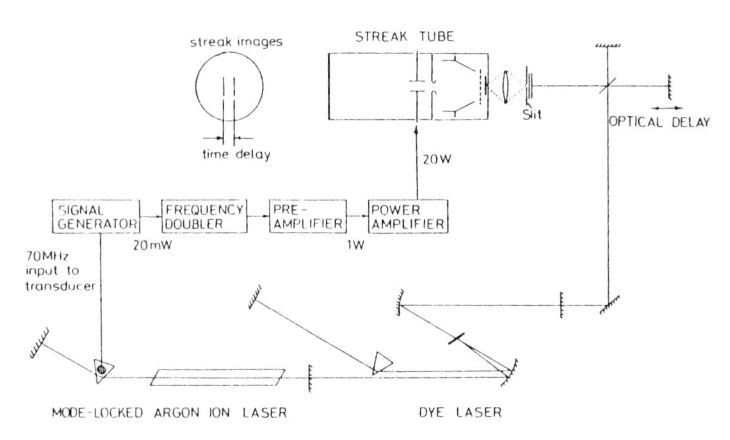

Fig.1 Experimental arrangement for synchronizing streak-camera to CW dye laser pulse train (repetition rate 140 MHz).

The experimental arrangement is shown schematically in Fig.1. The master RF oscillator provides a 1 W signal for the transducer of the acousto-optic mode-locking prism of the Argon-ion laser (Spectra-Physics Model 165), and a 20 mW signal for frequency-doubling and amplification (to 20 W) to synchronously drive the streak-tube (Photochron I) deflection plates. Alternatively synchronous driving is achieved with the signal from a photodiode detector of the mode-locked laser pulse train. The sinusoidal driving voltage (Fig.2), ∿ 3.5 kV, is adjusted to ensure that there is a linear deflection of ∿ 5 cm in the center of the image tube phosphor. Calibration of the deflection is easily obtained by varying the separation of the two pulses generated (from each incoming pulse) in the optical delay "interferometer" arrangement. The synchronously pumped [3] Rhodamine 6 dye laser can generate pulses as short as ∿ 1 ps at an average power of 40 mW (300 W peak power) when the optimum cavity length is maintained to an accuracy of ∿ 1 μm and the Argon laser is stably mode-locked [8]. The streak-camera tube (Photochron I, SII photocathode) has been previously shown [4] to have a time resolution limit of 2.5 ps at the dye laser wavelength, set by photo-electron time-dispersion [1]. In the present work, with a writing speed of 5×10^9 cm s^{-1} at the phosphor and a dynamic

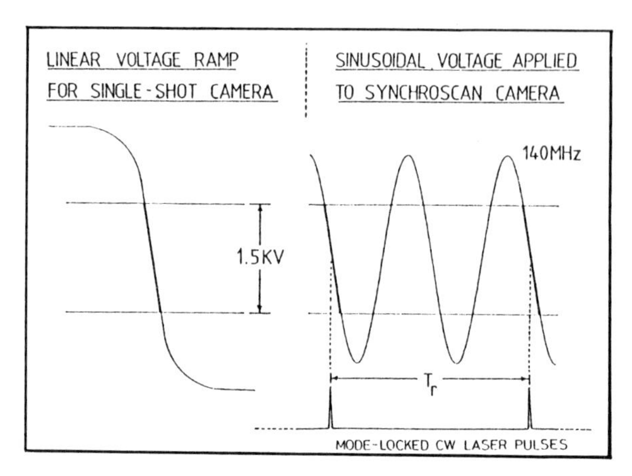

Fig.2 Sinusoidal driving voltage for synchronous streak-camera.

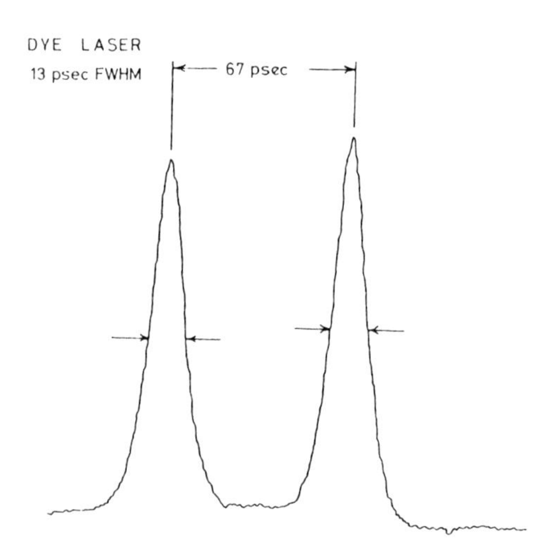

Fig.3 Linear intensity record using O.M.A. of two dye laser pulses generated in optical delay line. To obtain this record 4×10^7 pulses were integrated for 400 milliseconds. The streak-camera resolution is $\sim$ 5 ps. The recorded pulse width of 13 ps is increased by laser pulse jitter and R.F. drift and interference.

spatial resolution of 5 lp/mm the overall streak-tube resolution limit becomes $\sim$ 4.5 ps. The OMA (PAR Model 12050) was optically coupled to the streak-tube phosphor with a F/1.5, 8 cm focus lens. The spatial resolution of the OMA (15 lp/mm) did not limit the overall time-resolution. Despite this the recorded pulse width was 13 picoseconds (Fig.3).

At present it is difficult to determine what factors limit the time-resolution. The most obvious parameters are drift in the

R.F. driving electronics, jitter in the repetition rate of the mode-locked CW lasers and perhaps variation in the durations of the dye laser pulses. A jitter of 10 ps has been reported[10] in the synchronization of the pulse trains of two CW dye lasers synchronously pumped by a mode-locked Argon-ion laser. Interference between the high-power R.F. circuit driving the streak tube and the circuit of the acousto-optical modulator also could contribute to the instability of the overall system. The various factors, at present broadening the record by $\sim$ 10 ps, are being investigated using a passively mode-locked dye laser, which gives bandwidth limited pulses of durations 0.3 ps [9].

As the dye laser must be affected by the quality of the Argon-ion laser mode-locking, we studied the performance of this laser. Earlier investigations using as detector a fast photodiode [10] and a single shot streak-camera [11] had given distinctly different results. When the acousto-optic modulator was driven by a Wankel and Gottelman (Model LM 568) oscillator a pulse duration of 87 ps (FWMH) was obtained as shown in Fig.4. With a Harris Intertype oscillator [12] the pulses were at best 95 ps and were not reproducible, showing the importance of a stable oscillator. In both cases the average output power was $\sim$ 600 mW (514.5 nm line). With the continuously driven streak-camera optimization of the Argon-ion laser cavity length, the period of the RF oscillator and the acoustic resonance is readily achieved. Our results are in very good agreement with those reported in [10].

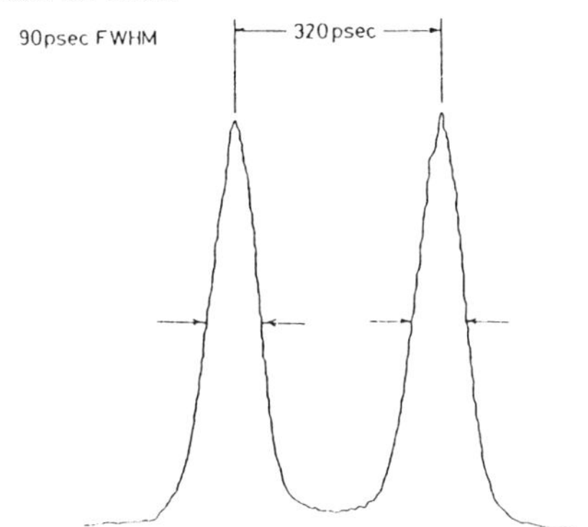

Fig.4 O.M.A. record (as in Fig.2) of argon-ion laser pulses.

The arrangement used for measuring fluorescence lifetimes is shown in Fig.5. While excited state lifetimes have been directly measured with picosecond streak cameras, including emission from dyes under mode-locking conditions [13,14] and fluorescence from DODCI [15], from photosynthetic samples [16] and from scintillation materials [17], the measurements have previously been carried out on a single-shot basis. With the repetitively scanning streak-camera these experiments become remarkably simple and the effects of varying the laser pulse parameters are immediately seen on the monitoring oscilloscope. When the dye laser was tuned

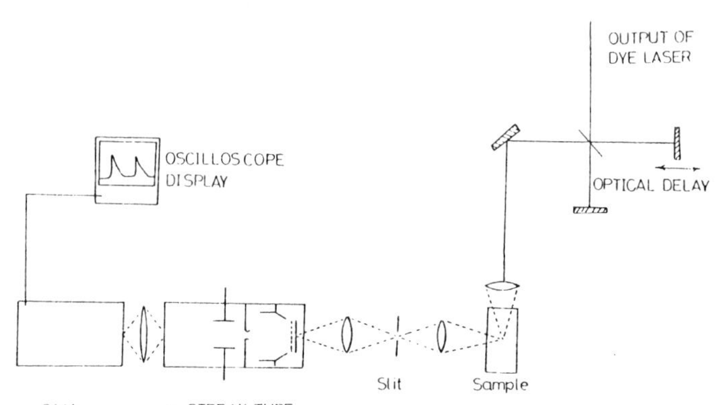

Fig.5 Arrangement for measuring fluorescence lifetimes with mode-locked CW dye laser and synchronously driven streak-camera.

to 590 nm and the output beam was focussed (5 cm focal length, F/2 lens) into a cell containing the mode-locking dye DODCI (10^{-3}M in ethanol) a 1/e relaxation time of $\sim$ 1 ns (Fig.6) was obtained in good agreement with the generally accepted value of 1.2 ns [3]. The fluorescence lifetime of Erythrosin B (Fig.6) was measured in a similar manner. Variation of decay times with wavelength by passing the fluorescence through a scanning monochromator, is at present under study.

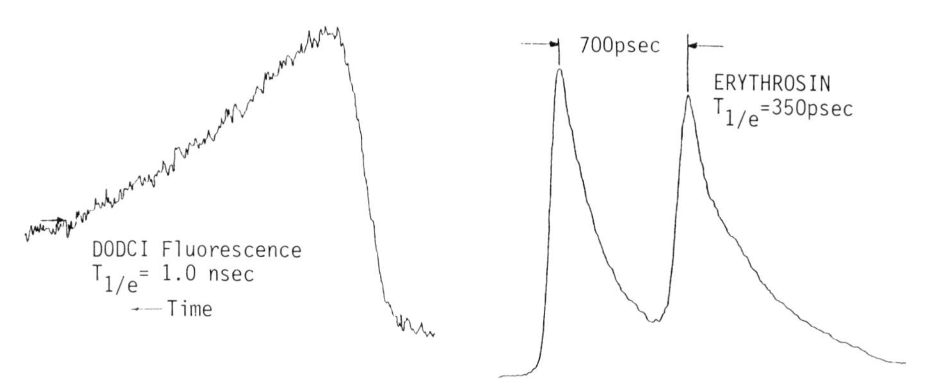

Fig.6 Measurement of fluorescence lifetimes using arrangement of Fig.4. (a) DODCI and (b) Erythrosin B. Each record represents 4 × 10^7 measurements integrated over 400 milliseconds.

From these preliminary results it is clear that the combination of a C.W. mode-locked laser and a repetitively driven synchronous streak-camera provides a very convenient method for the selective excitation and direct measurement of picosecond phenomena in materials of interest in photochemistry, biology and condensed matter physics. With further improvements in the jitter of the CW dye laser pulse repetition rate and of the streak-camera RF driving electronics stability, the time-resolution limit (< 1 ps) set by streak-tube photoelectron dispersion should be achievable.

The authors wish to thank S. F. Bryant, J. P. Ryan and M. M. Salour for many helpful discussions. Financial support from the Science Research Council and the Paul Instrument Fund is gladly acknowledged.

References

1. D. J. Bradley and G. H. C. New: Proc. IEEE, 62, 313 (1974) and references therein.
2. D. J. Bradley and W. Sibbett: Appl. Phys. Lett., 27, 382 (1975).
3. D. J. Bradley: Topics in Applied Physics, Vol. 18, Ultrashort Light Pulses, Ed. S. L. Shapiro (Springer Verlag, New York, 1977) pp. 17-81 and references therein.
4. E. G. Arthurs, D. J. Bradley, B. Liddy, F. O'Neill, A. G. Roddie, W. Sibbett and W. E. Sleat: Proc. of 10th International Congress on High Speed Photography (A.N.R.T. Paris, 1972) p. 117.
5. D. J. Bradley: UK Patent Application No. 34544/72 (US Patent 3973117 1976).
6. R. Hadland, K. Helbrough and A. E. Houston: Proc. 11th International Congress on High Speed Photography, Ed. P. J. Rolls (Chapman and Hall, London, 1975) p. 107.
7. D. J. Bradley: "Methods of generation and measurement of picosecond pulses in the VUV to infrared spectral regions", Proc. of 32nd Symposium on Molecular Spectroscopy, Ohio, 1977. (To be published in Journal of Physical Chemistry).
8. J. P. Ryan and D. J. Bradley: unpublished.
9. I. S. Ruddock and D. J. Bradley: Appl. Phys. Lett., 26, 296 (1976).
10. J. P. Heritage and R. K. Jain: Appl. Phys. Lett., 32, 41 (1978).
11. E. D. Jones and M. A. Palmer: Optical and Quantum Electronics, 9, 451 (1977).
12. The Harris oscillator was found to have a strong 50 Hz modulation of its output signal.
13. E. G. Arthurs, D. J. Bradley, A. G. Roddie: Chem. Phys. Lett., 22, 230 (1973).
14. E. G. Arthurs, D. J. Bradley, P. N. Puntambekar, I. S. Ruddock and T. J. Glynn: Opt. Commun., 12, 360 (1975).
15. J. C. Mialocq, A. W. Boyd, J. Jaraudias and J. Sutton: Chem. Phys. Lett., 37, 236 (1976).
16. S. L. Shapiro, V. H. Kollman and A. J. Campillo: FEBS Lett., 54, 358 (1975).
17. S. L. Shapiro, R. C. Hyer and A. J. Campillo: Phys. Rev. Lett., 33, 513 (1974).

Picosecond Pulse Generation with a cw GaAlAs Laser Diode

P.-T. Ho, L.A. Glasser, E.P. Ippen[1], H.A. Haus

Department of Electrical Engineering and Computer Science
and Research Laboratory of Electronics
Massachusetts Institute of Technology
Cambridge, MA 02139, USA

We report the operation of a compact picosecond pulse generator based on the CW GaAℓAs double heterostructure diode laser. Pulse durations of 23 psec have been obtained with a pulse repetition rate of 3 GHz. The diode operates CW at room temperature in an external resonator whose modes are actively locked by direct modulation of the diode current. The output pulses are measured by auto-correlation using second harmonic generation in $LiIO_3$.

The system is shown in Fig. 1. The commercial laser diode is uncoated on both output surfaces and is placed at the center of curvature of the external aluminum mirror. The mirror is spherical and has a radius of 5 cm. The composite resonator consists of three reflecting surfaces: the two diode cleavage planes and the external mirror.

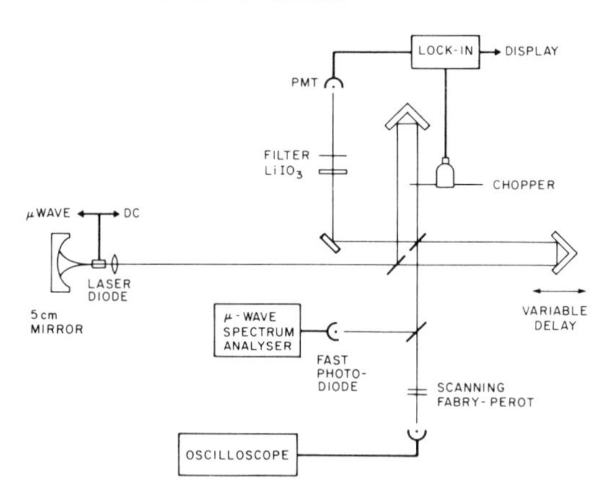

Fig.1 Experimental arrangement for generating picosecond pulses with a laser diode, measuring the pulse durations by autocorrelation, and monitoring both the microwave and optical spectra.

The isolated laser diode has a threshold current of 190 mA; the output wavelength is centered at 810 nm and is divided between several longitudinal modes 3.17 Å apart. The composite resonator reduces the threshold to 145 mA and the emission linewidth to no more than 0.5 Å under CW operation, as measured

[1]Permanent Address-- Bell Laboratories, Holmdel, New Jersey 07733

by a grating spectrometer. There are, however, multiple external resonator modes present as evidenced by a 3 GHz = c/2L beat signal on the microwave spectrum analyzer. The DC drive current of the laser is modulated with several milliwatts of microwave power at the observed 3 GHz beat frequency. The effect of tuning the modulation frequency may be observed via the spectrum analyzer. There is a dramatic spectral narrowing of the beat as the drive is tuned to resonance. The optical spectrum on the other hand is broadened.

The output of the laser is directed to a modified interferometer where it is divided into two, approximately colinear, beams with variable relative delay. Intensity autocorrelation is performed by phase-matched SHG in $LiIO_3$. The two infrared beams have relative intensities of 8:5; the weaker beam is chopped. Fig. 2 shows an experimental curve. The pulse duration is 23 psec assuming a Gaussian pulse shape. Some contrast is lost in the autocorrelation measurement because of spatial inhomogeneities in the beams. The contrast ratio is still greater than 3:1 because only one of the beams is chopped. In the absence of active modelocking the autocorrelation trace broadens and decreases dramatically in contrast and amplitude.

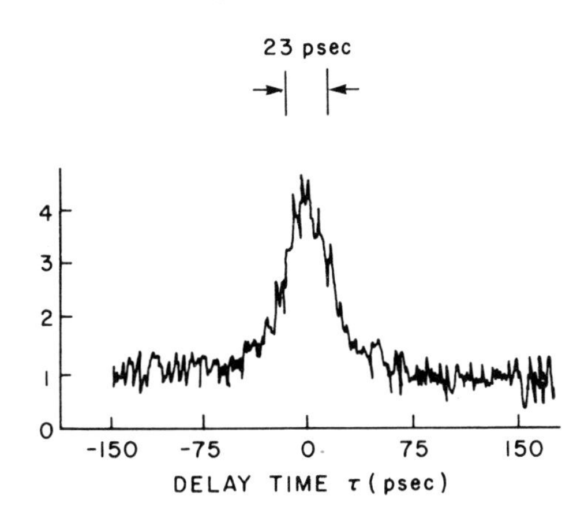

Fig.2 Pulse correlation by second harmonic generation. The duration of 23 psec (FWHM) assumes a gaussian pulse shape.

The fine structure on top of the autocorrelation trace indicates the presence of substructure with a period $2\ell_d/c$ corresponding to the (short) roundtrip time in the diode of optical length ℓ_d. This is an indication that the modes of the combined resonator are locked in clusters spaced by $c/2\ell_d$ along the frequency axis. Several interesting conclusions and interpretations follow from this picture. First of all it is clear that the composite-resonator loss has minima spaced apart by $c/2\ell_d$. Theory shows[1] that the peak gain versus frequency of a modelocked homogeneously broadened laser has to be higher than the peak gain of the free-running laser in order to accomodate more neighboring modes. The increase of the gain in the presence of a modelocking drive may lead to operation of modes in other loss minima along the frequency axis which do not operate when no modelocking drive is present. This explains the broadening of the optical spectrum when the modelocking drive is turned on. Further, the clustering of the modelocked modes around the loss minima presents coherent locking of different mode clusters and leads to a nondeterministic substructure underneath the pulse envelope with a temporal separation of $2\ell_d/c$. Finally, the picture suggests what should happen when an anti-reflection coated diode is used in the resonator. The local loss minima are removed and the loss versus mode frequency is now determined by accidental discrimination of the resonator modes by irregularities of the optical system. Modelocking in deterministically fixed clusters does not occur and in general one would expect worsening of the modelocking.

We have observed this to be true when a A.R. coated diode was substituted for the uncoated diode. The pulses were not less than 60 psec.

On the other hand, the above picture suggests that a careful control of the loss versus frequency characteristic (by e.g. gratings, prisms) in an external resonator could lead to better control of the loss characteristic and hence shorter pulses with no substructure. Our future work will explore this possibility.

The 23 psec pulses are the shortest ever reported from a CW diode laser. An additional important feature of this device is that these pulses are available in a controlled way at a microwave repetition rate.

This work was supported in part by the Joint Services Electronic Program (Contract DAAB07-76-C-1400).

1. H. A. Haus: J. Quantum Electron., QE-11 323 (1975)

Investigation of the Parameters Affecting Subpicosecond Pulse Durations in Passively Mode Locked Dye Lasers

J.-C. Diels, E.W. Van Stryland, and D. Gold

Center for Laser Studies, University of Southern California
Los Angeles, CA 90007, USA

We report the shortest pulses generated to date using a simple dispersionless three mirror cavity. In addition, for the first time, accurate fringe resolved autocorrelation measurements with a peak to background ratio of 8 to 1 are presented. These fringes enable us to determine in a single recording whether or not the pulses are bandwidth limited.

Passive mode locking has been made previously in a dispersive cavity containing either two jets (7 optical elements) or one jet of active medium and one short cell of saturable absorber (8 optical elements in the cavity). The shortest pulse durations measured with these cavities were respectively 0.5ps [1] and 0.3ps [2] . Pulses as short as 0.17ps have been generated using a simple dispersionless cavity of only 4 optical components (2 focusing mirrors and a flat ouput mirror). The saturable absorbers (dioxadicarbocyanine-iodide or DODCI and malachite green) were mixed with the Rhodamine 6G-ethylene glycol solution in the single jet stream of the laser. There are no prisms, etalons or dye cells which may act as a pulse bandwidth limiting element. After all, dye cells cannot be shortened indefinitely. The shortest pulses were obtained using a mirror having a reflectivity of 99.7% between 600 and 620 nm. Within that range, the dispersive properties of the dye solution select the wavelength.

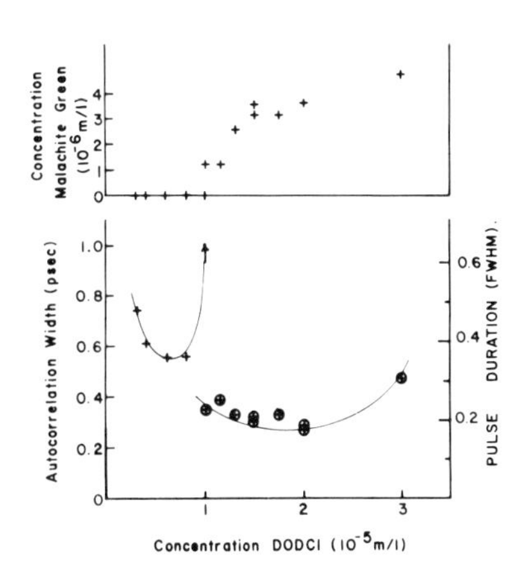

Fig. 1 Concentration dependence of the pulse duration

The dye was excited by cw argon ion laser with stabilized output power (Control Co, model 552). The pulse duration, as measured at threshold by a standard second order autocorrelation method, is plotted in Fig. 1, as a function of the composition of the dye solution. The left ordinate indicates the FWHM of the autocorrelation trace, while the actual pulse duration (FWHM) is shown on the right ordinate. A scale ratio of 1.55 is used, corresponding to

$sech^2$ shaped pulses, as justified by autocorrelation fits presented below. The concentration of Rhodamine 6G was 2×10^{-2} M/l. As DODCI alone is added to the solution, the pulse duration drops rapidly to a minimum. These data points (no malachite green, DODCI concentration $\times 10^{-5}$ M/l) are extremely dependent on particular cavity configuration and losses. At a concentration of 10^{-5} M/l of DODCI, the laser pulses have low coherence and long duration, as observed from the fringe resolved autocorrelation measurement. Good mode locking is restored by the addition of malachite green, the concentration of which is plotted in the upper part of the figure.

To measure pulse durations by second order autocorrelation methods, the pulse train is split into two beams. Both beams are recombined with a known differential delay. They are then sent through a frequency doubling crystal. The signal recorded is the second harmonic pulse energy as function of the delay between trains. All the methods used differ in the type of nonlinear mixing. In the zero background methods |3|, one detects the product of both fields, while for methods including background [2], the fields of both pulse trains are added. If $E(t) = \mathcal{E}(t)\cos(\omega t + \phi t)$ is the electric field of a light pulse, the function of delay measured by the latter method is the time average:

$$\int |\{E(t) + E(t-\tau)\}^2|^2 dt = 3/8 \int \{\underline{\mathcal{E}^4(t) + \mathcal{E}^4(t-\tau)} + \underline{4\mathcal{E}^2(t)\mathcal{E}^2(t-\tau)} + 4[\mathcal{E}^2(t) + \mathcal{E}^2(t-\tau)]\mathcal{E}(t)\mathcal{E}(t-\tau)\cos[\omega\tau - \phi(t) + \phi(t-\tau)] \mathcal{E}^2(t)\mathcal{E}^2(t-\tau)\cos 2[\omega\tau - \phi(t) + \phi(t-\tau)]\}dt \quad (1)$$

Only the underlined terms are measured in conventional techniques, which average out the fast variations at $\cos\omega\tau$ and $\cos 2\omega\tau$. The first two terms are constant factors independent of pulse overlap, and give a contribution of $2\int\mathcal{E}^4 dt$ in the wings of the autocorrelation trace. With the next term, (which is the only one measured in "zero background" methods) the autocorrelation function peaks at $6\int\mathcal{E}^4 dt$ in the center of the trace ($\tau=0$), giving the standard peak to background ratio of three to one. We have constructed an accurate autocorrelator that is able to resolve the next interference terms, which are the only ones that contain phase information. The contributions of all terms at $\tau=0$ add up to $16\int\mathcal{E}^4 dt$, yielding a peak to background ratio of eight to one, which is exactly what was observed. A sample of an experimental recording near the center of the trace is shown in Fig. 2. The solid line is the experimental recording while the circles indicate the functional fitting of (1), assuming the phase factor is constant within the range. While the sample

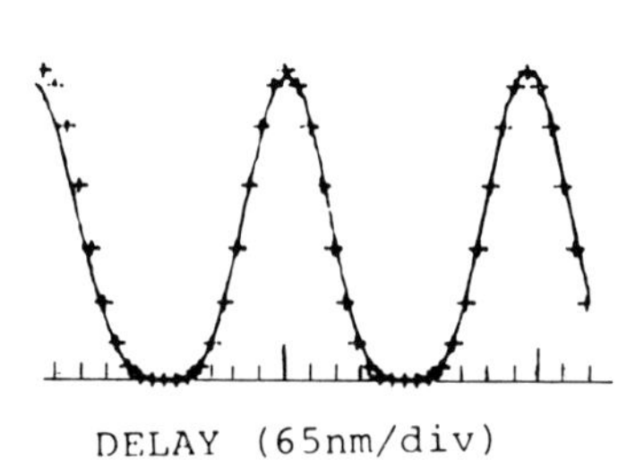

Fig. 2

of Fig. 2 covers only a delay of two light periods, our measurements scan continuously across the full autocorrelation width.

The advantage of recording the full autocorrelation function including the fast variation is that we keep phase or frequency information that is otherwise lost. Figure 3 shows a numerical comparison of the full autocorrelation traces (a and b) and conventional (c) for unchirped (a) against chirped (b) pulses. Figure 3c is the same for chirped and unchirped pulses. The pulses are gaussian shaped ($\mathcal{E}(t)=\exp\{-(t/W)^2\}$) with a linear chirp $\phi(t)=a(t/W)^2$. Even though the chirp is very small (a=0.16), its presence can be inferred at once from the shape of the corresponding autocorrelation curve. A quantitative study of the phase content of the pulse can be made by comparing the curve obtained by taking the average of the maximas and minimas of the complete autocorrelation function (white line in Fig. 3a and b) to the conventional autocorrelation. Such a comparison, with backgrounds subtracted and renormalized, is shown in Fig. 3d. From (1), it can be seen that the average (with background subtracted) has the shape $\int dt\,\mathcal{E}^2(t)\,\mathcal{E}^2(t-\tau)\{1 + \frac{1}{2}\cos 2[(\phi(t)-\phi(t-\tau)]\}$ When the phases are constant, this expression has the same shape as the zero background autocorrelation function.

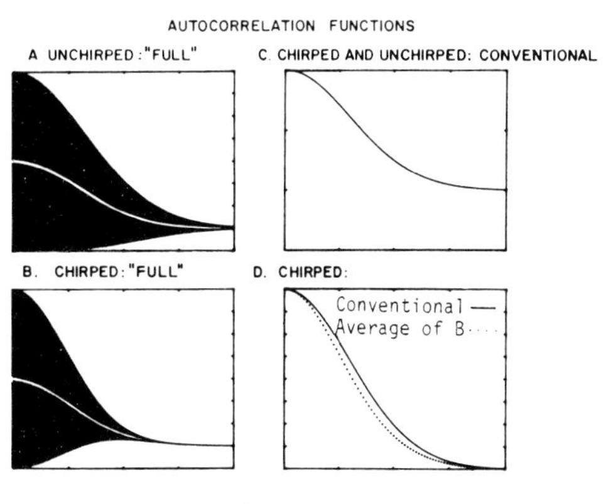

Fig. 3

An experimental comparison is made in Fig. 4. On the upper left corner of Fig. 4, a full experimental recording of the complete autocorrelation function is shown (only the envelope of the maximas and minimas of the interferences is shown). Note that the peak to background ratio is exactly eight to one. The energy average of that measurement shown in the upper right corner is the conventional autocorrelation measurement, with peak to background ratio of exactly 3. The solid line is the experimental recording, while the crosses are a fit to the autocorrelation of a superposition of two sech shaped (amplitude of the electric field (+) pulses. The shorter pulse has a width of 0.19ps and contributes to the central portion of the trace. The broader pulse only contributes to the wings of the autocorrelation function.

The 3/1 autocorrelation experimental curve is retraced on the bottom of Fig. 4 (solid line), and compared to the average of the top and bottom traces of the full autocorrelation measurement (dashed line). The fact that both curves are different, indicates that our signal is not pure in phase. Analysis of oscilloscope traces, pulse spectrum and autocorrelation measurements indicate that the pulse train oscillates regularly between two different regimes. Thus, the two sech pulses of

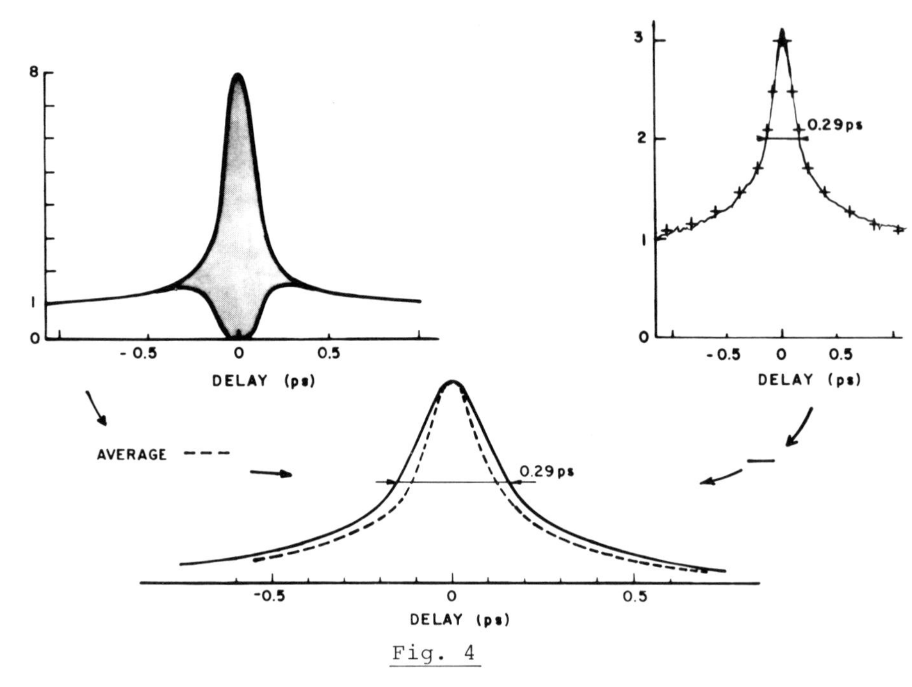

Fig. 4

the aforementioned fit have different frequencies. Measurements indicate that closest to threshold, we have the narrowest pulses centered at 615 nm. As the pump slightly increases, the pulses broaden while the average frequency shifts upward.

It should be noted that the autocorrelator, being made of BK7 glass, does affect the measurement itself. According to linear dispersion theory, [4] a measured pulse of 0.19ps FWHM had a width of 0.17ps before entering the interferometer.

References

1. C.V. Shank, E.P. Ippen: Appl. Phys. Lett 24, 373 (1974).
2. I.S. Ruddock, D.J. Bradley: Appl. Phys. Lett. 29, 296 (1976).
3. E.P. Ippen and C.V. Shank, Appl. Phys. Lett. 27, 488 (1975).
4. D. von der Linde, Appl. Phys. 2, 281 (1973).

Work supported by NSS, number ENG 76-06231.

Measurements of Polarization Relaxations in Materials Using Synchronously Mode-Locked Dye Lasers

Chi H. Lee

Department of Electrical Engineering
University of Maryland, College Park, MD 20742, USA

D. Ricard

Laboratoire d'Optique Quantique du C.N.R.S., Ecole Polytechnique
F-91128 Palaiseau, France

T.R. Royt

Naval Research Laboratory, Washington, DC 20375, USA

Abstract

Synchronous mode-locking allows picosecond pulse manipulation for materials studies involving excitation and probing of free decay. Recent results of relaxation measurements in calcite are discussed.

Important experimental advantage can be gained in studies of nonlinear optical effects and fast physical and chemical reactions if one is able to generate two or more independently tunable light pulses in the picosecond domain that are either precisely synchronized or separated by controllable delay times. It has been shown that such pulse manipulation can be achieved with synchronously mode-locked dye lasers [1]. We present here our measurements of ultra-fast materials relaxation processes obtained with these lasers. Very recent experiments have resulted in direct measurements of the picosecond relaxation times of optical phonons in calcite. Excitation of these vibrations by means of the synchronously mode-locked dye laser obviated the need for the transient stimulated Raman scattering approach with its concomitant high intensity effects and difficulties in exciting vibrations of low relative integrated cross-section.

The synchronously mode-locked dye laser has been studied by a number of groups but little has been published on physical studies that employ this technology. The device is basically a laser that becomes mode-locked when an instantaneous excitation of the active medium occurs at the same time during each cavity oscillation period, which can be accomplished by pumping with the output from a mode-locked laser of equal cavity length. This was demonstrated as early as 1967 by GLENN, BRIENZA and De MARIA [2] and is fact was the first dye laser of any sort to be mode-locked. Other early work has been detailed by SHANK and IPPEN in their 1973 review [3] of mode-locked dye lasers. These lasers were not termed "synchronously pumped" at the time and the interest was purely directed at obtaining a mode-locked dye laser. Other forms of mode-locking gained prominence subsequently, with the result that the approach lay relatively dormant until recently. Since 1974 interest in the technique has revived under the terminology of synchronous mode-locking. Work by many groups includes streak camera measurements of pulse structure and jitter, [1] demonstration of cw pumping, [4] pulse dynamics under cw pumping, [5] nonlinear mixing for IR and UV generation, [6] generation of tunable picosecond pulses in the VUV using a resonantly enhanced 4-wave mixing process in metal vapors made possible by overlapped and tunable pulses from two synchronously mode-locked dye lasers that are pumped in common, [7] synchronous pumping of a jet-stream-dye with a mode-locked kryton laser, [8] and production of subpicosecond pulses by tandem cw pumping. [9] With the exception of almost prophetic work by TOPP and RENTZEPIS [10] in 1971, wherein pump and dye laser pulses were employed internal to a mode-locked dye laser cavity to two-photon pump and probe a laser dye, the work involving synchrous mode-locking has been primarily source development.

The physics of materials has been studied using synchronously mode-locked dye lasers in three recent investigations. The first two, performed by MATSUOKA et al. [11] and by ROYT and LEE [12], have contained relaxation measurements for the coherence of electronic polarizations in calcium and strontium respectively. The third investigation is by LEE and RICARD [13] and it extends this technique to measurement of phonon relaxation in calcite. The physics of interest is the dynamics of material polarizabilities. In particular, molecular and atomic environments may be studied by measuring the rates at which the energy and phase of coherent excitations are lost through interactions with that environment. Synchronously mode-locked dye lasers allow clean excitation of resonances at arbitrary frequencies, be they single-photon resonances requiring frequency conversion by nonlinear mixing or be they two-photon resonances at the sum- or difference-frequencies of two overlapped pulses. Synchronous mode-locking then allows the freely decaying polarization to be probed at any delay time by a pulse at a still different arbitrary frequency.

In our work on an electronic polarization a pulse at ν_1 was used to excite a two-photon resonance of strontium in the vapor phase at $\sim 700^{\circ}C$ and mixed with ~ 200 Torr of helium. The off-diagonal density matrix element connecting the $5s^2$ ground state with the 5s5d state identifies the polarization at $\nu_1 + \nu_2 = 2\nu_1$ that was excited via a nonlinear response to two photons at ν_1. The probing pulse at ν_3 follows the exciting pulse at ν_1 by a variable delay time, τ. A sum-frequency signal in the vacuum UV at $\nu_4 = 2\nu_1 + \nu_3$ results from anti-stokes scattering of the probe pulse off of the freely decaying $5s^2$-5s5d polarization. This signal was measured as a function of delay time, τ, of the pulse at ν_3 with respect to that at ν_1. Our measurements place the relaxation time in strontium at ~ 150 ps. This time is slightly less than the predicted value of 200 ps due to velocity dephasing. In calcium, it was found that relaxation occurred at exactly the rate predicted by velocity dephasing [11]. Strontium, however, differs in several respects from calcium. In addition to having a nonzero nuclear spin, strontium also has an energetically decending cascade of single quantum transitions out of the 5s5d state. Both spontaneous or, stimulated radiation and nonlinear processes involving the energetically lower states are possible. Such radiation has been observed from the 5s5d state of strontium during nanosecond pumping. [14] Strong radiation of this type leads to longitudinal relaxation of the polarization which can reduce the polarization in a time dependent fashion beyond that caused by doppler or transverse stochastic dephasing. The extent to which such processes contribute to our measured polarization relaxation in strontium is a subject of current investigation.

We have investigated a generically related phenomenon in calcite. The optical phonon at $\nu_p = 1086\ cm^{-1}$ was coherently excited by irradiation with two synchronized picosecond pulses at ν_1 and ν_2, from the pump and the synchronously mode-locked dye laser respectively, such that $\nu_p = \nu_1 - \nu_2$. A beamsplit portion of a 530 nm pump pulse was used as the probe to generate a signal at $\nu_4 = \nu_1 - \nu_2 + \nu_3$, where $\nu_3 = \nu_1$. Care was taken to insure that all the relevant beam intensities were kept below the threshold level for transient stimulated Raman scattering.

In making measurements of this kind it is particularly important to avoid unnecessary contributions to the nonresonant signal by high intensity broadband excitations. While our tuned coherent excitation approach is helpful in this respect, further precautions were taken as follows. To check that we coherently excited the 1086 cm^{-1} optical phonon of calcite we measured the energy generated in calcite as the frequency $2\nu_1 - \nu_2$ was tuned. When $\nu_1 - \nu_2$ is equal to ν_p (the proper frequency of the phonon) the signal should be greatly enhanced. Such an experiment has already been performed with nanosecond pulses by LEVENSON and BLOEMBERGEN [15]. Their dye lasers were quasi-monochromatic compared to the Raman linewidth. The signal at $2\nu_1 - \nu_2$ is due to the nonlinear susceptibility $\chi^{(3)}(\nu_1, \nu_1, -\nu_2)$ which is made of two parts: a Raman resonant term, $\chi_R^{(3)}$, proportional to $(\nu_1 - \nu_2 - \nu_p + i\Gamma)^{-1}$ where Γ is the halfwidth of the resonance and a non-resonant (mainly electronic) term $\chi_{NR}^{(3)}$ which can be considered as constant in this frequency range. When $\nu_1 - \nu_2$ is far from ν_p, the non-resonant term $\chi_{NR}^{(3)}$ gives rise to a background signal. When $\nu_1 - \nu_2$ hits the resonance the signal is strongly increased. In this way we minimize the possibility that our polarization measurements contain inaccuracies introduced by nonresonant parametric effects. For the broadband green pulse excitation, off resonant spectral components give rise to a large non-resonant signal and the resonance is washed out even though the center frequencies of the ν_1 and ν_2 hit the resonance. To increase the resonant enhancement factor we passed the ν_1 beam through a Fabry-Perot etalon of free spectral range 22 cm^{-1} and finesse 8. The ν_1 pulses then had a linewidth of 2.75 cm^{-1} comparable to the dye laser linewidth, leading to an enhancement factor larger than 10.

Having set $\nu_1 - \nu_2$ at resonance and checked that we were coherently exciting the optical phonon, we were able to proceed with the measurement of dephasing time τ. To this end we sent a probe pulse train with a variable delay through the calcite crystal and measured the coherently scattered beam at the anti-Stokes frequency $\nu_3 + \nu_p$, ν_3 being the frequency of the probe pulse. The probe pulse train was beamsplit from the pump pulse train, ν_1, such that $\nu_3 = \nu_1$. To distinguish the probe beam from the excitation beam we sent it at a small angle (4°) to the latter and selected the light emitted in the direction of the probe beam by means of an aperture. The coherent anti-Stokes signal (we were close to phase-matching) then passed through a spectrograph and was detected by a photomultiplier.

Coherent anti-Stokes signals were measured as a function of the delay between excitation and probe pulses. Our analysis of the data yields a dephasing time (at room temperature) of $\tau = 7.1 \pm 1$ ps which is in reasonable agreement with that measured by ALFANO and SHAPIRO [16] of 8.5 ± 2 ps. On the contrary it disagrees with the time measured by LAUBEREAU et al. [17] (4.4 ps) and the spontaneous Raman data given by PARK [18] (3.8 to 4.8 ps). These discrepancies led us to measure the spontaneous Raman linewidth of our calcite sample. The argon-ion laser line at 4880Å was used as the excitation source. The scattered Raman signals near 5153Å were recorded. After correction for instrumental width, we deduced the width of the Raman line to be 0.8 cm^{-1}. This corresponds to a dephasing time of 6.6 ps, in reasonably good agreement with our directly measured value of 7.1 ps. We find, therefore, that when both methods were applied to the same sample, there was no discrepancy between the results.

The main objective of this work was to demonstrate two general applications of synchronously mode-locked dye lasers. In particular, we have studied coherent excitation of electronic and vibrational states and made subsequent measurements of the relaxation of these coherent polarizations.

Picosecond CARS allows a wide variety of experiments to be performed with time resolved spectroscopy of Raman active modes. Although in the simplest case the width of a Lorentzian line is sufficient to determine its dephasing time, in more complicated situations, knowledge of the dephasing time gives complementary information. Care has also to be exercised to minimize the non-resonant electronic contributions to the dephasing time. This point has not been taken into account in discussions of dephasing times of molecular and lattice vibrations in the literature. It would be of great interest to extend this technique to include the measurement of the energy lifetime of vibrational modes.

Note added: After submission of this paper, Dr. Laubereau informed us that they have recently measured the Raman linewidth of the A_{1g} mode of their calcite specimen. Their result is 1.20 ± 0.05 cm^{-1} (FWHM), in good agreement with their direct measurements reported in [17].

References

1. T.R. Royt, W.L. Faust, L.S. Goldberg, and C.H. Lee, Appl. Phys. Lett. 25, 514 (1974).
2. W.H. Glenn, J.J. Brienza and A.J. DeMaria, Appl. Phys. Lett. 12, 54 (1968).
3. C.V. Shank and E.P. Ippen, "Mode-Locking of Dye Lasers", Dye Lasers, ed. F.P. Schafer, Springer, New York, 1973.
4. C.K. Chan and S.O. Sari, Appl. Phys. Lett. 25, 403 (1974).
5. N.J. Frigo, T. Daly and H. Mahr, J. Quantum Electron., QE-13, 101 (1977).
6. C.A. Moore and L.S. Goldberg, Optics Commun. 16, 21 (1976).
7. T.R. Royt, Ph.D. Thesis (University of Maryland) 1976.
8. J. Kuhl, R. Lambrich and D. von der Linde, Appl. Phys. Lett. 31, 657 (1977).
9. J.P. Heritage and R.K. Jain, Appl. Phys Lett. 32, 101 (1978).
10. M.R. Topp and P.M. Rentzepis, Phys. Rev. A3, 358, (1971).
11. M. Matsuoka, H. Nakatsuka and J. Okada, Phys. Rev. A12, 1062 (1975).
12. T.R. Royt and C.H. Lee, Appl. Phys. Lett. 30, 332, (1977).
13. C.H. Lee and D. Ricard, Appl. Phys. Lett. 32, 168 (1978).
14. J. Reif and H. Walther, Appl. Phys. 15, 361 (1978).
15. M.D. Levenson and N. Bloembergen, Phys. Rev. B10, 4447 (1974).
16. R.R. Alfano and S.L. Shapiro, Phys. Rev. Lett. 26, 1247 (1971).
17. A. Laubereau, G. Wochner and W. Kaiser, Optics. Commun. 14, 75 (1975).
18. K. Park, Phys. Lett. 22, 39 (1966); 25A, 490 (1967).

IV. Biological Processes

Picosecond Studies of Electron Transfer in Bacterial Photosynthesis and in Model Svstems

D. Holten and M.W. Windsor

Department of Chemistry, Washington State University
Pullman, WA 99164, USA

W.W. Parson

Department of Biochemistry SJ-70, University of Washington
Seattle, WA 98195, USA

1. Bacterial Photosynthesis

Green plants have two photosystems, called PS1 and PS2, both of which absorb a photon and cooperate in the conversion of light energy into the chemical potential needed for the oxidation of water and the reduction of CO_2 to carbohydrate. Photosynthetic bacteria have only a single photosystem, which more closely resembles PS1 in plants. The primary oxidant has a potential of about +0.45 V. This is not enough to oxidize water, as occurs in green plants. Consequently, bacteria require a supply of more readily oxidized chemical species, e.g., sulfide, succinate, and thiosulfate. The primary reductant has a potential of about -0.15 V and, as in plants, is coupled both to CO_2 reduction and to the synthesis of ATP. Photosynthetic bacteria, like green plants, also contain an antenna system that harvests the incident photons and transfers the excitation to a reaction center (RC). By suitable detergent treatments of bacterial chromatophores, RC preparations essentially free of antenna pigments can be obtained. This possibility, coupled with the presence of only a single photosystem, has made bacterial RCs the preferred choice of most investigators for detailed studies of the charge separation process, although a few picosecond studies have been made with intact chromatophores. For recent reviews see [1-4].

On reaching the RC the excitation first produces an excited state of a pigment complex containing a special pair of bacteriochlorophyll (BChl) molecules. Charge separation then occurs within the RC, which results in the transfer of an electron to a "primary" acceptor, often called X, which consists of a quinone complexed to a non-heme iron. At the same time the primary donor, the special pair, is oxidized to the corresponding cation. Recent work has shown that this primary charge separation occurs in about 200 psec [5-7]. Thus Nature captures solar energy and stores it as chemical potential in a time appreciably shorter than 1 nsec.

In the past 2 or 3 years, much new information concerning the kinetics of this primary charge separation and the identity of the intermediate species involved has been obtained with the aid of picosecond spectroscopy. Rhodopseudomonas sphaeroides has been the main experimental subject, but additional insights have been gained recently from studies on Rhodopseudomonas viridis. Our recent results on this species will be emphasized in this report. Figure 1 summarizes our present knowledge of the primary charge separation, giving the structures of the intermediate species and the kinetics of the steps that connect them. Only Figure 1A corresponds to the physiological conditions, with the acceptor, X, unreduced. The other two schemes show what happens with prior reduction of X (Fig.1B) or both X and

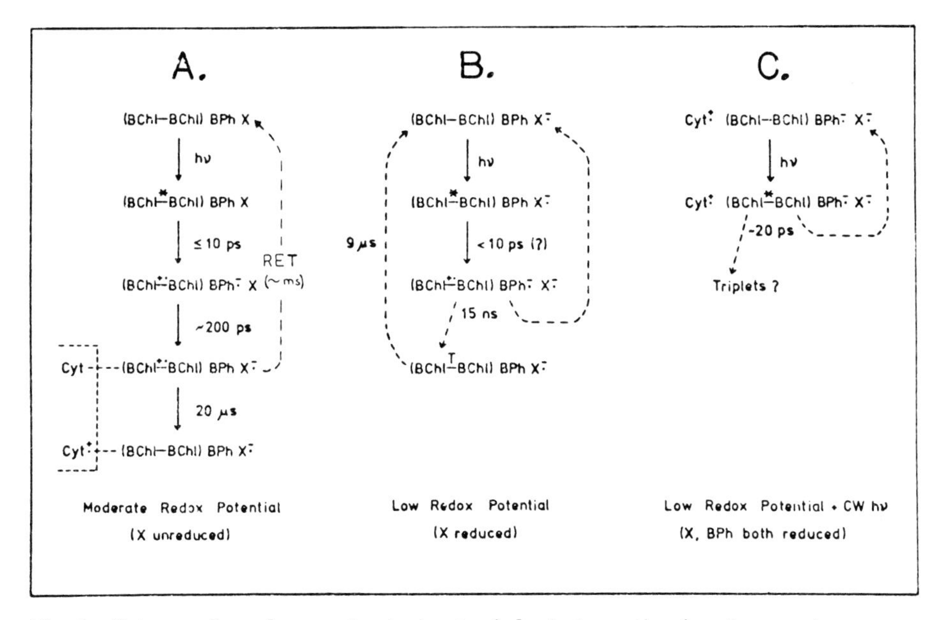

Fig.1 Scheme of early events in bacterial photosynthesis at room temperature. Scheme (a) corresponds to physiological conditions. The abbreviations P ≡ (BChl-BChl) and I ≡ BPh are used by some authors. Note, though, that I is usually taken to denote BPh in close association with other pigment molecules in the RC (see text). The intermediate P^F is BChl$\overset{+\cdot}{—}$BChl)BPh$\bar{\cdot}$X. RET denotes reverse electron transfer.

BPh (Fig.1C). Experiments under these artificial conditions are helpful in unraveling what goes on under physiological conditions. You will find frequent reference to Figure 1 helpful in digesting the rather large amount of material we now present.

RC preparations of both R. sphaeroides and R. viridis contain four molecules of BChl, two of bacteriopheophytin (BPh), one quinone complexed to a non-heme iron atom and several proteins. All these components are packaged in a lipid bilayer membrane. The UQ (ubiquinone in R. sphaeroides) complex is probably situated at or near the membrane surface with ready access to the aqueous medium. Figure 2 shows an "artist's conception" of this arrangement, together with additional data to which reference will be made later. Unlike R. sphaeroides, R. viridis RC's contain bound cytochrome-c, an advantage for more detailed studies of the primary charge separation process, as described below: RC's of R. viridis contain BChl b and BPh b instead of BChl a and BChl b as in R. sphaeroides RC's. As a result the major absorption bands of R. viridis are shifted to the red compared to those of R. sphaeroides.

The results of our picosecond and nanosecond studies [7] of R. viridis RCs may be summarized as follows. At moderate redox potentials (Fig.1A), which correspond to physiological conditions, flash excitation produces the excited state (BChl*BChl) or P*, and this transfers an electron to BPh (or I) in less than 20 psec to produce the radical pair state P^+I^- (also called P^F). [Picosecond studies [8] at 1310 nm, a wavelength at which an absorbance increase uniquely reflects the formation of P^+, indicate a risetime of <10 psec, in good agreement with this result.] The state P^F transfers an electron to X in about 230 psec. An additional step with 35-psec kinetics occurs in the region 800-810 nm only. At low redox potential (X reduced as in Fig.1B), P^F lives for about 15 nsec and decays, in part, to a longer-lived triplet state, P^R. All of the above results closely parallel those in R. sphaeroides. Spectroscopic data on the various intermediates support the view that the intermediate I is BPh b interacting strongly with one or more molecules of BChl b in the RC and that P involves two or more molecules of BChl b. Under conditions of superreduction (Fig.1C), i.e. continual illumination at low redox potential so as to place the RC in the state PI^-X^-, excitation with an 8-psec flash produces absorption changes that reflect the formation of the first excited singlet state, P^*. Under these special conditions, P^* decays with a time constant of 20 psec.

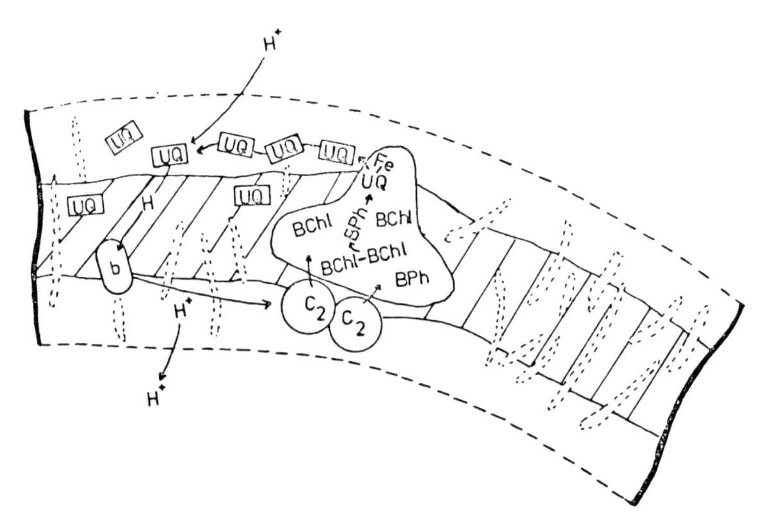

Fig.2 Artist's conception of the bacterial reaction center, showing primary electron transfer in the RC, secondary cytochromes b and C_2, and pathway (short arrows) of charge separation across the membrane. Elongated dashed shapes denote carotenoids. Adapted from an original by P. L. Dutton [15].

The absorbance changes that arise from the conversion of P to P^* are shown in Figure 3. These data represent the first observation of the first excited singlet state of the BChl b complex. The predominant feature of the spectrum is a bleaching of the 960-nm absorption band. The bleaching extends to shorter wavelengths, with a distinct shoulder near 850 nm, which we take to reflect the 850-nm component of the 830-nm absorption band, but drops to zero at 830 nm (Fig.3).

At still shorter wavelengths, a very broad absorbance increase extends into the visible region, broken by a trough at 610 nm. Judging from studies on BPh a and related compounds in vitro [9,10], the absorbance changes are consistent with those that one expects to accompany the formation of the excited singlet state, P^*. The absorbance increase at 1310 nm, which is characteristic of the formation of P^+, does not occur under these conditions [8]. The absorbance changes attributed to P^* decay essentially to completion by 1 nsec (filled circle at 960 nm in Fig.3). Most of the decay occurs with an exponential time of about 20 psec, but there is a residual component that decays more slowly with a time constant of several hundred picoseconds.

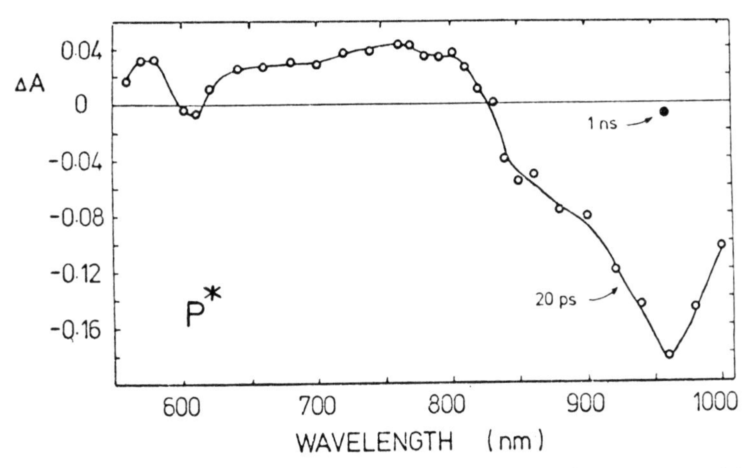

Fig.3 Difference spectrum for the absorbance changes caused by excitation of *Rps. viridis* RC's with 8 ps flashes, after reduction of I by continuous illumination at low redox potential. The spectrum reflects the formation of the excited singlet state, P^*, of the RC [7].

The difference spectrum in Fig. 3 also provides some guidance to assigning the different components of the 830-nm absorption band in the RC spectrum. The component near 850 nm on the red side of the 830-nm band is bleached along with the 960-nm band when P goes to P^*. This result agrees with prior work that shows that these bands also bleach in concert when P is converted to P^+. Since exciton interactions between two or more BChl *b* molecules could give rise to the 850- and 960-nm bands, these results support the idea that the primary donor in the RC is the excited singlet state P^* of (most likely) a pair of closely interacting BChl *b* molecules. However, additional interactions with one or both the remaining BChl *b* molecules are not excluded. Excitation of P to P^* or electron expulsion to give P^+ must then disrupt the interactions responsible for these bands. ESR and electron nuclear double resonance (ENDOR) studies on RCs from *R. viridis* suggest that the unpaired electron in P^+ is not shared equally by two identical molecules of BChl, as appears to be the case in *R. sphaeroides* [11]. Perhaps the two interacting molecules in *R. viridis* are situated in environments that differ significantly.

The spectral evidence also supports the idea that the intermediate acceptor I in *R. viridis* involves both BChl *b* and BPh *b* and that in the reduced I^- state, it is the BPh *b* component that is reduced, just as is believed to occur in *R. sphaeroides*. *R. viridis* RCs also show a fast (within 20 psec of excitation) absorbance change at 810 nm when PI goes to P^+I^-. The decay of this change at 810 nm takes place with a time constant of ∿35 psec. This result closely parallels the corresponding changes that occur in *R. sphaeroides* at 790 nm [6]. However, the possibility that these changes are an artifact that results from the absorption of a second photon has not been completely ruled out.

The conversion of $P^{*}IX$ to $P^{+}I^{-}X$ incurs a loss of about 30% of the original energy, and a further 30% is lost in the second step, $P^{+}I^{-}X$ to $P^{+}IX^{-}$, in which the electron continues its migration from P^{F} to the primary acceptor, which gives an overall energy conversion efficiency of 35-44%. It appears unlikely that such a low efficiency would survive the pressure of natural selection. The spectral changes discussed above raise the intriguing possibility that Nature has learned how to harness a portion of these apparent losses to attendant structural changes in the photosynthetic membrane [11] that would be coupled to proton trnasport and ATP synthesis.

2. Model Systems

As a model for the primary reactions of photosynthesis, we have studied electron transfer from photoexcited bacteriopheophytin (BPh) to p-benzoquinone (Q), m-dinitrobenzene (m-DNB) and methylviologen (4,4'-dimethyldipyridilium chloride, MV) in solution. Electron transfer from the excited singlet state (BPh^{*}) to all three acceptors is thermodynamically favorable; however, only Q has a redox potential low enough to accept electrons from the triplet state (BPh^{T}). Following diffusional encounter, the immediate products of the electron transfer reactions are short-lived radical pairs involving the BPh cation radical ($BPh^{+}\cdot$) and the reduced form of the quencher. The radical pair states can then decay rapidly to the ground state by reverse electron transfer (RET) or by dissociating to give free radicals (charge separation). The competition between these two pathways might depend on several factors, including the energy of the radical pair state, the sizes and net charges of the donor and acceptor molecules, the solvent polarity and viscosity and whether the electron donor was a singlet or triplet. The studies we present here have focused on elucidating the relative importance of such factors.

All three quenchers cause reductions in the lifetime of BPh^{*}, the fluorescence quantum yield and the yield of BPh^{T}. We found that p-benzoquinone quenches both BPh^{*} and BPh^{T} in solution very effectively, but that the formation of $BPh^{+}\cdot$ was detectable only when BPh reacted in the triplet state [10]. This type of behavior appears to be common place for the quenching of chlorophylls and pheophytins by electron transfer in solution. These observations can be reconciled by assuming that electron transfer to Q occurs from both BPh^{*} and BPh^{T}, but that the radical pair [$BPh^{+}\cdot Q^{-}\cdot$] formed via the singlet reaction deactivates rapidly to the ground state by reverse electron transfer. This would have to occur before $BPh^{+}\cdot$ and $Q^{-}\cdot$ could separate, which we have estimated to require about 230 ps [12]. The radical pair state formed by electron transfer from BPh^{T} to Q apparently preserves triplet character during this time, so that spin selectivity reduces the rate of deactivation to the ground (singlet) state, affording a greater opportunity for the radicals to separate.

The radical pair states formed by the interaction of BPh^{*} and MV or m-DNB lie at an appreciably higher energy (∿1.3 ev) than in the case of p-benzoquinone (∿0.8 ev). Methyl viologen (MV^{2+}) also differs from both m-DNB and Q in having an electrical charge of +2, and the radical that is formed by its one-electron reduction ($MV^{+}\cdot$) is cationic. Therefore coulombic repulsion causes the charge transfer state [$BPh^{+}\cdot MV^{+}\cdot$] to dissociate more rapidly than do the other radical pairs. The dissociation lifetime is estimated to be ∿2 psec. In this case, it was found that one can detect the formation of $BPh^{+}\cdot$ in a reaction in which BPh^{*} is the electron donor, although the quantum yield is only about 10%. In the case of m-DNB, BPh^{*} was quenched, but the formation of $BPh^{+}\cdot$ was undetectable [9]. Simply raising the energy of the radical pair (which might be expected to decrease the rate of the back reaction) does not appear to be sufficient to make the model reaction approach the remarkable yield of essentially 100% that the photosynthetic apparatus achieves.

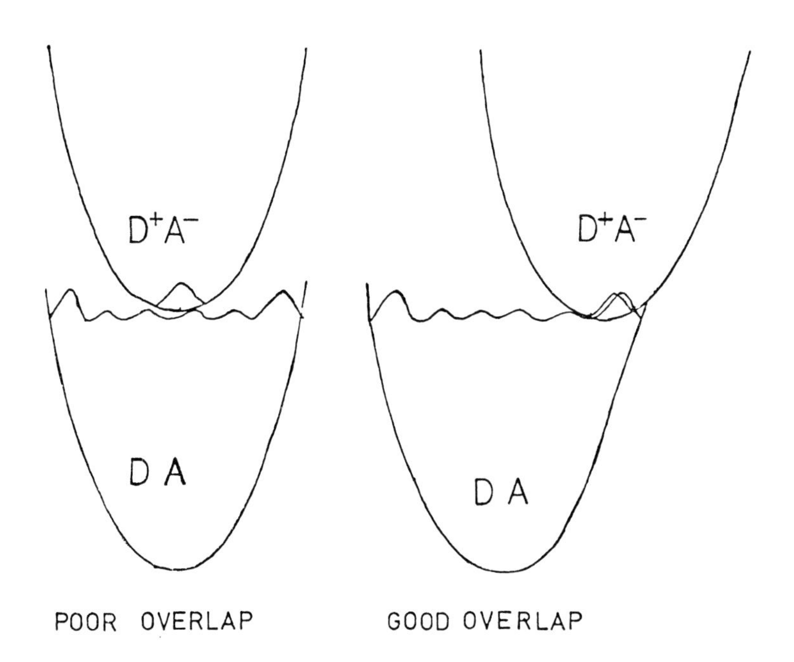

Fig.4 Potential energy diagrams illustrating conditions for good or poor Franck-Condon overlap between the radical pair state (D^+A^-) the ground state (DA). A horizontal displacement between the upper and lower curves could represent a change in a nuclear coordinate for either D or A.

The rate of the forward electron transfer reactions involving BPh *in vitro* ($\sim 10^{11}$ sec^{-1}) are very close to that calculated from fluorescence measurements for the initial electron transfer from P^* to I *in vivo*. However the rate of reverse electron transfer reaction from I^- to $P^{\dot{+}}$ is about 10^3 times slower than the corresponding reactions *in vitro*. This difference in reactivity cannot be explained simply in terms of the amount of energy to be dissipated in the back reaction. The free energy gap of about 1 ev separating [P^+I^-] from the ground state [11] is somewhat greater than that for [$BPh^{\dot{+}}Q^{\dot{-}}$] and smaller than that for [$BPh^{\dot{+}}$m-$DNB^{\dot{-}}$] or [$BPh^{\dot{+}}MV^{\dot{+}}$]. Transformation of this amount of electronic energy to vibrational degrees of freedom *in vitro* with such a large rate constant appears to be possible if there is a rather large shape change upon losing or gaining an electron in at least one of the participants. This provides good vibrational overlap (Franck-Condon factors) between the radical pair and ground states. Figure (4) illustrates the conditions for good overlap (large shape change) or poor overlap (small shape change) between the radical pair state and a high vibrational level of the ground state. A large shape change would correspond to a horizontal displacement of the upper curve with respect to the lower one.

X-ray crystallographic measurements on molecules similar to quinone, indicate that a reasonably large shape change could occur for molecules the size of Q, MV or m-DNB, thus accelerating the rate of the back reaction in these cases. However, X-ray crystallographic data on some porphyrins and recent resonance raman measurements on BChl and BPh indicate that the addition or removal of an electron from larger molecules such as BPh or BChl causes much smaller changes in bond lengths and bond angles, thus reducing the rate of any back reaction involving only these molecules. This offers one explanation as to why the reverse reaction from I^- to P^+ [i.e. BPh^- to $(BChl)_2^+$] is so much slower than the processes in vitro. Another possible explanation involves the symmetries of the different orbitals involved in the forward and reverse electron transfer reactions along with some geometrical considerations [9]. For additional discussion of these topics see [2, 9,12,13,14].

These concepts provide some useful insights into the mechanism of electron transfer in the reaction center. The initial electron transfer step, $(BChl)_2^*BPh \rightarrow (BChl)_2^+BPh^-$, occurs in <10 psec. This step must be extremely fast to prevent the excitation energy from "hopping" back out of the RC into the antenna pool. The observed time of <10 psec is reasonable because of at least two factors. First, there is the possibility for excellent orbital overlap between BChl and BPh since the π-electron systems of the molecules are quite similar, and the rate of electron transfer depends strongly on orbital overlap of the participants. Second, this step incurs a loss of a rather small amount of electronic energy, $\sim$0.4 eV, which can be readily dissipated by activation of a small number of vibrational modes within the RC [14]. Based on the reasoning given above, one would expect the back reaction, $(BChl)_2^+BPh^- \rightarrow (BChl)_2\ BPh$, with $\Delta E \sim 1.0$ eV, to be relatively slow. This is so because $(BChl)_2$ and BPh are both large molecules and we therefore have the "poor overlap" situation shown in Fig. 4. The subsequent forward transfer of the electron from BPh to UQ, which takes about 200 psec, incurs a loss of another 0.4 eV. After this step, the electron is sufficiently far from $(BChl)_2^+$ that quenching by fast reverse electron transfer is not expected to be a problem.

REFERENCES

1. A. J. Campillo, S. L. Shapiro: In *Ultrashort Light Pulses*, ed. by S. L. Shapiro, Topics in Applied Physics, Vol. 18 (Springer, Berlin, Heidelberg, New York 1977).

2. D. Holten and M. W. Windsor, Ann. Rev. Biophys. Bioeng. 7, 189 (1978).

3. R. E. Blankenship and W. W. Parson, Ann. Rev. Biochem. 47, 1978 in press.

4. W. W. Parson and R. J. Cogdell, Biochem. Biophys. Acta 416, 105 (1975).

5. K. J. Kaufmann, P. L. Dutton, T. L. Netzel, J. S. Leigh, and P. M. Rentzepis, Science 188, 1301 (1975).

6. M. G. Rockley, M. W. Windsor, R. J. Cogdell, and W. W. Parson, Proc. Natl. Acad. Sci. USA 72, 2251 (1975).

7. D. Holten, M. W. Windsor, W. W. Parson, and J. P. Thornber, Biochim. Biophys. Acta 501, 112 (1978).

8. T. L. Netzel, P. M. Rentzepis, D. M. Tiede, R. C. Prince, and P. L. Dutton, Biochim. Biophys. Acta 460, 467 (1977).

9. D. Holten, M. Gouterman, W. W. Parson, and M. W. Windsor, Photochem. Photobiol. in press.

10. D. Holten, M. Gouterman, W. W. Parson, M. W. Windsor, and M. G. Rockley, Photochem. Photobiol. 23, 415 (1976).

11. J. Fajer, M. S. Davis, D. C. Brune, L. D. Spaulding, D. C. Borg, and A. Forman, Brookhaven Symp. Biol. 28, 74 (1977).

12. M. Gouterman and D. Holten, Photochem. Photobiol. 25, 85 (1977).

13. R. E. Blankenship and W. W. Parson, Vol. 3 of Topics in Photosynthesis edited by J. Barber (Elsevier, Amsterdam, 1978).

14. R. H. Felton in The Porphyrins edited by D. Dolphin (Academic, New York, 1978).

15. P. L. Dutton, K. M. Petty, H. S. Bonner, and S. D. Morse, Biochim. Biophys. Acta 387, 536 (1975).

Primary Photosynthesis Selectively Excited by Tunable Picosecond Parametric Oscillator

S.A. Akhmanov, A.Yu. Borisov, R.V. Danielius, V.S. Kozlovskij, A.S. Piskarskas, and A.P. Razjivin

Moscow State University, Moscow, USSR, and

Vilnius University, Vilnius, USSR

Several projects for solar energy conversion have been described recently (like methane production from chlorella etc.) but they hardly can contribute substantially to the solution of our energy problem. Indeed, the optimal energy storage in photosynthetic organisms is not more than 2-3% of the overall solar radiation [1,2] while the minimal energy yield (ϕ_w) for industrial solar energetics should not be less than 5-8%. Such ϕ_w can be only realized in photoelectric systems with quantum yield (ϕ_e) not too much different from unity.

At present only semiconductor photocells made on the basis of superpure inorganic crystals are known to meet this requirement. But one more efficient photoelectric system appears to exist - this is photosynthesis. Indeed, it has been shown that for the primary stages leading to the stabilization and separation in space of the charges of opposite sign, the process is characterized by $\phi_w \sim 15\%$ [2,3]. This implies high values ($\gtrsim 0.9$) of the corresponding ϕ_e, according to recent data obtained with various photosynthetic objects [4].

It is established that chlorophyllous molecules in vivo lose their photoinduced singlet excitations via trivial intramolecular mechanisms in about 10^{-9} s [5]. Therefore, photophysical stages of active photosynthesis (energy migration and trapping) should be completed in time interval at least ten times shorter to provide $\phi_e \gtrsim 0.9$. It evidently means that chlorophyll fluorescence lifetimes should be less than 100 ps for active photosynthesis and in fact such values were measured recently for some purple bacteria [6,7] and photosystem-1 of plants [8].

Thus by arranging these primary stages (the delivery of absorbed light energy from hundreds of antenna molecules to RC) on a picosecond time scale Nature succeeded in realizing high ϕ_e and consequently reasonable ϕ_w. And what is important, it is achieved without superpurity or nearly ideal cristallinity. This is the greatest advantage of this type of photoelectric system which is based on organic dye molecules operating with a current-free mechanism of energy channeling. It promises that the technology of model systems (if we succeed in their modeling) should be reasonable and cheap. Of course both the dye and the matrix should be much more stable in direct sunlight (up to several years) than the natural proteins and chlorophyllous molecules. These considerations give very strong incentive for investigating the main principles of picosecond photosynthetic phenomena in potential model systems.

The investigation of mechanisms of the primary photophysical stages of photosynthesis requires spectrometers with picosecond time resolution and with some of their features dictated by the specific spectral properties of photosynthetic objects. Many picosecond spectrometers have satisfactory time resolution, but are not optimal in sensitivity, accuracy and especially spectral range available. We have described an absorption picosecond spectrometer [9], built up specially for investigations in the field of photosynthesis and dye model systems, that, we believe, is adequate pretty much for this purpose.

Two picosecond parametric oscillators (PPO) pumped by the second harmonic of the monopulse picosecond YAG:Nd^{3+} laser were used in this spectrometer (Fig. 1) for selective excitation and probing of a sample. Pulse duration of the driving generator was 32 ± 4 ps; pulse energy amplification was 1.2×10^{-2}J. The excitation beam PPO (two KDP crystal scheme with angular tuning) provided coverage of the wavelength range 780-1400 nm, and an output pulse energy up to 10^{-3} J. Varying the temperature of the $LiNbO_3$ crystals used in the probing beam PPO permitted a wavelength tuning range of 700 - 2100 nm.

The excitation beam traverses an adjustable optical delay line controlled by a step motor. The excitation and probing beams cross the 1 mm sample cell from opposite sides, the angle between their direction being 175°. The probing beam was apertured to a 0.3 mm diameter spot at the sample cell, probing a uniformly excited part of the sample. Three fast amplitude analyzers were employed to register each point of every curve, each point being the average of 32 pulses. The probing pulse reflection from the front surface of the cell was used as a reference signal to exclude (avoid) the effect of probing pulse energy fluctuations.

PPO pulse duration measurements were based on noncollinear generation of the second harmonic. Both the excitation and probing pulse durations were determined to be 25 ± 5 ps. The approximate similarity of the crosscorrelation function and the autocorrelation function points out the stability of synchronization between the two PPO pulses. Although a 25 Hz pulse repetition frequency was accessible, as low as 1 Hz frequency was used in experiments to avoid photoinduced damage of the RC preparation.

Primary photophysical stages of bacterial photosynthesis involve the act of absorption of light quantum ($\sim 10^{-15}$s) by chlorophyll molecules of the light harvesting antenna, excitation energy migration to the specific energy trap - the reaction center (RC) ($\lesssim 10^{-10}$s), conversion of excitation energy to the energy of separated charges of opposite signs (10^{-11} - 10^{-10}s) and energy stabilization up to 10^{-3} - 10^{-1}s. RC is an ordered structural complex which includes 3 protein subunits,

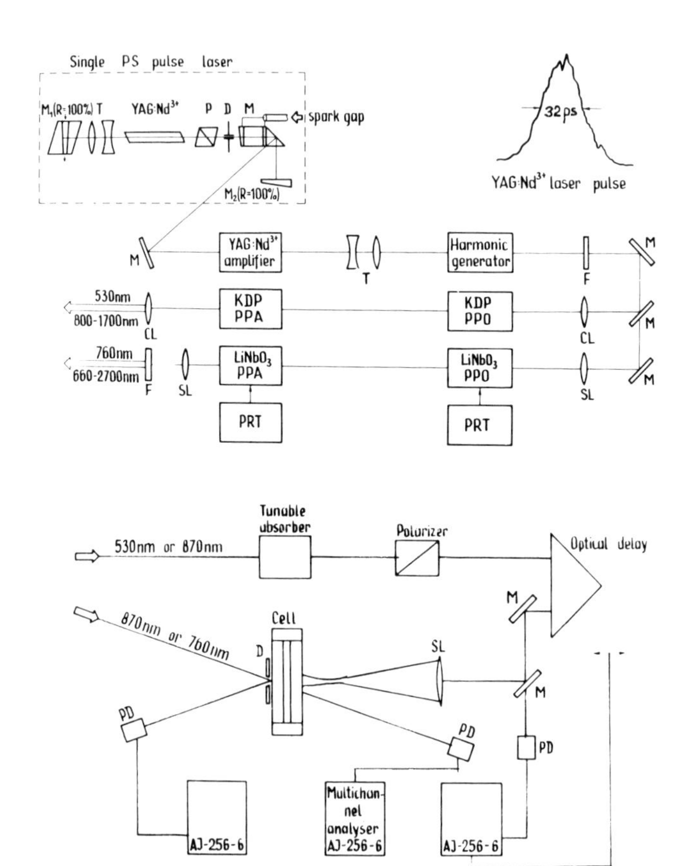

Fig.1 Block-diagram of picosecond absorption spectrometer. PPO - picosecond parametric oscillator, PPA - picosecond parametric amplifier, PRT - precision regulator of temperature, PD - photodiode, P - polarizer, M - modulator, D - Aperture, F - optical filter, CL - cylindrical lens, SL - spherical lens, T - telescope.

2 molecules of bacteriopheophytine (BPh), 4 molecules of bacteriochlorophyl (BChl) two of which are attributed to the spectral from P800 and two of which are complexed in a dimer P870. RC also includes some other molecules, in particular, ubiquinone.

Conversion of excitation energy, channeled to P870, consists of the oxidation of P870 and the reduction of the primary acceptor. Oxidation of P870 appears as a bleaching of the RC absorption band with maximum at 870 nm.

We first estimated the absolute value of the quantum yield of this process ϕ_e in picosecond region on the basis of the dependence of the RC preparation transmittance (T_{870},%) on $h\nu$ in excitation pulse ($h\nu/cm^2$). As seen from Fig.2, ϕ_e calculated from the experimental dependence of T_{870} on $h\nu/cm^2$ tends to unity as the number of quanta per RC becomes less than 1.0 - 0.1.

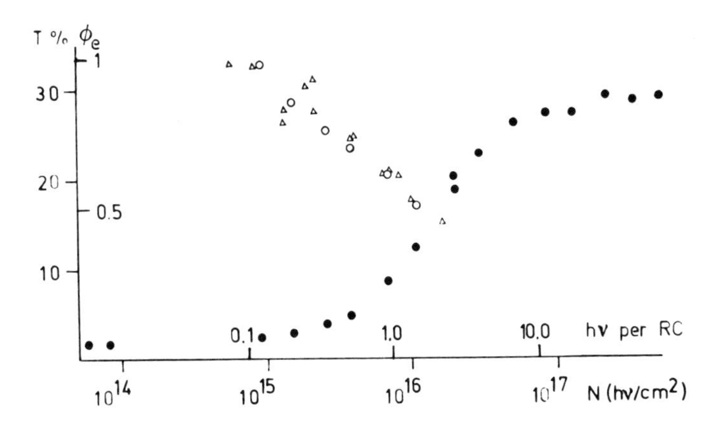

Fig.2 Dependences of transmittance (T_{870},%) of RC preparation (filled circles) and quantum yield (ϕ_e) of charge separation in RC (open circles and triangles) on number of quanta in excitation 870 nm pulses ($h\nu/cm^2$) or per RC. Values of ϕ_e presented by triangles were calculated on basis of transmittance curves not shown here.

Kinetics of the absorption changes at 870, 810, 796 and 748 nm, induced in the RC spectra by 532 nm pulses (BRh absorption band with maximum at 535 nm) and by 870 pulses (long wavelength absorption band of BChl dimer P870) are presented in Fig.3. We put into practice selective excitation of RC at 870 nm instead of often-used excitation at 532 nm because there is absorption of some other pigments of RC at 532 nm in addition to BPh at this wavelength, the proportion of their extinction coefficients being uncertain. This uncertainty creates a problem for quantitative measurement and interpretation of results with 532 nm pulse excitation. The kinetics of the absorption change at 748 and 796 nm induced by 532 nm pulses show an additional bleaching during the excitation pulse in contrast to the ones induced by 870 nm pulses. This additional bleaching is due to excitation energy transfer via the spectral forms of RC pigments. Disappearance of the additional bleaching simultaneously with excitation pulse points to the fact that the characteristic time of energy transfer process is considerably less than the pulse duration. The amplitude of the additional bleaching depends on excitation pulse intensity as well as energy transfer rates and its value may be used for estimation of rate constants. Mathematical simulation of experimental bleaching kinetics permitted us to estimate the rate constants of energy transfer in RC from Bph to P800 and from P800 to P870 to be $3 \times 10^{12} s^{-1}$ and

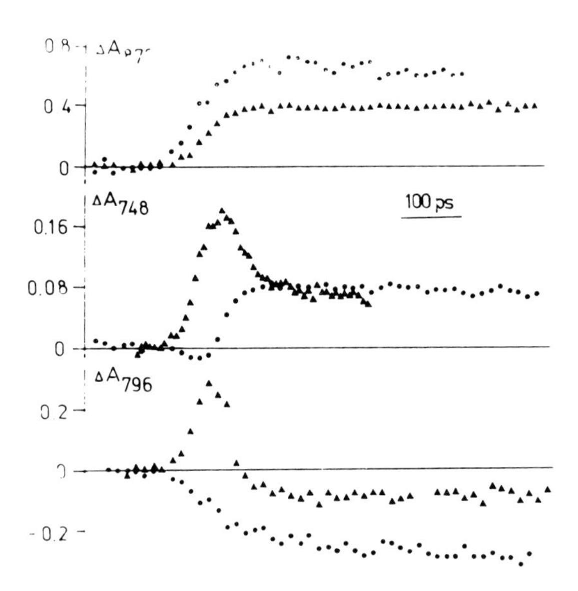

Fig.3 Kinetics of absorption changes of RC at 870, 748 and 796 nm induced by 870 nm (circles) or 532 nm (triangles) excitation pulses.

$2 \times 10^{12} s^{-1}$ respectively. Some details of the calculation procedure make us consider these values as upper limits and the actual ones may be two or three times lower.

Taking this into account one may consider characteristic times of energy transfer between spectral forms of RC pigments to be about 0.2 - 1.0 ps.

Figure 3 shows that the kinetics at 796 nm and especially the one at 748 nm, induced by 870 nm pulses, are delayed in comparison with the respective kinetics at 870 nm. The shape of kinetics and the value of the delay are consistent with the interpretation, proposed by V. A. SHUVALOV, *et al.*, that P800 is the primary electron acceptor, and Bph - the secondary one.

Interesting kinetics of RC absorption changes at 810 nm induced by 870 nm pulses are presented at Fig.4. The relaxation time of the short-lived component was dependent on intensity of excitation. At comparatively low levels of excitation (approximately 10^{16} $h\nu/cm^2$), bleaching with a long relaxation time was observed. As the intensity of 870 nm pulses increased the relaxation time of short-lived component decreased. This phenomenon is evidently related to stimulated processes in RC and enables one to determine the level of intensity which does not cause nonlinear processes in RC.

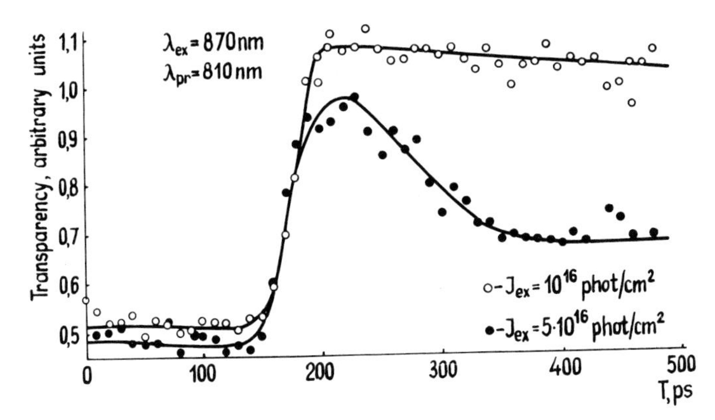

Fig.4 Kinetics of absorption changes of RC on intensity of excitation pulses.

References

1. Calvin, M. (1976) Photochem. Photobiol. 23, 425-444.
2. Boardman, N. K. (1978) Proc. 4th Int. Congress Photosynthesis, Reading 1977 (Hall, D. O., Coombs, J. and Goodwin, T. W., eds.) pp. 635-644.
3. Bolton, J. R. (1978) Proc. 4th Int. Congr. Photosynthesis, Reading 1977 (Hall, D. O., Coombs, J., and Goodwin, T. W., eds.) pp. 621-634.
4. Barsky, E. L., Borisov, A. Yu., Il'ina, M. D., Samuilos, V. D., Fetisova, Z. G. (1975) Molecular Biol. (Russian) 9, p. 275-282.
5. Godic, V. I., Borisov, A. Yu., (1977) FEBS Lett. 82, p. 355-358.
6. Borisov, A. Yu., Godic, V. I. (1972) I. Bioenerg. 3, 211-220.
7. Campillo, A., Shapiro, S.: In *Ultrashort Light Pulses*, ed. by S. L. Shapiro, Topics in Applied Physics, Vol. 18 (Springer, Berlin, Heidelberg, New York 1977) p. 316.
8. Il'ina, M. D., Borisov, A. Yu. (1973) Biochim Biophys. Acta 305, p. 364-371.
9. Akhmanov, S. A., Borisov, A. Yu., Danielius, R. V., Piskarskas, A. S., Razjivin, A. P., Samuilov, V. D. (1977) Pisma v JETF 26, p. 655-658.

Light Collection and Exciton Dynamics in Photosynthetic Membranes[1]

A.J. Campillo and S.L. Shapiro

University of California, Los Alamos Scientific Laboratory
Los Alamos, NM 87545, USA

1. Introduction

The first steps in photosynthesis involve light absorption by specialized antenna pigments followed by efficient energy migration to a reaction center complex whereupon the excitation is utilized to drive the photosynthetic process. Figure 1 illustrates how the harvesting is thought to occur. The dots represent chlorophyll a (Chl a) molecules and RC represents a reaction center. Not shown are accessory pigment molecules which assist in absorbing light. There are a variety of accessory pigment species [1] with complimentary overlapping absorptions that together span the visible spectrum. Once excited, the accessory pigment molecules very quickly (∿ 1 ps) transfer their energy via a nonradiative dipole-dipole interaction to a Chl a molecule. The Chl a's first excited singlet state has a lower energy than the singlet states of the other antenna pigments and so the excitation cannot easily jump back. Possible, however, are further similar Förster transitions between neighboring Chl a molecules and consequently the excited state, or Frenkel exciton, begins a random walk through the Chl a antenna matrix until it wanders near the reaction center complex whereupon it is quickly captured. The entire process, from initial pigment absorption to reaction center capture, although involving many hundreds of separate energy transfer steps, occurs in less than one nanosecond and has quantum yields for photon utilization approaching 100%.

The hypothetical model of the antenna system of a plant shown in Fig. 1 is highly simplified. The actual situation is far more complicated[1] and the exact organization of the antenna system is unknown. Photosynthesis in green plants takes place on a thin (~10 nm) membrane called the thylakoid which is shaped like a flattened hollow sac. The major constituent of the thylakoid is chlorophyll which exists in a variety of chlorophyll-protein complexes [1]. The various protein complexes, each composed of a dozen or so chlorophyll molecules, accessory pigments or other specialized photosynthetic material (cytochromes, ferredoxin etc.), are the building blocks of the photosynthetic apparatus. The dimensions of these building blocks are unfortunately too small to discern their precise organization using electron microscopy. Thus we must rely on indirect information such as that obtained with picosecond spectroscopic techniques. With the highest magnification available in electron microscopy repetitive structures, called quantasomes, measuring about 20 by 15 by 10 nm and arranged much like cobblestones, can be seen on the thylakoid membranes. These quantasomes are often identified with the photosynthetic unit (PSU), a hypothetical entity which contains all the functional

[1]Work performed under the auspices of the U. S. DOE

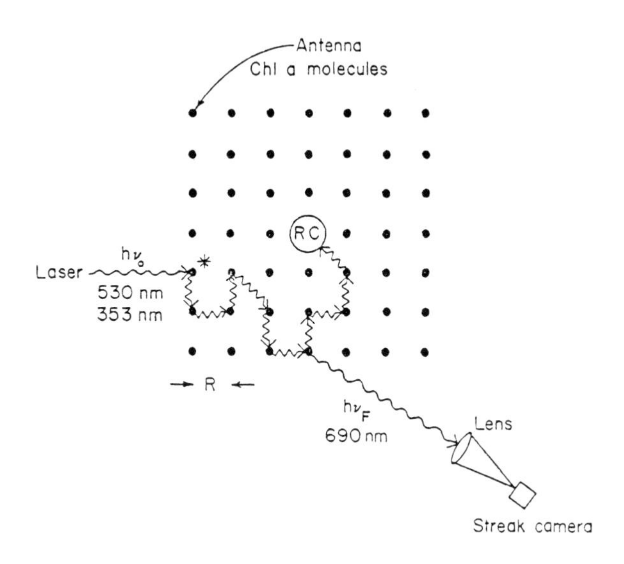

Fig. 1 Simplified representation of the antenna system of a plant. Light is absorbed by accessory pigments (not shown) and transferred to Chlorophyll a molecules (black dots) which funnel the excitation to a reaction center (RC)

material necessary to perform photosynthesis independently. There is evidence for and against the existence of such units [1,2]. If the units are independent, each would contain one reaction center having its own antenna system as depicted in Fig. 1. This model is often referred to as the "puddle" model. An alternative model does not recognize individual PSU's and postulates a common antenna feeding all reaction centers. This model is often referred to as the "lake" model. In this model an exciton upon encountering a filled or closed reaction center would continue to migrate to another RC. This would not be possible in the puddle model where the quanta might eventually be lost to fluorescence. The question as to whether the lake or puddle description is most appropriate is one which is still to be satisfactorily resolved in the literature.

An additional complication is introduced because in green plants there are thought to be at least two photosystems [2], PSI and PSII, which are in close proximity, each having their own distinct cooperative photochemistry. PSI, the smaller of the two units, upon red light absorption produces a strong reductant and a weak oxidant, whereas PSII, upon absorption of 'bluer' photons produces a strong oxidant and a weak reductant. Electron flow from the weak reductant to the weak oxidant is coupled to the conversion of adenosine diphosphate and inorganic phosphate to adenosine triphosphate (ATP). ATP assists the strong reductant to reduce CO_2 to a carbohydrate (CH_2O), and the strong oxidant oxidizes H_2O to O_2. PSI and PSII are thought to be located on the outer and inner surfaces of the thylakoid, respectively.

The various techniques of picosecond spectroscopy can be readily employed to deduce information about the antenna apparatus of plants that is not presently obtainable by other means. One such scheme, see Fig. 1, would involve excitation by a picosecond laser pulse at an appropriate wavelength and simultaneous ultrafast streak camera observation of the fluorescence from various molecular species. The observed fluorescence would accurately probe the excited state density of the species under observation. In particular, in green plants, the fluorescence decay from Chl a is predominently determined by the "exciton" migration time to the reaction center, which effectively quenches the Chl a excited state population. This rate can in turn be utilized to test various models under consideration. Further information may be obtained by studying the additional quenching mechanisms that come into play at higher excitation intensities. For example, by flooding the antenna network shown in Fig. 1 with additional excitons, CAMPILLO et. al. [3-4] have shown that the excitons have a high probability of interacting with each other via a singlet fusion scheme, $S_1 + S_1 \rightarrow S_n + S_0 \rightarrow S_1 + S_0$ + heat, that effectively annihilates one of the excitons and leads to lower observed quantum efficiencies and shorter fluorescence decay times. Here S_1 is the first excited singlet state of Chl a, S_0 its ground state and S_n a higher excited state. The manner in which excitons interact *in vivo* should strongly reflect the existence or non-existence of boundary conditions (puddle model), provide information on the coupling of the two photosystems and also assist in obtaining values for various rate coefficients. In this paper we will review fluorescence studies our group has performed in recent months to unravel the complex organization of the antenna system of green plants [5,6].

2. Experimental Setup

The experimental apparatus utilized in these studies for fluorescence lifetime and quantum yield measurements was described previously [7] and is similar to that shown in [8], which appears elsewhere in this volume. Briefly, a 1060 nm pulse was selected from the pulse train emitted by a mode-locked Nd:YAG laser, frequency shifted to 530 nm in a KDP crystal, and then allowed to excite the samples, which in this report were isolated spinach chloroplasts at 77 K. The sample was illuminated uniformly by aperturing the gaussian beam from the laser and by imaging the aperture with a lens onto the sample cuvette. A pickoff plate directed a portion of the beam into a Laser Precision energy meter, so that the laser energy was measured on each shot. The fluorescence from the sample was collected and imaged onto the slit of an Electrophotonics ICC-512 (S-20 response) streak camera. Appropriate narrow band interference filters were used to isolate the fluorescence band of interest. Streaks were imaged onto a PAR silicon vidicon optical multichannel analyzer and then displayed on an oscilloscope after each shot. Shots could be accumulated on a Nicolet signal averager, thereby allowing lower excitation intensities and also greatly improving the signal to noise ratio. Because there is considerable variation from shot to shot in the start of the streak due to the electrical jitter, a reference point in time was established by allowing a weak green pulse, which bypasses the sample, to enter the streak camera a few hundred picoseconds before the arrival of the fluorescence. The zero time for the fluorescence onset and the excitation pulse-shape characteristics about the zero time position could be determined by examining 530 nm light scattered off the sample. Besides direct lifetime measurements, total time integrated fluorescence yields were obtained by diverting the fluorescence with a movable mirror onto a photomultiplier.

3. The Singlet-Singlet Fusion Process

CAMPILLO et. al. [3,4] proposed and demonstrated that Chl a singlet fusion processes occurred readily in photosynthetic membranes which were excited with relatively low energy ps pulses. SWENBERG et. al. [9] and CAMPILLO et. al. [10] were able to fit early observations with a surprisingly simple formulation reproduced here. A Stern-Volmer rate equation which includes the effect of exciton annihilation is given by

$$\frac{dn_s}{dt} = C\, I(t) - kn_s - \gamma\, n_s^2 \qquad (1)$$

In (1), n_s is the population density of excited Chl a singlets, k is the reciprocal of the observed lifetime at low intensities, γ is the appropriate bimolecular rate constant, I(t) is the laser intensity and C is a constant which includes the absorption cross section. If I(t) is approximated by a delta function, (1) is easily solved

$$n_s(t) = \frac{1}{-\frac{\gamma}{k} + \left[\frac{1}{n(o)} + \frac{\gamma}{k}\right] e^{kt}} \qquad (2)$$

In (2), n(0) is the initial density after laser excitation. An expression for the total time integrated fluorescence yield ϕ (I) can be obtained straight-forwardly and is given by

$$\frac{\phi(I)}{\phi_o} = \frac{k}{\Gamma I} \ln\left[1 + \Gamma I/k\right] \qquad (3)$$

Here ϕ_o is the yield at low I and $\Gamma \equiv \gamma n(o)/I$. This simple formalism has been shown to completely describe the exciton dynamics of mutants of the photosynthetic bacteria, Rhodopseudomonas sphaeroides [10]. However, because this model assumes no boundary conditions (i.e. lake model) and does not take into account the presence of two photosystems, it remained until the work described here to demonstrate the usefulness of this formalism in the case of green plants.

We should mention that to include the possibility of incorporating the puddle model into the formalism would require the use of Poisson statistics and the expression for $\phi(I)$ would be dramatically different from (3). Thus experimentally it is straightforward to differentiate between the two cases.

4. Constructing an Appropriate Model of the Green Plant Antenna Apparatus

The most general model we might construct would consist of separate PSI and PSII reaction center complexes, each possessing its own antenna system. Although this is reasonable, there exists evidence which would seem to question this general model. For example, when the photosynthetic membranes of the higher plants are fractionated using detergents, several different chlorophyll protein complexes can be identified. Ten to eighteen per cent of the total chlorophyll occurs in a type of protein-chlorophyll complex which is enriched with the P700 reaction centers of PSI [11]. Another 40-60% of the chlorophyll occurs as protein complexes containing equal amounts of chlorophyll a and b. These complexes have been designated as light-harvesting (LH) chlorophyll-protein complexes [11]. The LH pigment-protein complexes act as antenna, and it has been proposed that they can feed excitation energy to either of the two photosystems. A third type of chlorophyll-protein complex is believed to be closely associated with the photosystem II reaction centers [12]. Based on this tripartite organization of chlorophyll *in-vivo*, BUTLER and KITAJIMA have proposed that the low temperature fluorescence bands of chloroplasts at 685, 695 and 735 nm can be assigned to the LH, PSII and PSI chlorophyll-protein complexes, respectively [12,13]. Various fluorescence induction and quenching experiments at 77 K using cw sources have provided some support for this tripartite fluorescence model [13-15].

We have therefore tested the validity of the tripartite model by examining the fluorescence from the 685, 695 and 735 bands at 77 K in spinach chloroplasts at various excitation intensities. The major portion of the experiments and conclusions presented in this section were previously published in two papers [5,6] and were performed in collaboration with N. E. GEACINTOV and C. E. SWENBERG at N.Y.U., J. BRETON at the Centre d'Etudes Nucleaires de Saclay, and R. C. HYER at the Los Alamos Scientific Laboratory.

The results of an intensity dependent quenching experiment are shown in Fig. 2. The dots represent but a few of the hundreds of data points determined at the emission wavelength of 735 nm while the solid line summarizes numerous data points observed at 685 nm. Both curves are normalized by dividing by their respective low intensity quantum yield ϕ_0. Surprisingly, both curves overlap. The functional roll-off is characteristic of singlet-singlet annihilation processes, and indeed the solid line in Fig. 2 is also exactly (2) with the appropriate rate constants. If we assume a stoichiometric ratio of 300 Chl a molecules to one reaction center then X=1 indicated in Fig. 2 represents one 530 nm photon absorbed per reaction center. Using this assumed absorption cross-section and the observed lifetime of PSII emission at low intensities (800 ps), (3) can be fit to our data with $\gamma_{ss} \sim 5 \times 10^{-9} cm^3/sec$.

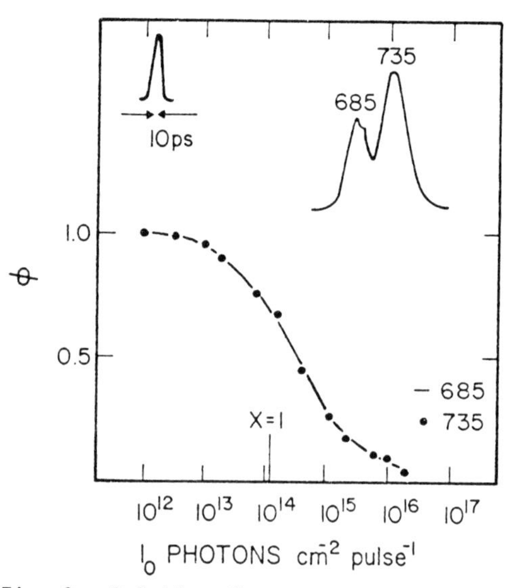

Fig. 2. Relative fluorescence yield ϕ as a function of the incident intensity I_0 determined at the emission wavelengths of 685 and 735 nm (spectrum shown in insert)

If we assume that the 685 and 735 nm emission originates from different antenna pigment systems and that singlet-singlet annihilation occurs within both systems, it is difficult to explain why the two quenching curves in Fig. 2 are identical. The reasons against this are the following: (A) the lifetimes of the singlet excitons are different in the two photosystems, the decay at 735 nm measured by us being 1.5 ns, while the decay time at 685 nm is 0.8 ns. A more efficient quenching at 735 nm is thus expected. (B) There is a greater pigment heterogeneity in the light harvesting pigments than within the PSI pigments and consequently one would expect different γ s. (C) The distribution of energy and thus the density of singlet excitons may be different within the PSI and II light harvesting systems. All these factors should give rise to a difference in the quenching curves observed at 685 and 735 nm. No such differences are observed. It is highly unlikely that all of the above mentioned factors should fortuitously combine to cancel all these differences and to produce identical quenching factors at 685 and 735 nm. If however, both PSI and PSII are coupled to a common light harvesting system, as proposed in the tripartite model, and the exciton annihilation occurs in the LH system, then the data is consistent. Furthermore, the excellent fit of (3) to the data of Fig. 2 implies that the "lake" model is the appropriate description for this common light harvesting pool. A value for the exciton diffusion length (ℓ_d >20nm) can be derived [16] using the value of γ_{ss} observed here and is consistent with a picture of the exciton ranging over several reaction centers (lake model).

The lifetime data is equally informative. Fig. 3 shows signal averaged emission data from PS II and I at an excitation intensity of 2×10^{14} photons/cm^2 and represents one of a series of such curves taken over a range of intensities. The data can be summarized:

I. No difference was observed between the 685 nm and 695 nm emission, both displaying a 800 ps exponential decay at excitation intensities of 5×10^{13} photons/cm^2 or less and shorter nonexponential decays, consistent with (2) at higher intensities.

II. The risetime of the 685 nm fluorescence is less than the resolution of our ps fluoroscope (20 ps).

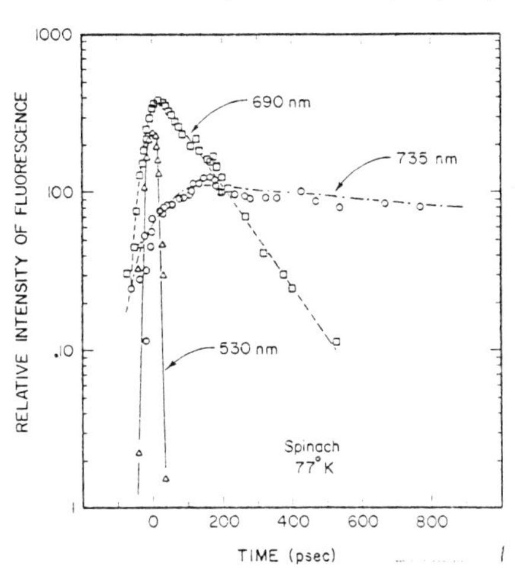

Fig. 3 Time resolved fluorescence from spinach chloroplasts at a temperature of 77 K and an incident intensity of 2×10^{14} photons cm^{-2}. Emission at 690 nm rises promptly, following 530 nm excitation pulse, whereas emission at 735 nm rises much more slowly (from Ref. [6])

III. The decay time of the 735 nm emission was 1.5 ns and showed no variation with intensity over the range 5×10^{12} to 8×10^{15} photons/cm^2.

IV. A risetime of 140 ± 50 ps (10-90%) was observed for the 735 nm emission at intensities below 10^{15}photons/cm^2 and rapidly decreased above that value.

The different dependence of the observed lifetimes of PSI and PSII emission with intensity despite the fact that their total time integrated fluorescence quantum efficiency quenching curves were identical further supports the picture of a tripartite organization with the exciton annihilation process occurring in the LH system. The observations I and II above, indicate that PSII is strongly coupled (forwards and backwards) to the LH system. The quenching curves shown in Fig. 2 and the observed risetime in IV above indicate that PSI is coupled to LH with a rate constant $K \sim (140 \text{ ps})^{-1}$. The lack of variation of the 735 nm lifetime with intensity would imply little or no back reaction with the LH system. Also implied in the data is that the cross section for PSI absorption must be 1 Å^2 or less due to the observed lack of evidence of multiple excitation, even at intensities of 8×10^{15} photons/cm^2. This cross section appears to be too small to represent Chl *a* emission from PSI. Recently, SATOH and BUTLER [17] proposed that the 735 nm fluorescence is due to the formation at low temperature of a Chl *a* species within PSI called C705. The proposed C705 species would have a cross section of the appropriate size to account for our data. This and previous data presented in [17], we believe, confirms the assignment of the C705 species as the source of the 735 nm emission. Since C705 exists only in PSI, its emission provides an accurate probe of the exciton dynamics within PSI and does not affect our previous conclusions concerning the validity of the tripartite model.

There is yet an inconsistency in the data if we treat the LH, PSII and PSI aggregates as lumped systems with the following assumptions: 1) The LH and PSII are so closely coupled that they are indistinguishable and have a common fluorescence decay rate of β_{II} ($\beta_{II}^{-1} \sim 800$ps); 2) The rate of energy transfer from LH to PSI is K ($K^{-1} \sim 140$ps);and 3) PSI fluorescence decays at a rate β_I ($\beta_I^{-1} \sim 1.5$ ns). The solution of the rate equations for such coupled systems predicts that the PSII fluorescence decay rate and PSI risetime should both equal $(\beta_{II} + K)^{-1} \sim 140$ ps at low intensities and $(\beta_{II} + K + C\gamma_{ss} \alpha I)^{-1}$ at high intensities. This is not observed except at intensities greater than 10^{15} photons/cm^2. The difficulty probably arises because the LH aggregate is a distributed network of molecules and not a lumped element. However, by postulating two types of LH systems, LH(1) and LH(2), the data appears consistent once again. The two types of systems are coupled together in a common LH pool, but LH(1) is distinguishable in that its chlorophyll molecules are tightly coupled to the PSI pigment system with $K > \beta_{II}$. It is possible that LH(1) represents a relatively small fraction of the total light harvesting pigments, perhaps consisting of LH pigments which are in close proximity with PSI. LH(2) consists of chlorophyll molecules which are less tightly coupled to PSI (possibly because of a large physical separation) and which decay mainly by the rate constant, $\beta_{II} \sim (800\text{ps})^{-1}$. The fluorescence yield from this system is relatively higher than that of PSI. The 735 nm risetime is thus unaffected by exciton annihilation within LH(1) up to intensities of 10^{15} photons/cm^2 when the annihilation rate begins to compete with the rapid LH(1) to PSI transfer rate, K. Fig. 4 depicts schematically this modified tripartite fluorescence model.

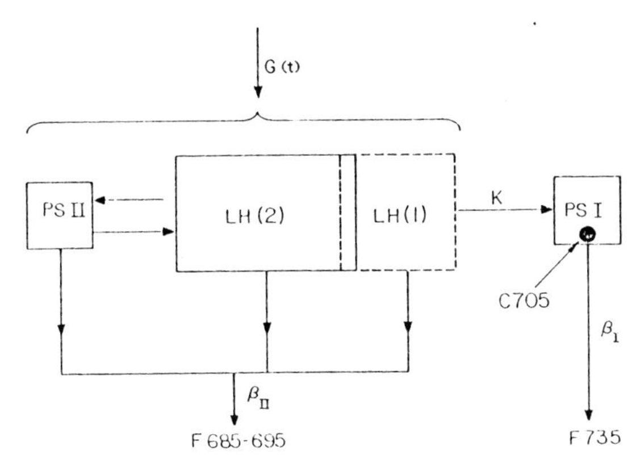

Fig. 4 Modified Tripartite Fluorescence Model (from Ref. [6])

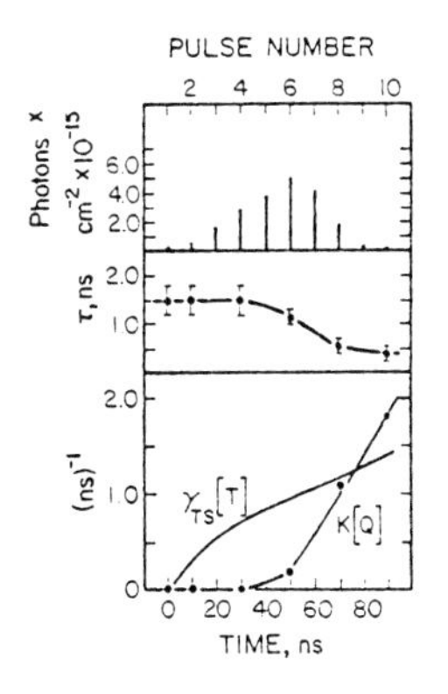

Fig. 5 (A) Intensity profile of a typical pulse sequence. (B) Fluorescence (PSI) decay times measured at 77 K, 735 nm emission band. (C) K[Q], quenching constant x quencher concentration within PSI calculated from the experimental data in (B). $\gamma_{TS}[T]$ is a theoretical curve representing the predicted buildup of triplets, [T], in the LH system x γ_{TS}, the singlet-triplet exciton annihilation rate constant. (from Ref. [18])

5. Quenching of PSI Fluorescence by Triplet Excitons

Figure 5 summarizes a recent experiment performed in collaboration with N. E. GEACINTOV and C. E. SWENBERG of N.Y.U. and K. WINN and R. C. HYER of Los Alamos [18]. Unlike previous experiments which were performed using single picosecond excitation pulses, here the excitation source, as shown in Fig. 5A, is a short train of 530 nm pulses, with a peak intensity of 5 x 10^{15} photons/cm^2 pulse. When isolated chloroplasts of spinach at 77 K are excited, it is found that the decay of the 735 nm fluorescence is initally 1.5 ns but fluorescence from later pulses display shorter exponential decays. The 1/e points of the observed decays are plotted in Fig. 5B. It can be seen later pulses in the train produce successively shorter 735 nm fluorescence lifetimes. Presumably, a long lived intermediate state, most likely a triplet state, is created by intersystem crossing from singlets formed by earlier pulses in the train. The triplets, in turn, quench the fluorescence via a singlet-triplet fusion process (i.e. $S_1 + T_1 \rightarrow S_0 + T_n \rightarrow S_0 + T_1$ + heat) and have an expected decay rate, $(\beta_I + KQ)$. Here Q is the population of the quencher and K the appropriate bimolecular rate constant. This effect partially accounts for the anomalously short Chl a lifetimes observed by early picosecond researchers when they excited their samples with pulse trains. In Fig. 5C, the experimentally determined value of K[Q] based on Fig. 5B is plotted. Also plotted is a calculation predicting the expected value of γ_{ST}[T] based on the anticipated buildup of triplets in the LH system. In this calculation we assumed that triplets are formed from singlets via intersystem crossing at a rate in-vivo equal to that experimentally observed in-vitro and also assume a value of $\gamma_{ST} \sim 6 \times 10^{-9}$ cm^3/sec predicted by RAHMAN and KNOX [19]. Considering the nature of the assumptions made in the calculation, there is good agreement between the absolute magnitude of γ_{ST}[T] and K[Q]. If the magnitude of the intersystem crossing rate or γ_{ST} is assumed to be different, this will affect the magnitude but not the shape of the predicted curve. We interpret the difference in the shapes of the two curves to arise as a consequence of triplets requiring 50 ns to migrate from the LH aggregate to PSI. This result appears to be consistent with previous studies performed by our group in organic solid photosynthetic analogues where it was observed that the diffusion rates of singlet excitons were two orders of magnitude greater than those of triplet excitons.

References

1. K. Sauer: In Bienergetics of Photosynthesis, ed. by Govindjee (Academic Press, New York, San Francisco, London 1975) pp. 115-181
2. Govindjee, R. Govindjee: In Bioenergetics of Photosynthesis, ed. by Govindjee (Academic Press, New York, San Francisco, London 1975) pp 1-50
3. A. J. Campillo, S. L. Shapiro, V. H. Kollman, K. R. Winn and R. C. Hyer, Biophys. J. 16, 93 (1975)
4. A. J. Campillo, V. H. Kollman and S. L. Shapiro, Science 193, 227 (1976)
5. N. E. Geacintov, J. Breton, C. E. Swenberg, A. J. Campillo, R. C. Hyer and S. L. Shapiro, Biochim. Biophys. Acta 461, 306 (1977)
6. A. J. Campillo, S. L. Shapiro, N. E. Geacintov and C. E. Swenberg, FEBS Lett. 83, 316 (1977)
7. A. J. Campillo, S. L. Shapiro and C. E. Swenberg, Chem. Phys. Lett. 52, 11 (1977)
8. A. J. Campillo, J. H. Clark, S. L. Shapiro and K. R. Winn, appears elsewhere in this volume
9. C. E. Swenberg, N. E. Geacintov and M. Pope, Biophys. J. 16, 1447 (1976)
10. A. J. Campillo, R. C. Hyer, T. G. Monger, W. W. Parson and S. L. Shapiro, Proc. Nat. Acad. Sci. USA 74, 1997 (1977)
11. J. P. Thornber, Ann. Rev. Plant Physiol. 26, 127 (1975)
12. W. L. Butler and M. Kitajima, Proceed. Third Internat, Congress Photsyn., Rehovot, ed. by M. Avron (Elsevier, Amsterdam 1974), pp 13-24
13. W. L. Butler and M. Kitajima, Biochim. Biophys. Acta 396, 72 (1975)
14. M. Kitajima and W. L. Butler, Biochim. Biophys. Acta 376, 105 (1975)
15. M. Kitajima and W. L. Butler, Biochim. Biophys. Acta 408, 297 (1975)
16. N. E. Geacintov, J. Breton, C. E. Swenberg and G. Paillotin, Photochem. Photobiol. (IN PRESS)
17. K. Satoh and W. L. Butler, Biochim. Biophys. Acta, 502, 103 (1978)
18. N. E. Geacintov, C. E. Swenberg, A. J. Campillo, R. C. Hyer, S. L. Shapiro and K. R. Winn, Biophys. J. (IN PRESS)
19. T. S. Rahman and R. S. Knox, Phys. Stat. Sol. (b) 58, 715 (1973)

Fluorescence and Energy Transfer in Photosynthesis

G.S. Beddard, G.R. Fleming, G. Porter, and C.J. Tredwell

The Royal Institution of Great Britain, 21 Albemarle Street
London W1X 4BS, England

An accurate determination of the kinetic decay law governing the excited state decay of *in vivo* chlorophyll and associated light harvesting pigments is of fundamental importance to an understanding of the excitation energy transfer process in photosynthesis (1). We have used time resolved fluorescence decay techniques based around ultrashort pulse lasers to follow the temporal characteristics of light emission from photosynthetic systems.

Experimental

The results reported here have been obtained on two separate instruments, (1) a picosecond Nd-glass laser streak camera system and (2) a CW mode-locked dye laser with single photon counting detection. These two systems are described separately below.

1. Streak Camera System

The system has been described in detail elsewhere (2); the passively mode-locked Nd glass laser produces 6ps pulses, the second harmonic at 530nm is used for excitation. Intensities of $\sim 10^{14}$ photons/cm^2/pulse were used for the measurements. Fluorescence from the samples was time resolved and stored digitally by a streak camera (Imacon 600) and optical multichannel analyser combination. The linearity of the detection system is better than 3% over a dynamic range of about 100.

2. Synchronously Pumped System

Picosecond tuneable pulses are obtained from a Rhodamine 6G dye laser synchronously pumped by a mode-locked argon ion laser (CR 12).

A stable r.f. source was used to drive the mode locker on the ion laser resulting in pulses of 90ps duration. The duration of the dye laser pulses are 2.4ps and 4.7ps using a reduction factor of 1.8 (3) from the auto-correlation width for 3 plate and 2 plate Lyot filters respectively.

For the photon counting measurements the pulse repetition rate was reduced from 75MHz to 33kHz using a Pockels cell between crossed polarisers. A contrast ratio of better than 500:1 between the transmitted and rejected pulses was achieved. The subsequent laser output was divided along two paths by a beamsplitter. One was attenuated and used to excite the sample while the other was incident upon a Texas Instruments TI XL 56 silicon avalancne photodiode which provided the start signal for the time-to-amplitude converter. Fluorescence emitted at right angles to the excitation beam was detected through appropriate filters by a Mullard 56 TUVP photo-multiplier tube. Temporal linearity of the photomultiplier tube was obtained by reducing the light sensitive area of the photocathode to 3mm diameter. Time calibrations were carried out by monitoring the excitation pulses through suitable optical delays. The laser power at the sample cell was measured using an Alphametrics photometer. Experiments were performed with this instrument with incident laser intensities within the range 10^9 to 10^{11} photons/cm^2/pulse. As a check on the instrument's performance the dye molecule Rose Bengal was measured using 580nm excitation and the same emission filters as used with the photosynthetic systems. The fluorescence had an exponential decay over three decades decrease in fluorescence intensity of 597ps in methanol and of 122ps in water, which compares well with previous measurements (4).

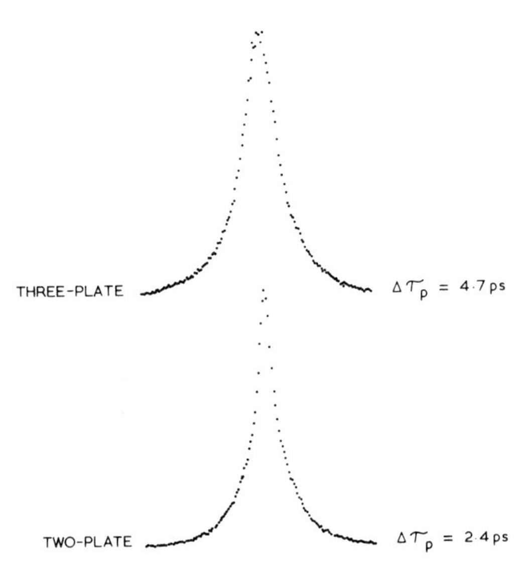

Fig.1 Background free autocorrelation traces of the mode-locked pulse train from the Rhodamine G laser. Upper trace 3 plate birefringent filter, lower trace 2 plate filter

The algae Chlorella pyrenoidosa and Porphyridium cruentum were cultured as described previously (5,6). Pea (Pisum cativum) chloroplasts were isolated with the outer envelope intact and hypotonically shocked immediately before additions were made and the fluorescence measured (6). A purified PS I preparation was obtained from pea chloroplasts by isolation of a stroma lamellae vesicle fraction using 0.2% digitonin (5).

Chlorella

The fluorescence from dark-adapted Chlorella, measured by photon counting was analysed with the assumption of a single exponential decay and gave reasonably good fits as judged by a chi-square criterion. However, the calculated best fit data revealed small systematic deviations from the actual data which indicated that the decay was probably non-exponential. This may be seen in Fig.2, where the fluorescence decay is close to, but not quite exponential over a 1000-fold decrease in intensity. The fit to the data could be considerably improved using two exponential terms, although the lifetimes varied slightly between different experiments. The two lifetimes obtained for dark-adapted Chlorella were found to be in the ranges 270 to 350ps and 530 to 650ps with the long component accounting for between 38% and 27% of the initial intensity.

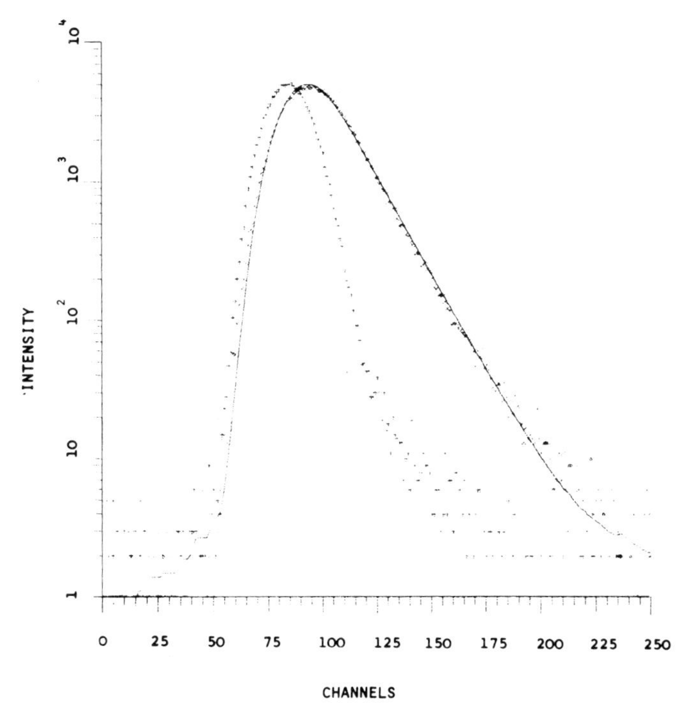

Fig.2 Time resolved fluorescence of dark-adapted Chlorella. 32.9ps/channel

It seems unlikely that the longer component is due to a proportion of the reaction centres being closed since a lifetime of about 1.5ns would then be expected, and attempts to fit the data with decays longer than 1ns proved unsuccessful. No effect upon the lifetime was discerned when the excitation wavelength was varied within the range 580nm to 640nm, i.e. in the region where the chlorophylls are directly excited. Excitation with the frequency doubled Nd laser at 530nm occurs where the carotenes absorb. Similarly, variation of the incident laser intensity from 10^9 to 10^{11} photons/cm^2/pulse did not affect the fluorescence decay.

The data we obtain for Chlorella and chloroplasts is in fair agreement with previous determinations using single laser pulses from a frequency doubled Nd^{3+} glass laser with streak camera detection where moderate excitation intensities ($\approx 10^{13}$- 10^{14} photons/cm^2/pulse) were used (2). We increased the incident light intensity up to 10^{11} photons/cm^2/pulse without observing any change in the decay characteristics. Typical measurements were taken with only 10^9 photons/cm^2/pulse which is at least four orders of magnitude less than the threshold for exciton annihilation processes (7).

The non-exponentially observed in the fluorescence decays can derive from several sources. These are (a) heterogeniety in the light collection pigment complexes between different organisms, (b) contributions to the total emission from PS I as well as PS II, (c) transient diffusional effects, and (d) different decay times from the Chl-a/b protein and Chl-a protein complexes. Recent work on mutants lacking Chl-a/b complexes (8) has shown that Chl-a complexes have decay times of ∿250ps similar to the 270ps component we observe.

Chloroplasts

The fluorescence decay for dark-adapted chloroplasts was fitted by a double exponential consisting of a major short component (413ps) and a long component (1463ps) comprising about 3.8% of the initial intensity (Table 1).

Table 1 Characteristics of the Fluorescence Decay of in vivo Chlorophyll-a

Sample	τ_1(ps)	τ_2(ps)	% $\tau/_2\tau_1$	ϕ_{calc}*	τ_{mean}(ps)
Chlorella Dark-adapted	492	-	-	0.024	-
Chloroplasts Dark-adapted	413	1463	3.8	0.022	453
Light + DCMU	453	1328	9.9	0.027	540
Light + DCMU + Mg^{2+}	462	1342	36.6	0.039	784
SLV Fraction	113	1192	3-9	0.0056	-

* ϕ_{calc} was calculated using the parameters for Chlorophyll-a of $\phi_o = 0.33$; $\tau_o = 6700$ps

In low salt buffer upon pre-illumination and the addition of DCMU, the proportion of the long component increased to about 10; the lifetime of the short component increased slightly. The addition of 5mM Mg^{2+} caused a further increase in the proportion of the long component to about 37% of the total. The lifetime of the longer component under the three different conditions is the same within experimental error. The decay kinetics of dark-adapted chloroplasts are presented in Fig.3.

We believe that our measurements of the dark-adapted state indicate that some PS II reaction centres are non-quenchers, i.e. effectively closed, on a statistical basis at any given time. This could possibly be due to some pre-illumination by scattered laser light (which seems unlikely since the effect was not observed in chlorella) or as a result of damage to some of the chloroplasts during preparation. This explains the 3.8% of ∿1.4ns decay in the dark-adapted samples. Even upon addition of DCMU many PS II reaction centres remain open, presumably because of the low level of illumination, but the proportion of the long lifetime is now increased to ∿10% (Table 1) and addition of Mg^{2+} further increases the proportion of long component present to ∿37%. The fluorescence yields and mean lifetimes are also consistent with this viewpoint. The weighted mean lifetimes calculated from our data can be compared to the lifetimes obtained by MOYA et al. [9] who, using a phase fluorimetric technique and assuming a simple exponential decay, obtained a fourfold increase in lifetime upon illumination, whereas our calculated mean lifetime increases by a factor of 1.7. The change in fluorescence yield also reflects this incomplete pre-illumination. Unfortunately, it was not possible to increase the level of light used for pre-illuminating the samples. Careful analysis of the decay curves showed that a sum of two exponentials fits the data well ($x^2 < 2$) whereas an equation of the form $exp(- at^{\frac{1}{2}} - bt)$ as was previously used [6] fits the data less well ($x^2 > 10$).

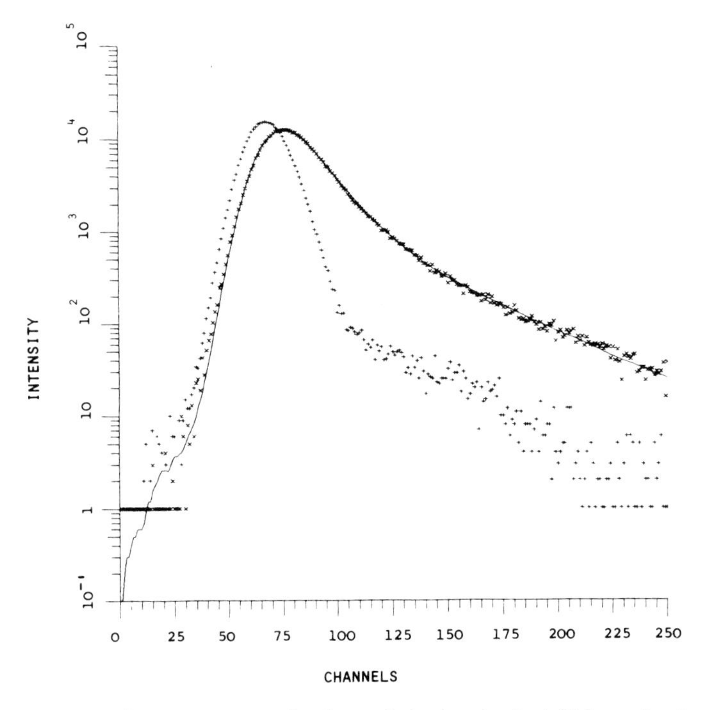

Fig.3 Fluorescence emission of dark-adapted Chloroplasts at 31.9ps/channel. The fluorescence is shown together with the calculated decay (solid line). The instrument function is also shown

The fluorescence decay of chloroplasts suggests that both open and closed traps may exist separately from, and independently of one another. If the excitation energy were able to move freely between light harvesting and antennae chlorophylls, such as suggested by BUTLER and STRASSER [10] in their tripartite model, then an exponential decay, varying continuously from ~400ps to ~1400ps, is expected as more traps are closed. The number of closed traps is proportional to the percentage of τ_2 the long decay component. We also find that the calculated quantum yield varies almost linearly with the number of traps closed.

Our data may be described by a model in which free migration is possible between the light harvesting complexes but once the energy reaches some point in the Chl-a antenna complex it is confined to this complex until trapping occurs. This has the effect of separating open and closed units from one another.

In the presence of cations and pre-illumination, the proportion of τ_2, the long decay, is increased from ∿10% to ∿37% over that in the absence of cations. If spillover from PS II to PS I occurs, then the short decay time, τ_1, should increase as PS I quenching of PS II is removed when cations are added (6). The fluorescence from PS I should also grow in as PS II decays, the exact decay being the convolution of PS II decay and the 113ps decay of PS I particles. As may be seen from Table 1, the decay time τ_1 is within experimental error the same both in the presence and absence of cations, contradicting that expected from spillover. The cation induced fluorescence increase may be explained by increased coupling between Chl-light harvesting and Chl-a_{II} antenna complex (10). The short decay τ_1 is then unaffected by cations, as observed, and PS I decays independently of PS II and may decay with ∿100ps lifetime as observed in SLV particles, Table 1.

SLV Fraction

The fluorescence decay was fitted by a double exponential, the major short component (113ps) being attributed to the PS I emission. The proportion of long component (approximately 1.2ns) varied for different preparations of the SLV fractions, being less the smaller the amount of photosystem II (PS II) left in the preparation, determined from the 77 K emission spectrum.

The PS I lifetime of 113ps (±10%) is longer than most previous determinations (as cited by SEARLE *et al.* (5)) but is in agreement with the value of about 100ps reported by SEARLE *et al.* (5). We have assigned the minor long component ($\gtrsim$1.2ns) to a residual amount of PS II still present in the preparation. Since the sample was not flowed, it was expected that the PS II reaction centres would be closed due to illumination by the laser, which resulted in this fluorescence lifetime being comparable to the long component seen in chloroplasts (Table 1). On the basis of this lifetime (113ps), and with the reasonable assumption that the natural radiative lifetime of chlorophyll is unchanged *in vivo*, we calculate a fluorescence yield of 0.0056. This is higher than the value of 0.003 measured by BOARDMAN *et al.* (11), who however also report that the fluorescence yield of PS II *is a* factor of 5 greater than that of PS I. The fluorescence yields of PS I and PS II reported by BROWN (12) for different organisms also show a difference of a factor of 4 to 5 between the two photosystems. The ratio of our calculated fluorescence yields for PS II and PS I is 3.9 which is in good agreement with the steady-state determinations.

Porphyridium Cruentum

The red alga, *Porphyridium cruentum*, possesses a series of accessory light harvesting pigments which absorb light in the spectral region between 500nm and 670nm (13). These pigments, phycobiliproteins, are contained within structures known as phycobilisomes attached to the thylakoid membrane (14). The phycobilisomes of *P.cruentum* contain three main pigments, namely B-phycoerythrin (λ_{abs} = 545nm, λ_{fl} = 575nm), R-phycoyanin (λ_{abs} = 617,

λ_{fl} = 636nm) and allophycocyanin (λ_{abs} = 650nm, λ_{fl} = 660nm) (13). Chlorophyll-a in P.cruentum is contained within the thylakoid membrane; as a result of the phycobiliprotein light harvesting complexes, the chlorophyll-a/b protein complex is not found in these algae.

Early steady-state fluorescence studies of energy transfer in P.cruentum (15) indicated that the phycobilisomes preferentially serve Photosystem II, and, subsequently, the energy transfer sequence has been proposed as :

B-phycoerythrin ⟶ R-phycocyanin ⟶ Allophycocyanin ⟶ chlorophyll-a

In collaboration with DR. BARBER and DR. SEARLE (Imperial College), we have studied the transfer of energy between these protein complexes using the mode-locked laser:streak camera system (16,17). The second harmonic of the neodymium:glass laser (530nm) provides an excellent excitation source for the first pigment in the transfer sequence, B-phycoerythrin. The fluorescence emitted by the sample was wavelength-resolved with interference filters at 578nm, 640nm, 660nm and 685nm, which correspond to the fluorescence maxima of the three phycobiliproteins and chlorophyll-a (16).

The time-resolved fluorescence emission from the four pigments, in the intact algae, are shown in Fig.4.

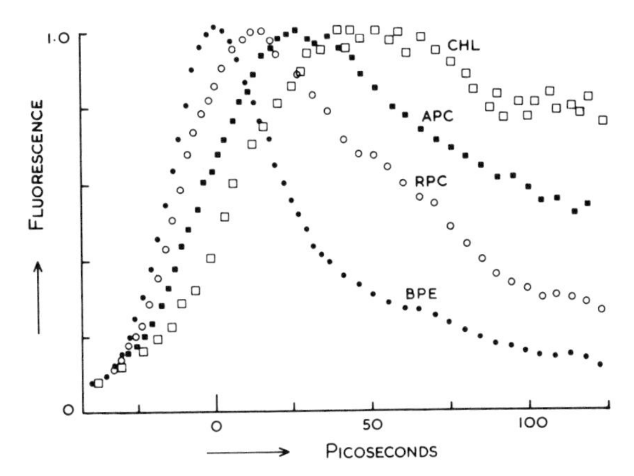

Fig.4 Risetimes of wavelength resolved fluorescence from P.cruentum. BPE is B-phycoerythrin; RPC is R-phycocyanin; APC is allophycocyanin; and Chl is chlorophyll-a

B-phycoerythrin fluorescence exhibits a risetime consistent with the time resolution of the streak camera and the duration of the excitation pulse (6ps), but the remaining pigments all show risetimes which are significantly longer. The risetimes to maximum intensity and the 1/e decay times for each of the four pigments are summarised in Table 2.

Table 2 Fluorescence of Porphyridium cruentum

	Wavelength nm	Risetime (ps)	$\tau_{1/e}$(decay) (ps)	
		Algae	Algae	Phycobilisomes
B-Phycoerythrin	578	0	70 ± 5	70 5
R-Phycocyanin	640	12	90 ± 10	-
Allophycocyanin	660	22	118 ± 8	4000
Chlorophyll-a	685	52	175 ± 10	-

All of the fluorescence decay curves were governed by an exponential decay law, with the exception of that from B-phycoerythrin which was non-exponential. Since the excitation intensity was kept below 10^{14} photons/cm^{-2}, the observed non-exponential decay law cannot be ascribed to exciton annihilation.

These results are consistent with a wave of excitons migrating rapidly through the various pigments towards the chlorophyll-a antennae of Photosystem II. The major part of the exciton's lifetime within the phycobilisome is spent in the relatively large B-phycoerythrin pigment bed (13); the progressive lengthening of the fluorescence lifetime, upon going from one phycobiliprotein to the next, reflects the time required for the exciton to cross that pigment bed. In comparison with chlorella or spinach chloroplasts, the Photosystem II chlorophyll-a emission has a relatively short lifetime which can be ascribed to the absence of the additional chlorophyll-a, normally contained in the chlorophyll-a/b protein complex of green photosynthetic species.

When the phycobilisomes are isolated from the thylakoid membranes of P.cruentum (17), energy transfer from allophycocyanin to chlorophyll-a is prevented. Consequently, the fluorescence lifetime of allophycocyanin increases to 4000ps, which is probably close to that of the pigment in dilute solution. As might be expected, the fluorescence lifetime of B-phycoerythrin remains unchanged, since energy transfer between the pigments in the phycobilisome is unaffected.

This system illustrates the efficiency of inter-pigment energy transfer in the photosynthetic unit. Quantum yields of fluorescence, calculated from the measured lifetimes are all below 1% implying an efficiency of energy transfer greater than 99%.

Acknowledgements

We are grateful to the Science Research Council, and the EEC Solar Energy Research and Development Programme for support of this work, and thank the Royal Society for the award of a John Jaffé Research Fellowship to G.S.B., and the Leverhulme Trust Fund and the Ministry of Defence for Fellowships to G.R.F. and C.J.T. respectively.

References

1. R.S. Knox in Primary Processes in Photosynthesis (J. Barber, ed.), p.55 Elsevier, Amsterdam
   For a comprehensive review, see Ultrafast Reactions in Photosynthesis, 1978, Photochem. Photobiol., in press.
2. G. Porter, J.A. Synowiec, and C.J. Tredwell, Biochim. Biophys. Acta, 459, 329, 1977
3. J. Heritage and R. Jain, Appl. Phys. Lett., 32, 101, 1978
4. G.R. Fleming, A.E.W. Knight, J.M. Morris, R.J.S. Morrison, and G.W. Robinson, J. Amer. Chem. Soc., 99, 4306, 1977
5. G.F.W. Searle, J. Barber, L. Harris, G. Porter, and C.J. Tredwell, Biochim. Biophys. Acta, 459, 390, 1977
6. J. Barber, G.F.W. Searle, and C.J. Tredwell, Biochim. Biophys. Acta, 501, 174, 1978
7. A.J. Campillo, S.L. Shapiro, V.H. Kollman, K.R. Winn, and R.C. Hyer, Biophys. J., 16, 93, 1976
8. G. Porter, C.J. Tredwell, J. Barber and G.F.W. Searle, (in preparation)
9. I. Moya, Govindjee, C. Vernotte, and J.-M. Briantais, FEBS Letters, 75, 13, 1977
10. W. Butler and R. Strasser, Proc. Natl. Acad. Sci., 74, 3382, 1977
11. N.K. Boardman, S.W. Thome, and J.M. Anderson, Proc. Natl. Acad. Sci. 56, 586, 1966
12. J.S. Brown, Biophys. J., 9, 1542, 1969
13. A.N. Glazer in Photochemical and Photobiological Reviews, Vol. 1, (ed. K.C. Smith), chapter 2, Plenum Press, 1976
14. E. Gantt and S.F. Conti, J. Cell Biol., 29, 423, 1966
15. L.N.M. Duysens, Ph.D. thesis, University of Utrecht, The Netherlands, 1952
16. G. Porter, C.J. Tredwell, G.F.W. Searle and J. Barber, Biochim. Biophys. Acta, 501, 232, 1978
17. G.F.W. Searle, J. Barber, G. Porter, and C.J. Tredwell, Biochim. Biophys. Acta, 501, 246, 1978

Picosecond Research of Some Biomolecules
(Bacteriorhodopsin, Bacteriochlorophyll and Bases of DNA)

P.G. Kryukov, V.S. Letokhov, Yu.A. Matveetz, D.N. Nikogosian, and A.V. Sharkov

Institute of Spectroscopy, Academy of Sciences, Moscow Region
Podol'skii Rayon, Troitzk 142092, USSR

1. Introduction

Our investigations in picosecond spectroscopy are developing in two main directions. The first one is the investigation of ultrafast processes in biological objects. The second one is the investigation of selective action of laser radiation on molecules in gaseous phase. The problem of the selective action of laser radiation on biological molecules seems to be very exciting since it promises an unique opportunity to direct biochemical processes by the action of laser light.

The characteristic absorption bands of large molecules or parts of large molecules are usually seen in the infrared absorption i.e. in the vibrational spectrum, but not in the electronic. To selectively excite the electronic state of a molecule in the most preferable way, a two step excitation via a selected vibrational intermediate state was used. This type of selective excitation was demonstrated for simple molecules in the gaseous phase [1] and very recently in a condensed medium (coumarin dye dissolved in CCl_4) [2].

The main problems to deal with in the case of condensed media are that the vibrational relaxation times lie in a picosecond time scale. Therefore to obtain noticeable population of vibrational levels in condensed media it is necessary to use powerful pulses with picosecond duration. Another difficulty which arises in experiments with biomolecules, especially in vivo is the infrared absorption of water which is used as a solvent. One way to overcome this difficulty is to excite the vibrational overtones [3]. Some time ago we demonstrated experimentally two step excitation through a vibrational overtone where there was no noticeable absorption of solvents as an intermediate state. This experiment was done with Rhodamine B in solution of D_2O [4].

Now we shall present some experimental results which reflect our efforts to solve some of the above-mentioned problems.

2. Experimental Techniques

The laser set-up emits ultrashort pulses of tunable frequency in parallel channels [5-7]. The master oscillator is Nd:YAG self-mode locked laser. A single pulse is selected with a Pockels shutter and amplified in two Nd:YAG amplifiers. Although the pulse duration (30 psec) exceeds the one obtainable from a neodymium glass laser, usage of Nd:YAG provides some advantages. The Nd:YAG laser has a high stability of generation without any satellites, it also has a higher repetition frequency. Our set-up operates with 1 pps repetition frequency and output energy is 50 ± 5 mJ. After amplification the laser beam is splitted into two parts for obtaining due to frequency conversion of two beams with different wavelengths. We used several methods for frequency conversion. Parameters of output pulses are presented in Table 1.

Table 1. Parameters of laser set-up with Nd:YAG self mode locked oscillator (pulse duration 30 psec, pulse repetition frequency 1 pps)

Mode of operation	Wave length[nm]	Pulse energy[mJ]	Crystal
Oscillator + amplifiers	1064	50 ± 5	-
Second harmonic generation	532	40	KDP
Fourth harmonic generation	266	20	KDP
Parametric emission with pump at 532 nm	800 + 1670	1 + 1.4	KDP
Second harmonic of parametric emission	400 + 835	0.1 + 0.4	$LiIO_3$
Fourth harmonic of parametric emission	220 + 400	0.001 + 0.02	KDP ADP $LiCOOH \cdot H_2O$
Parametric emission with pump at 1064 nm	1.46 + 4 μ	0.05 + 0.5	$LiNbO_3$
Raman emission in C_2H_5OH	630	2	-
Picosecond continuum in D_2O	400 + 1200	about 20% efficiency	-

3. Picosecond Spectroscopy of Bacteriorhodopsin at Room and Low Temperatures

During the time less than 6 psec a new absorption band appears in light adapted bacteriorhodopsin (BR) at room temperature under the ultrashort light pulse [8]. By its position this absorption band is close to the primary photoproduct (K) band at 635 nm in difference spectrum.

In our work the investigation of the primary stage of bacteriorhodopsin phototransformation at room and low temperatures has been carried out to assign reliably this band to the primary photoproduct but not to any other intermediate state (for instance, to the excited BR state). Making low temperature measurements we hoped to increase the formation time of K and therefore to resolve it (similar behavior was observed in visual pigment [9]).

The sample was excited by the pulse at 532 nm with energy 2.5×10^{-3} J/cm^2. Photoinduced absorption changes were registered by the delayed in time pulse of the picosecond continuum. Purple membranes from Halobacterium halobium were dissolved in ethylene glycol - water 2:1 mixture at 0.5 mM concentration. The 2 mm thick cell was placed into cryostat. It is well known [10] that photoproduct K can be fixed at the temperature lower -120°C. Due to this we were able to investigate the absorption spectrum with a probe pulse delay by several seconds (when product K was formed) and compare it with the spectrum obtained at picosecond delay.

Measurements at temperature -150°C did not show differences in the kinetics of formation of photoinduced absorption changes at 635 nm compared to the measurements at room temperature (Fig.1).

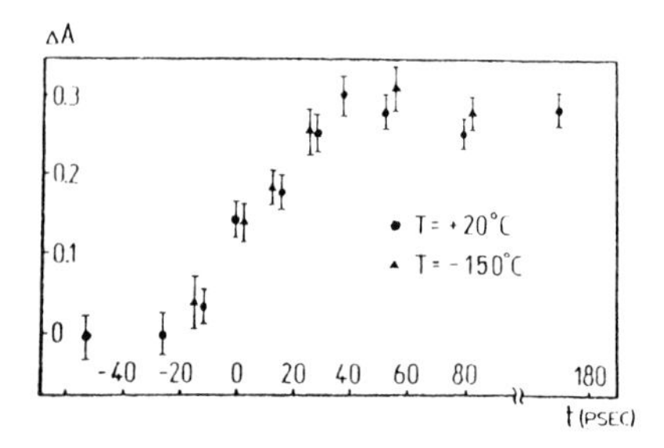

Fig.1 The kinetics of absorption changes in bacteriorhodopsin under excitation of laser pulse at 532 nm and measured at 635 nm (temperatures +20°C (●) and -150°C (▲)).

There were no subsequent changes in absorption at least during several seconds [11]. The usage of the deuterated sample [9] and lowering of the temperature up to 13°K did not change the character of the curve. However, the difference spectrum that was registered at this temperature with 40 psec delay in the region 550 ÷ 700 nm differs from the spectrum registered after several seconds and coinciding with the spectrum of K formation (Fig.2).

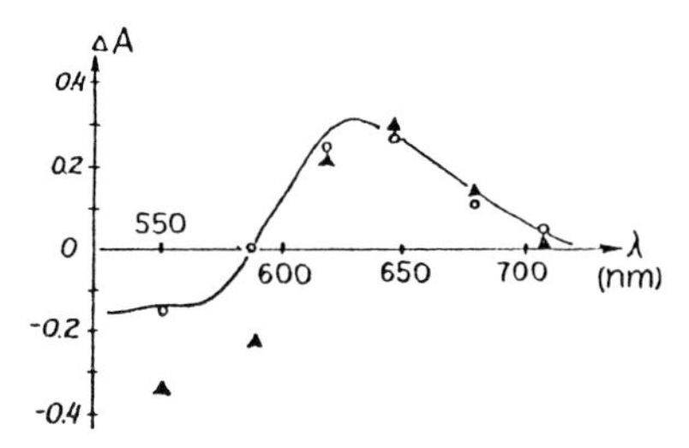

Fig.2 The difference spectra of formation of the primary photoproduct in bacteriorhodopsin at temperature 13°K. The measurements are conducted at the delay between the probing and exciting pulses 40 psec (▲) and 1 sec (O) (the solid line - the low temperature stationary measurements).

Measurements of kinetics of absorption changes at wavelengths 650 nm and 590 nm have about the same times (τ_{590} = 80 ± 10 psec, τ_{650} = 90 ± 20 psec) which do not change at temperature up to 77°K and become shorter than 25 psec at room temperature (Fig.3).

Thus we may state that the process of K formation occurs in a picosecond time scale even at temperatures close to the temperature of liquid helium. The intermediate state preceding to the formation of K (probably the excited state of bacteriorhodopsin) has the absorption spectrum close to the photoproduct K spectrum.

4. Investigation of Electron Transfer Reactions in Rodospirillum Rubrum Reaction Centers

In addition we attempted to study the primary photoreactions at bacterial photosynthesis [12] namely the electron transfer reactions in R. Rubrum reaction centers. It was shown [13] that the charge separation occurs between Bacteriochlorophyll dimer (BChℓ)P and Bacteriopheophytin (Bph) under the light pulse in a picosecond time scale. However, these measurements were

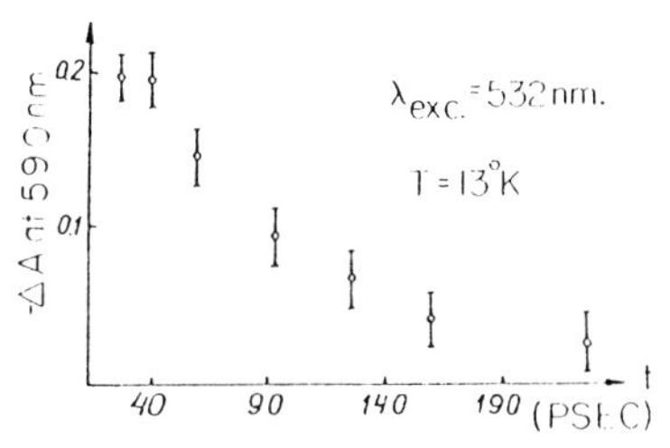

Fig.3 The kinetic of formation of the primary photoproduct in bacteriorhodopsin as measured at 13°K at 590 nm (λ exc=532 nm).

performed at 530 nm that did not exclude the possibility of formation of excited states of 2 BChl and 4 Bph molecules of reaction center as well as the charge separation reaction. Having a tunable laser of ultrashort pulses we were able to excite the primary electron donor P itself at 880 nm. Experimentally obtained differential absorption spectra of R. Rubrum reaction centers, registered with different delay relatively to the excited pulse are shown in Fig.4. The spectrum measured after 450 psec corresponds in the whole to the formation spectrum of cation radical P^+ with exception of 750 nm region where a shift under the action of anion radical of ubiquinone is observed. The absorption spectrum measured after 30 psec includes additional bands corresponding to the formation of anion radical of Bph^- beside the bands corresponding to the formation of P^+. The differential absorption spectrum measured with a zero delay relative to the excited pulse includes corresponding to the spectrum of formation of anion radical $BChl^-$ (in our case Bch (800). There are no bands corresponding to the formation of Bph^-.

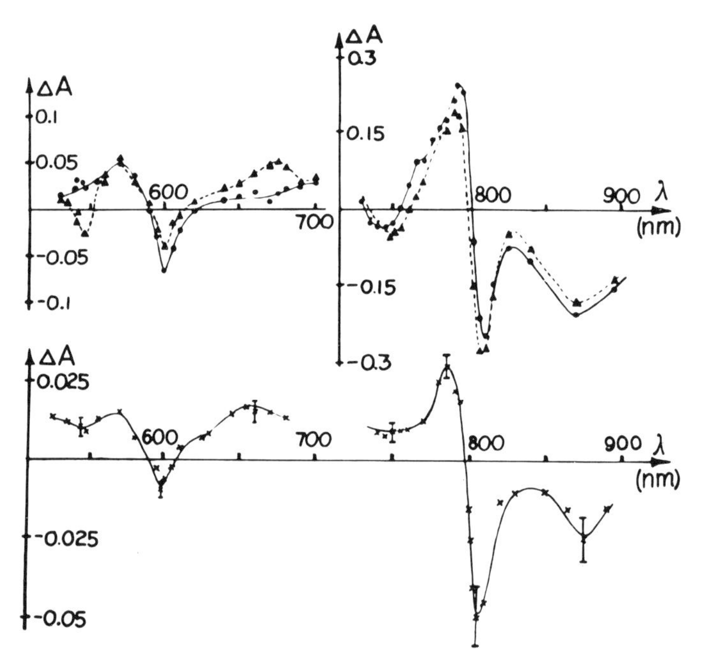

Fig.4 The difference absorption spectra of Rhodospirillum Rubrum reaction centers at the delay between the probing and exciting (λ=880 nm) pulses; 0 psec (**X**), 30 psec (▲), 450 psec (●)

Fig.5 illustrates the kinetics of changes on the wavelengths characteristic for P, BChℓ-800 and Bph. We interpret the data obtained experimentally in the following way. The electron transfers from P and P^+ is formed during the time shorter 15 psec

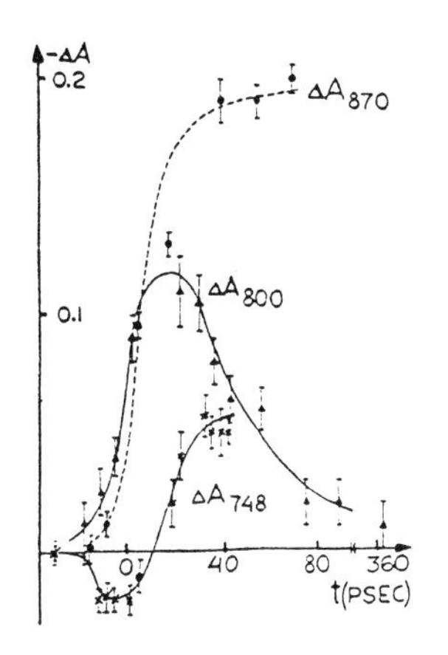

Fig.5 The kinetics of absorption changes in R.Rubrum reaction centers at 748 nm, 800 nm, 870 nm. The exciting pulse wavelength is 880 nm.

in accordance with 13. This electron is immediately (less than 15 psec) accepted by BChℓ-800 and $BCh\text{-}800^-$ is formed. Then during $\tau \sim 30$ psec the electron is transferred from $BCh\ell\text{-}800^-$ to Bph and Bph^- is formed. The Bph transfers the electron to ubiquinone with $\tau \sim 250$ psec in accordance with [13]. Such a sequence of electron transfer fully corresponds to the data about the displacement of chromophore interaction in reaction centers. These data were obtained in dichroism and circular dichroism study of reaction centers [14].

5. Photoreactions of Components of Nucleic Acids in Aqueous Solution Under UV Ultrashort Pulses Excitation

We have carried out the investigation on ultrashort laser radiation action on aqueous solutions of nucleic acid components. The UV laser radiation was tuned to the first electronical absorption band. The fourth harmonic of Nd:YAG laser (λ = 266 nm) coincides with this band of thymine and adenine. A noticeable bleaching of the first electronical band involved with irreversible photoreactions was observed when the aqueous solutions of thymine and adenine ($c=10^{-4}$M) were irradiated with ultrashort pulses of the fourth harmonic of Nd:YAG laser with intensity from 10^9 up 10^{10} W/cm^2. The dependence of this effect on the irradiation dose at various intensities is presented in Fig.6. The dependence of bleaching on the irradiation intensity is obviously nonlinear (Fig.7). This bleaching arose at simultaneous irradiation of both the fourth and the second harmonics (Fig.8). The rise has not been observed if the time delay between the second and the fourth harmonic pulses exceeded 100 psec. The effect has not been observed in case of the second harmonic only. Thus we conclude that photochemical reactions with participation of several quanta occur and their quantum yield achieving several percents. The thin layer cellulose chromatography and spectroscopic researches of the irradiation species have shown the absence of products of hydrate or dimer type which can be usually observed due to photolysis of thymine [15-16]. On

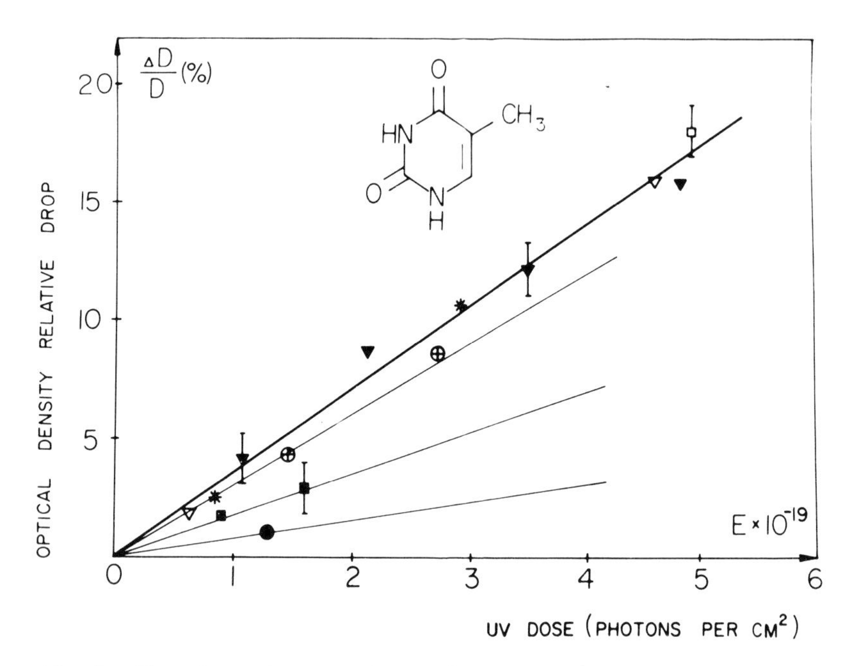

Fig.6 The dependence of a relative drop of optical density of thymine solution on UV irradiation dose. Irradiation intensity at λ = 266 nm in GW/cm^2: ● -0.18; ■ -0.35; ⊕ -0.63; ✱ -1.52; □ -2.28; ▽ -2.53; ▼ -2.58.

the base of these data we make a conclusion that photochemical reactions lead to the saturation of double bond $C_{(5)} = C_{(6)}$ and/or to the distraction of pirimidine ring in thymine case, and to the violation or the perturbation of purine nucleus conjugated system in case of adenine. We consider this investigation as a first step towards selective action of laser radiation of complex biological molecules.

Acknowledgments

We thank our colleagues for fruitful cooperation: Yu. A. Lazarev, E. L. Terpugov, and L. N. Chekulayeva from Institute of Biophysics, Putchino, Moscow area, 142292, USSR; V. A. Shuvalov, A. V. Klevanik from Institute of Photosynthesis, Putchino, Moscow area, 142292, USSR; A. V. Borodavkin, Yu. V. Morozov from Institute of Molecular Biology, Moscow, USSR; E. I. Budowsky, N. A. Simukova from Shemyakin Institute of Bioorganic Chemistry, Moscow, USSR.

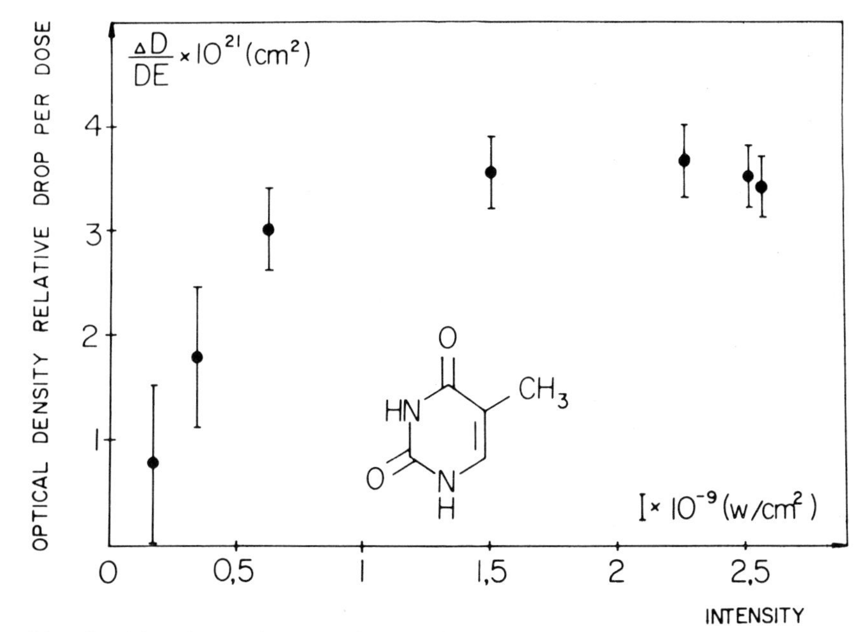

Fig.7 The dependence of a relative drop of optical density of thymine solution per dose on UV intensity.

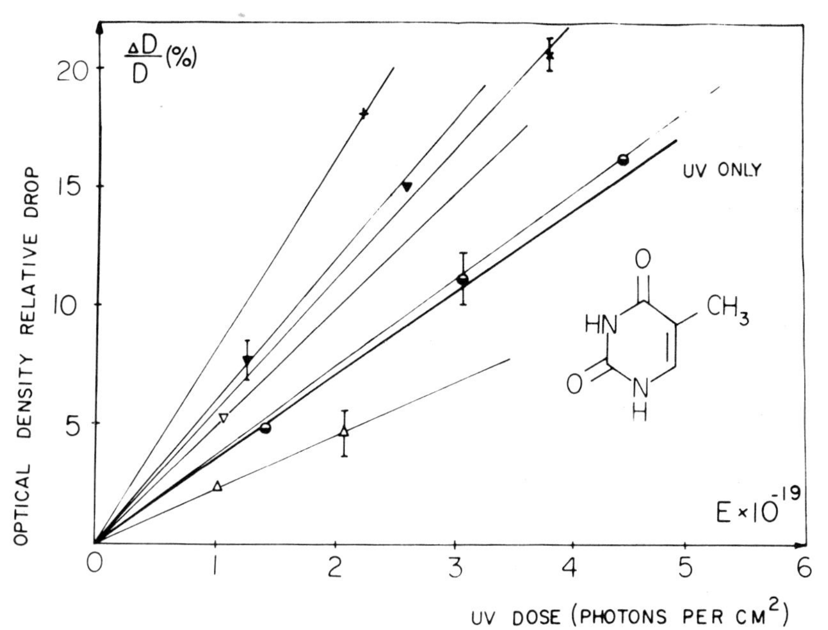

Fig.8 The same as Fig.1 but irradiation with λ = 266 nm and λ = 532 nm. Irradiation intensity in GW/cm^2 at λ = 266 nm and λ = 532 nm respectively: ▲ -0.52, 1.78; ● -1.34, 3.02; ▽-0.52, 16.53; ✕-2.28, 3.92; ▼-0.41, 15.17; +-4.91, 7.00.

References

1. R. V. Ambartzumian, V. S. Letokhov: IEEE Journ. of Quant. Electr. QE-7, 305 (1971), Appl. Opt. 11, 354 (1972).
2. A. Laubereau, A. Seilmeier, W. Kaiser: Chem. Phys. Lett., 36, 232, (1975).
3. I. P. Maier, A. Seilmeier, A. Laubereau, W. Kaiser: Chem. Phys. Lett., 46, 527 (1977).
4. P. G. Kryukov, V. S. Letokhov, Yu. A. Matveetz, D. N. Nikogosian, A. V. Sharkov: Kvantovaya Elektronika, 5, No. 8, (1978).
5. P. G. Kryukov, Yu. A. Matveetz, D. N. Nikogosian, A. V. Sharkov, E. M. Gordeev, S. D. Fanchenko: Kvantovaya Elektronika, 4, 211, (1977).
6. P. G. Kryukov, Yu. A. Matveetz, D. N. Nikogosian, A. V. Sharkov: Kvantovaya Elektronika, 5, No. 9, (1978).
7. Yu. A. Matveetz, D. N. Nikogosian, V. Kabelka, A. Piskarskas; Kvantovaya Elektronika, 5, 664, (1978).
8. K. J. Kaufmann, P. M. Rentzepis, W. Stoekenius, A. Lewis: Biophys. Biochem. Res. Comm. 68, 1109 (1976).
9. K. Peters, M. P. Applebury, P. M. Rentzepis: Proc. Nat. Acad. Sci. USA, 74, 3119 (1977).
10. W. Stoekenius, R. H. Loziek: J. Supramol. Structure, 2, 769, (1974).
11. P. G. Kryukov, V. S. Letokhov, Yu. A. Matveetz, E. L. Terpugov, Yu. A. Lazarev, L. N. Chekulayeva, A. V. Sharkov: Biophysika, 1, 171, (1978).
12. W. W. Parson, R. J. Cogdell: Biochem. Biophys. Acta, 416, 105, (1975).
13. M. G. Rockley, M. W. Windsor, R. G. Cogdell, W. W. Parson: Proc. Nat. Acad. Sci. USA, 72, 2251, (1975).
14. V. A. Shuvalov, A. A. Asadov, I. N. Krakhmaleva: FEBS Lett., 76 240, (1977).
15. C. J. Fisher, H. E. Johns: Photochem. Photobiol., 1b, 23, (1973).
16. H. S. Bagdasaryan: "Two-quantum photochemistry", "Nauka", Moscow, (1976).

The Use of Resonant Light Scattering as a Probe of Picosecond and Subpicosecond Relaxation and Dephasing Times in the Excited State

J.M. Friedman

Bell Laboratories, Murray Hill, NJ 07974, USA

Most biological macromolecules exhibit optical absorption lineshapes that are spectrally broad even at liquid He temperatures. Because line broadening mechanism might originate from relaxation and dephasing processes that are of biological and physical importance, it is of interest to deconvolute the various mechanisms that are contributing to a given spectrally broadened optical transition. When the line broadening is caused by relaxation and dephasing processes that are occurring on a subpicosecond timescale it is no longer routinely feasible to apply the elegant time resolved non-linear techniques in the determination of the dephasing and relaxation rates. Instead, these rates and mechanisms are more effectively determined by monitoring the spectrum of resonant scattered monochromatic light [1]. In this paper it will be shown that resonant light scattering can be used to expose the relaxation and dephasing origin of line broadening in the visible absorption spectrum of ferrocytochrome c at room temperature and at liquid He temperatures.

Ferrocytochrome c is a hemeprotein that functions as a mediator of electron transport in most living organisms. The functional properties of this molecule are determined by the protein matrix modified iron porphyion chromophore which gives rise to visible absorption spectrum of this protein. The visible absorption band at room temperature consists of two broad peaks separated by $\sim$1000 cm^{-1}s. The first peak called the α band corresponds to the nearly degenerate 0,0 transition of the lowest singlet $\Pi\Pi^*$ iron porphyrin transition. The second band is actually an envelope for several 0,1 transitions involving primarily non-totally symmetric vibrations that are active in vibronic coupling. At lower temperatures the two orthogonal components of the α band are resolvable. Both components of the α band have widths on the order of 100 cm^{-1} at 4.2K; at room temperature the α band has roughly twice the composite width of the 4.2K spectrum.

The observed spectral differences can originate from any of several possible line broadening mechanisms including uncertainty broadening due to fast non-radiative relaxation of the resonant excited state ($T_{rad} \approx 5\times10^{-8}$sec), thermal fluctuation induced dephasing processes, inhomogeneous broadening due to a distribution of porphyrin environments with an associated distribution of absorption lines and the trivial broadening mechanism due to level congestion. The last mechanism can be dismissed since high resolution resonance Raman excitation profiles [2] have been used to resolve the line shape of the pure 0,0 contribution. Consequently we focus on the first three possible line broadening mechanisms all of which are of interest biologically since they reflect interactions between either the porphyrin macrocycle and the protein matrix or the electronic interactions between the porphyrin and the iron [3,4]. We now consider how these line-broadening mechanisms affect the resonance Raman scattering spectrum [5].

When a chromophore is effectively isolated from any phase perturbing environment, excitation with monochromatic light generates monochromatic scattered light regardless of the resonance conditions. In the monochromatic excitation limit the excited state contributions to the scattering state, which is an exact eigenstate of the system defined by photon plus chromophore, can be regarded as resonances with the individual fourier components of the Lorentzian lineshapes associated with the non-stationary Born-Oppenheimer type states. The time interval over which a given B-O state can contribute to the scattering state is given by the reciprocal of the associate resonant fourier component. This time dependence is of course not observable in a stationary state experiment. On resonance, i.e. when the monochromatic photon energy corresponds to the peak transition frequency associated with a given B-O state, the scattering time associated with that state is the reciprocal of the Lorentzian damping constant which is in this limit the lifetime ($T_1 = \Gamma_i^{-1}$) of the B-O state (assuming statistical limit, etc.). The oscillator strength of the B-O state can be thought of as having been diluted over a Lorentzian lineshape determined by the lifetime broadening. If there is an inhomogeneous distribution of peak transition frequencies resulting from a distribution of non-communicating (on the scattering time scale) chromophoric environments, then the scattered radiation is still monochromatic but the composite scattering cross section is a sum of cross sections with differing contributing fourier components.

In the discussion so far we have been treating the energy conserving system as consisting of molecule and radiation. However in most condensed phase experiments the molecule interacts with the environment which must now be considered part of the new isolated system interacting with the radiation. If the molecule interacts with the environment then there can occur emission as a result of either dephasing between the resonant state and the ground state or relaxation out of the resonant state. The latter typically generates, in an incoherent fashion, a population of non-resonant levels with respect to the system defined by incident light plus molecule.

Dephasing of the resonant contribution to the scattering process can occur when the molecules interacting with the light source experience random perturbations that cause the energy of the resonant state to fluctuate in time with respect to other electronic or vibronic states. The time scale of the fluctuation relative to the scattering time determines the extent to which each fluctuation will dephase the resonant process. For example, if the resonant state energy fluctuates on a time scale that is slow relative to the resonant scattering time i.e. $(E_p-E_i+i\Gamma_i/2)^{-1}$, then a monochromatic photon still essentially excites a single fourier component in each molecule, but the members of the ensemble of molecules will have different fourier components excited. The result nonetheless is monochromatic re-emission with a cross section that is determined by the average distribution of energy denominators. If, however, the fluctuation occurs on a time scale that is fast compared to the scattering time then the monochromatic photon will simultaneously excite at random intervals a distribution of fourier components of the individual perturbed molecules. The resulting re-emission will be broadened relative to the isolated molecule. It would seem that in many molecular systems it should be possible to switch from one scattering limit to the other by shifting on and off resonance with a long lived adiabatic state. Far off resonance, the scattering time is essentially instantaneous, $(E_p-E_i)^{-1}$, and consequently the resulting re-emission will still be monochromatic. On resonance if the fluctuations are fast enough the full Lorentzian distribution of fluctuation renormalized fourier components of i will be excited resulting in an additional resonant re-emission that is broad. The susceptibility of a resonant photon molecular system to the introduction of incoherence will strongly depend upon the lifetime of the resonant state. For very short lived system the re-emission will be damped $(1/\Gamma_i)$ but monochromatic even in the condensed phase whereas for a relatively long lived resonant state, the re-emission can be expected to be dephased hence broadened in all but the most unreactive of environments. The ratio of the total integrated incoherent scattering to coherent scattering integrated overall energy has been shown [7] to be equal to $\frac{2T_1-T_2}{T_2} = \frac{\Gamma_2^*}{\Gamma_i/2}$ when the impact and separation approximations hold [8] (near or at resonance). From this ratio we clearly see that the dephased re-emission becomes a progressively smaller fraction of re-emission for shorter scattering times i.e. larger values of Γ_i for a fixed proper dephasing rate $= \Gamma_2^*$.

In Figures 1 and 2 [9] are shown the spectrum of scattered light generated by monochromatic light that has been tuned over the visible absorption spectrum of ferrocytochrome c at room temperature and 6K respectively. In addition to the well characterized sharp resonance Raman lines [10] some of the spectra display broad spectral features occurring at fixed absolute frequency. The broad features have been assigned as incoherent re-emission primarily from the zero vibrational level [2,9]. For excitations into the $\beta(0,1)$ band this broad re-emission is considered non-resonant.

The observation that at 6K [9] the broad emission from the zero vibrational occurs only for excitations resonant with $\beta(0,1)$ transitions and not for the $\alpha(0,0)$ excitation excludes the possibility that the broadening is due to the sampling of or the exchanging of different spectrally sharp inhomogeneous sites on the time scale of the excitation and emission lifetimes. This result follows since we would expect that the dephasing processes

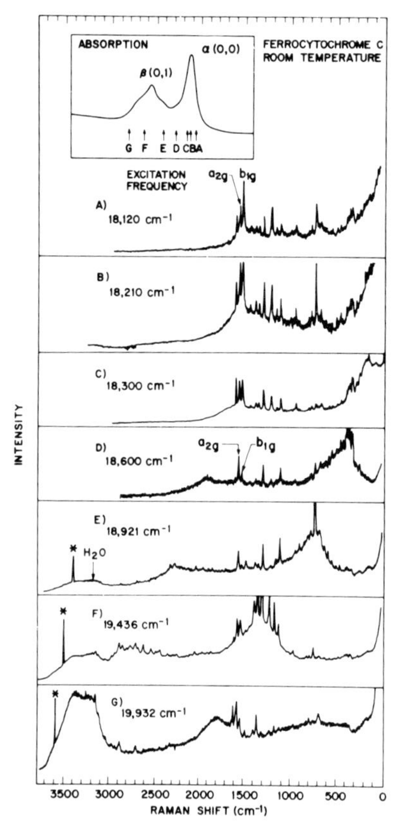

Fig.1 Room temperature re-emission spectra from cytochrome c obtained at several excitation frequencies throughout the α (0,0) and β (0,1) absorption band (see insert). The line marked with an asterisk is an instrumental artifact. The broad band between 3000 and 3500 cm^{-1} results from H_2O scattering.

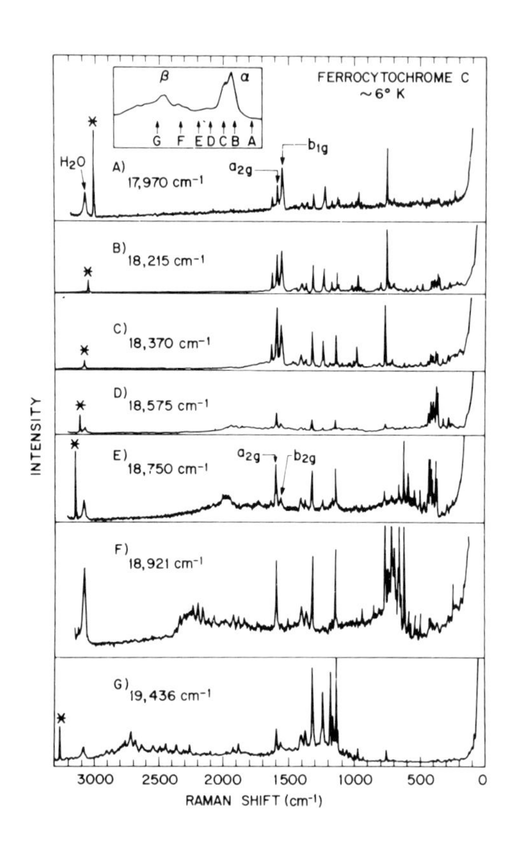

Fig.2 Re-emission spectrum of cytochrome c obtained at 6°K. The line marked with an asterisk is an instrumental artifact and the band near 3100 cm^{-1} results from Raman scattering from H_2O in the ice.

that are responsible for that effect would also generate multiple site emission (hence broad) when the excitation is resonant with the α band. The absence of the broad emission for α band excitations at 6K indicates that there is minimal resonant dephasing on the time scale of the reciprocal of the width of the absorption. Furthermore from Raman excitation profile data [2], the damping constant of the 0,0 and 0,1 transitions are comparable. Consequently the broad emission from this level subsequent to vibrational relaxation from a $\beta(0,1)$ resonance excitation must reflect the approximate inherent homogeneous width of the emitting level. From Raman excitation profile data [2] it has been shown that this homogeneous width corresponds to a lifetime of tens of femtoseconds which is consistent with quantum yield measurements [11,12].

In contrast to the spectra resulting from α band resonances at 6K where there was no observable incoherent contribution (hence $\Gamma_2^* << \Gamma_2$), the corresponding room temperature α band excitations result in spectra with roughly equal contributions from incoherent (broad) and coherent (Raman) scattering. Consequently, from the previously discussed incoherent to coherent ratio we have $\Gamma_2^* \approx \Gamma_i/2$. From the 6K Raman excitation profile,[2], where $\Gamma_i >> \Gamma_2^*$ we have a value of Γ_i^{-1} equal to 45 femtoseconds which results in a proper dephasing time at room temperature of $\sim$ .1 psec. It follows that whereas the 6K absorption spectrum is lifetime broadened the lineshape of the room temperature absorption spectrum is determined by comparable contributions from $\Gamma_i/2$ and Γ_2^*. As expected from the value of Γ_2^*, the r.t. absorption spectrum is roughly twice the width of the 6K spectrum.

References

1. J. M. Friedman: In *Advances in Laser Chemistry*, ed. by A. Zewail, Springer Series in Chemical Physics, Vol. 3 (Springer, Berlin, Heidelberg, New York 1978)
2. J. M. Friedman, D. L. Rousseau and F. Adar: Proc. Nat. Acad. Sci. 74, 2607-2611 (1977).
3. R. M. Hochstrasser: in Probes of Structure and Function of Macromolecules and Membranes, eds. B. Chance, C. Lee and J. K. Blaise, Vol. 1 (Academic Press, New York, 1971) pp.57-62.
4. D. Huppert, K. D. Straub and P. M. Rentzepis: Proc. Nat. Acad. Sci. 74, 4139-4143 (1977).
5. For a recent review, see F. Novak, J. M. Friedman and R. M. Hochstrasser: Resonant Scattering of Light by Molecules: Time Dependent and Coherence Effects; in Lasers and Coherence Spectroscopy, ed. J. I. Steinfeld (Plenum Press, New York, 1977).
6. S. Mukamel and A. Nitzan: J. Chem. Phys. 66, 2462-2479(1977).
7. R. M. Hochstrasser and F. A. Novak: Chem. Phys. Letts. 53, 3-7 (1978).
8. A. Nitzan, S. Mukamel and A. Ben-Reuven: Chem. Phys. 24, 37-41 (1977).
9. J. M. Friedman and D. L. Rousseau: Chem. Phys. Letts., in press.
10. J. M. Friedman and R. M. Hochstrasser: J. Am. Chem. Soc.

98, 4045-4052 (1976).

11. F. Adar, M. Gouterman and S. Aronowitz: J. Phys. Chem. 80, 2184-2191 (1976).

12. J. M. Friedman and R. M. Hochstrasser, Chem. Phys. 6, 1551-1561(1974).

V. Poster Session

Picosecond Optical Absorption at 1.06 μm and 1.55 μm in Thin Germanium Samples at High Optically-Created Carrier Densities

A.L. Smirl, J.R. Lindle, and S.C. Moss

Department of Physics, North Texas State University
Denton, TX 76203, USA

Recently, the enhanced transmission of single, ultrashort optical pulses at 1.06 μm through germanium as a function of incident pulse energy has been measured [1]. In addition, the temporal evolution of this enhanced transmission has been determined on a picosecond time scale using the excite and probe technique [1,2]. The latter measurements reveal that the probe transmission increases for 50-100 psec following excitation with an intense optical pulse. ELCI *et al.* [3] have attributed this rise in the probe transmission to a cooling of the hot electron-hole plasma created by the excitation pulse. In sharp contrast to this interpretation, AUSTON *et al.* [4] have stated that they expect the energy relaxation time to be too short to account for the rise in the probe transmission. AUSTON *et al.* also suggest that enhanced free-hole [5] and Coulomb-assisted indirect absorption [6] effects can be significant at the high optically-created carrier densities encountered in the excitation-probe measurements at 1.06 μm. Indeed, they suggest that these processes could introduce a minimum in the absorption versus carrier density relationship. AUSTON and MCAFEE [7] have proposed an alternative explanation for the temporal evolution of the probe transmission by combining the details of the way the absorption saturates as a function of carrier density with a monotonic decrease in carrier density with time due to Auger recombination [8]. The rise in probe transmission with time can then be explained in the following manner. We denote the density at which the minimum absorption occurs as n_{min}. The absorption of the excitation pulse creates an initial carrier density greater than n_{min}. As the carrier density is decreased by Auger recombination, the absorption coefficient will decrease in time until the carrier density reaches n_{min}, then increase. Thus, the probe transmission will increase then decrease if the initial, optically-created carrier density is greater than n_{min}. We stress that the success of the second model as it now stands depends on the absorption decreasing then increasing with carrier density: *there must be an absorption minimum*.

Here, we report measurements of the *combined* free-carrier, intervalence band free-hole, and indirect absorbance *in thin* germanium samples at a wavelength of 1.55 μm during excite and probe experiments at a wavelength of 1.06 μm. Our interests in these measurements are twofold. First, we want to ascertain whether or not free-carrier, free-hole, and indirect absorption effects are important in excite-probe experiments at 1.06 μm. Second, if these effects are important, can they, together with Auger effects, account for the rise in probe transmission.

The experimental configuration is depicted in Fig.1. This arrangement is similar to the arrangement utilized by AUSTON *et al.* [8]. In this application of the excite and probe technique, a high density plasma is created by direct absorption of an intense excitation pulse, and the evolution of the plasma is monitored by a second probe pulse. The excitation pulses were selected by a laser-triggered spark gap and a Pockel's cell from trains of pulses produced by a mode-locked Nd:glass laser. The pulses were 5 to 10 psec in duration and had peak powers of approximately 10^8 watts at a wavelength of 1.06 μm, and they produced a measured irradiance of approximately 4×10^{-3} J/cm^2 when focused on the crystal surface. The plasma produced by the absorption of the excitation pulse was probed using weak pulses of two types: one had an energy greater than the direct energy band gap for germanium and the other had an energy less than the direct gap but greater than the indirect gap. The former was derived from the excitation pulse using a beam splitter as shown in Fig.1. The latter,

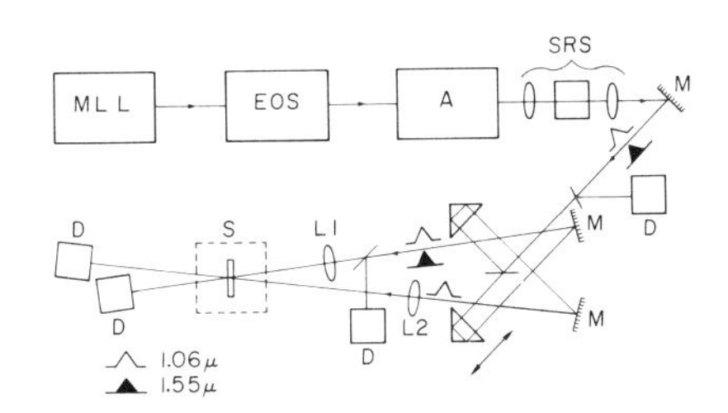

Fig.1 Block diagram of the experimental configuration for excite and probe measurements at 1.06 μm and 1.55 μm, where MLL denotes the mode-locked laser, EOS the electro-optic switch, A the laser amplifier, SRS the Stimulated-Raman-Scattering cell, M a mirror, D a detector, L1 and L2 lens, and S the sample

having a wavelength of 1.55 μm, was produced by stimulated Raman scattering in benzene. We emphasize that the energy of a quanta at 1.06 μm (1.17 ev) is sufficient to excite direct band-to-band transitions in germanium as well as free-carrier, free-hole, and indirect transitions; whereas, the energy of a quanta at 1.55 μm (0.80 ev) falls below the direct band gap but above the indirect gap and is, thus, only a measure of the combined free-carrier, free-hole, and indirect processes. The incident excitation pulse irradiance was measured, and the overlap of excitation and probe pulses was ensured, employing techniques described in [8].

The germanium sample was a high purity ($\rho_{min} = 40\,\Omega$ cm) single crystal cut with the (111) plane as face. The sample was polished and etched with Syton to a thickness of 6 μm as determined by interferometric techniques.

The results of the measurement of the change in absorbance of the thin germanium crystal as a function of increasing carrier number (incident excitation pulse energy at 1.06 μm) are shown in Fig.2 for photon energies of 1.17 and

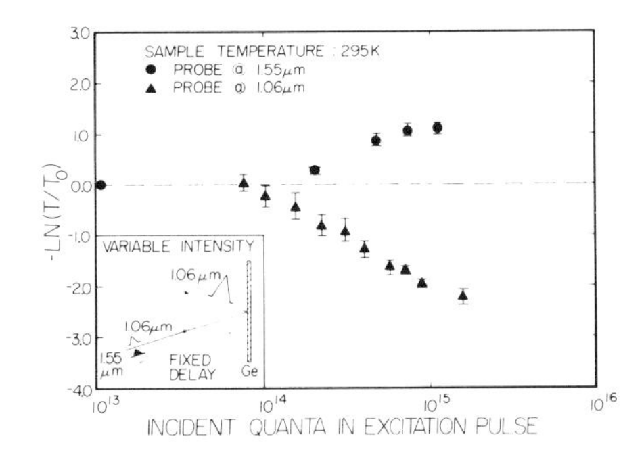

Fig.2 Change in absorbance, $-\ln(T/T_0)$, of the germanium sample at 1.06 μm and 1.55 μm as a function of incident excitation pulse energy at 1.06 μm, where T_0 is the linear transmission of the sample at the wavelength under consideration

0.8 ev. The data were obtained in the following manner. The crystal was illuminated by variable energy optical pulses of wavelength 1.06 μm, and the transmission of each pulse was measured. Each pulse at 1.06 μm was followed immediately (at a fixed delay of 15 psec) by pulses that monitored the absorbance of the crystal at wavelengths of 1.55 μm and 1.06 μm. The optical absorbance at 1.17 ev is seen to decrease by approximately 2.3 as the carrier number increases. This corresponds to a transmission increase of an order of magnitude. By contrast, the absorbance at 0.8 ev increases roughly by 1.0, corresponding to a decrease in transmission of a factor of 2.7. Each data point shown is the average of at least ten separate observations. The data were very reproducible within the error bars.

A striking feature of the data presented in Fig.2 is that the absorbance of the crystal at 1.06 μm does not decrease then increase as required by the model suggested by AUSTON and MCAFEE [7]. In fact, as can be seen from Fig.2, any decrease in carrier density with time caused by carrier recombination will be accompanied by an increase in the total absorbance at 1.06 μm. Thus, a temporal decay of carrier density alone can not be combined with the absorption versus density relationship to account for the rise in probe transmission at 1.06 μm. In addition, the change in absorbance at 0.8 ev, which is sensitive to free-carrier, free-hole, and indirect absorption effects, is smaller in magnitude and opposite in sign to that measured at 1.17 ev, which is sensitive to direct absorption effects as well. Although care must be taken when extrapolating absorbance measurements at 0.8 ev to 1.17 ev, we believe that the sign and magnitude of this absorbance change at 0.8 ev and the observed monotonic decrease in overall absorbance at 1.17 ev suggests that the saturation of the absorbance at 1.17 ev is dominated by changes in the direct absorption coefficient.

The results of excite-probe experiments that measure the temporal evolution of these changes in absorbance are presented in Fig.3. In this experiment the sample was irradiated by an optical pulse at 1.06 μm containing roughly 10^{15} quanta and was probed at various delays by weak pulses having wavelengths of 1.55 μm and 1.06 μm. The transmission of the probe pulse at 1.06 μm initially increases as the delay between the two pulses increases, reaching a peak value at a delay of approximately 50 psec. Meanwhile, the absorbance of the sample

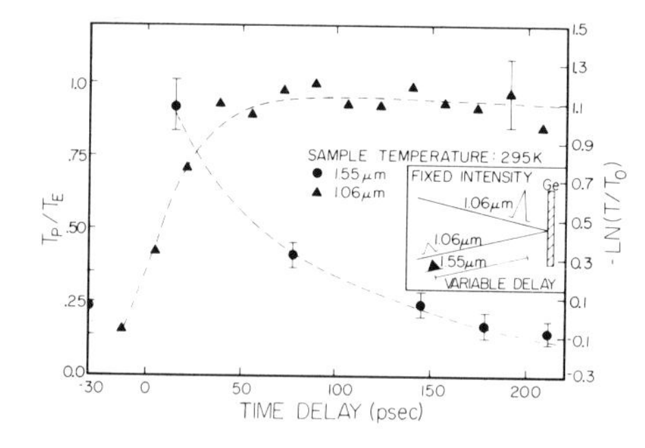

Fig.3 Probe pulse absorption versus delay between the excitation pulse at 1.06 μm and 1.55 μm. The probe data at 1.06 μm is plotted as the normalized ratio of probe pulse transmission to excitation pulse transmission, T_P/T_E, and the data at 1.55 μm as the change in absorbance, $-\ell n(T_P/T_0)$, where T_0 is the linear transmission of the probe pulse at this wavelength

at 1.55 μm decreases by roughly 1.0 corresponding to a transmission increase of approximately 2.7. Different units were chosen for the ordinates of the 1.06 μm and 1.55 μm curves to facilitate presentation of both curves in the same figure. The probe measurements at 1.06 μm are identical to those performed by SMIRL *et al.* [1]. Also, the results of the probe measurements at

1.55 μm are similar to those obtained by AUSTON et al. [8]. However, these authors stated that they performed their measurements at excitation intensities such that the absorption of the excitation pulse was linear. Our experiments are clearly performed in the nonlinear region. In addition, the measurements of AUSTON et al. were performed on a 300 μm-thick sample; our sample was 6 μm-thick. Thus, our experimental configuration is almost identical to that employed in our original excite and probe experiments at 1.06 μm.

The measurements presented in Fig.3 indicate that free-carrier, free-hole, and indirect absorption can be significant at the carrier densities encountered during the excite and probe experiments at 1.06 μm presented here. Note that the increase with time of the probe transmission at 0.8 ev caused by the decrease in free-carrier, free-hole, and indirect absorbance is of the proper magnitude (if extrapolated unchanged to 1.17 ev) to account for a major portion of the probe rise at 1.17 ev. AUSTON et al. [8] attribute this decrease of the probe pulse absorbance at 1.55 μm with delay to a decrease in free-carrier absorption caused by a temporal decay in carrier density due to Auger recombination. The present experiments only allow the measurement of the change in the combined free-carrier, free-hole, and indirect absorbance, and they do not provide for a convenient separation of their individual contributions. In addition, we feel that the decrease in absorbance at 1.55 μm with delay could be accounted for by a number of mechanisms. One of these possible mechanisms is carrier reduction by Auger recombination. Another possibility, however, is that the decrease in absorbance at 1.55 μm is caused by a cooling of a hot carrier distribution created by the excitation pulse at 1.06 μm. This decrease in absorbance could be attributed to a decrease in free-carrier absorption and a decrease in the available states for indirect absorption as the large, hot carrier distribution cools.

In summary, the measurement of the transmission of thin germanium samples at 1.55 μm and 1.06 μm as a function of optically-created carrier densities indicates, over the range of densities encountered in these experiments, that the absorption versus density relationship at 1.17 ev does not exhibit a minimum and suggests that the combined free-carrier, free-hole, and indirect absorbance changes are opposite in sign and smaller in magnitude than the changes caused by saturation of the direct absorption. Second, the excite and probe measurements at these two wavelengths indicate that free-carrier, free-hole, and indirect absorption processes could be significant in interpreting excite and probe measurements at 1.06 μm.

The authors wish to acknowledge their debt to Dave Auston and Sigrid McAfee of Bell Laboratories for suggesting and sharing the model presented in this paper and for their helpful discussions.

This work was supported by the Office of Naval Research and the North Texas State University Faculty Research Fund.

[1]A.L. Smirl, J.C. Matter, A. Elci, and M.O. Scully, Op. Commun. 16, 118 (1976).
[2]C.V. Shank and D.H. Auston, Phys. Rev. Lett. 34, 479 (1975).
[3]A. Elci, M.O. Scully, A.L. Smirl, and J.C. Matter, Phys. Rev. B16, 191 (1977).
[4]D.H. Auston, S. McAfee, C.V. Shank, E.P. Ippen, and O. Teschke, Solid State Electron. 21, 147 (1978).
[5]R. Newman and W.W. Tyler, Phys. Rev. 105, 885 (1957).
[6]C. Haas, Phys. Rev. 125, 1965 (1972).
[7]S. McAfee and D.H. Auston, private communication, July 28, 1977.
[8]D.H. Auston, C.V. Shank, and P. LeFur, Phys. Rev. Lett. 35, 1022 (1975).

Advantages of the Planar Streak Camera in Picosecond Recording

A. Lieber

University of California, Los Alamos Scientific Laboratory
Los Alamos, NM 87545, USA

Availability of picosecond and sub-picosecond light pulses from lasers has opened entirely new fields of study of picosecond phenomena. In spite of new, ultra-fast, channelplate oscilloscopes and fast detectors, the only recording system capable of yielding the picosecond temporal history of one spatial dimension remains the ultra-fast streak camera. Streak cameras with fiberoptics inputs are capable of normalizing and recording a variety of optical signals with bandwidths far exceeding any present or projected oscilloscope.

Until a few years ago all commercial streak cameras used a streak tube based upon the RCA C73435. This tube was first intended as a shutter tube and was later adapted for streak use [1-2]. Engineering improvement programs on this basic design have transpired at many laboratories through out the world [3-6]. Serious problems have emerged as design limits were approached and have been the subject of much controversy [7].

The fact remains that cameras based upon the old design are flawed mainly because of problems associated with the internal photoelectron optics of the streak tube. For tube picosecond recording, these cameras suffer from blooming, spatial resolution loss, low dynamic range, pulse temporal distortion, and produce streaks of very low statistical quality. Usage of pinhole optics in the streak tube serves as a bottleneck to passage of the extremely high photoelectron pulses required to register high quality picosecond streaks. To maintain resolution, the gain of the tube and hence the peak current must be limited to control spacecharge defocusing effects at the pinhole. Problems with the streak tube are compounded by the fact that it is basically a soft field device. A grid structure must be added to increase the photocathode extraction field for picosecond response. This structure results in unwanted image magnification which effectively lowers the tube's sensitivity by spreading the already limited number of photoelectrons over a greater output area. Magnification requires more expensive large diameter follow-on image intensifiers, and distorts spatial linearity.

Inherent design weaknesses of the old streak tube are amplified when the design is applied as an x-ray sensitive streak tube [8]. The design limits at about 20 ps, which is insufficient to record instabilities encountered in laser-fusion experiments that are as short as several picoseconds. For this reason another streak tube based upon another family of intensifiers, the planar or proximity-focused tube, was developed as an outgrowth of the laser-fusion program at Los Alamos Scientific Laboratory several years ago [9]. In this device, cross-over and pinhole electron optics are eliminated. Electrons flow from cathode to phosphor in a ribbon or sheet removing the limitations on peak currents and associated problems. Since this type of tube inherently utilizes ultra-high electric fields to operate, grid structures are unnecessary. Unity magnification is provided without distortion. High electric fields yield the added bonus of mapping the photoelectrons from cathode to phosphor without blooming over a large range of signals. Problems of pulse temporal distortion and after pulsing are not present in this device--the streak profile represents an accurate history of the incident light. In the old tube spatial distortions in picosecond operation depend upon photocathode illumination profile--instrumental errors are a function of the experiment. Planar optics eliminates this difficulty.

The ability of a camera to record features of the incident light profile over a large range of signals is as important to picosecond photography as ultimate obtainable temporal resolution itself. A practical definition of dynamic range remains the signal range over which the camera can record features of a given temporal resolution. While operating with picosecond resolution, cameras based upon the old tube generally display a range of less than 50 X.

The dynamic range of a new commercial camera based upon the planar streak tube has been measured and the results are shown in Fig. 1. A series of images of a Nd:YAG laser beam were formed along the camera slit using an etalon. Each image was degraded in intensity over its predecessor by 0.23. Complex pulse trains were generated to determine over how many steps the camera could track. For these measurements an avalanche-krytron sweep generator module was used which was triggered from a photodiode. Sweep rate was chosen to produce a modest resolution of 8 ps. The images were recorded on film which was calibrated and densitometerized. In the figure the output values are shown along the ordinate. These levels very nearly track the $(0.23)^n$ inputs. It is evident from the figure that the camera is capable of resolving the double pulse with better than 10 ps resolution over steps corresponding to a dynamic range in excess of 1000X. Confirmation of this range is also evident in the upper trace. In this streak switched out pulse and bleedthrough are recorded in a single trace. The manufacturer's quoted shutter ratio for the switch of 800:1 confirms the other measurement. The distortionless nature of the format as well as high statistical quality of the traces is also evident. Densitometer scans across streak of varying light levels yield similar FWHM values. These features make the camera an ideal companion for operation with an OMA or other digital readout system.

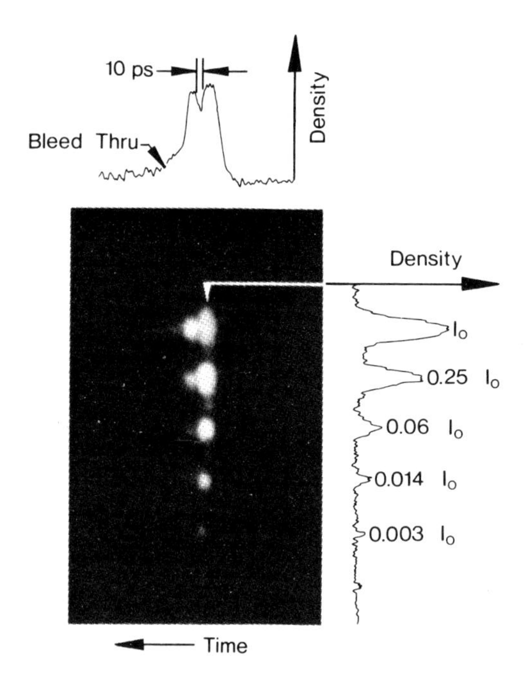

Fig. 1 Dynamic range test of planar streak camera. Output values of scan across streaks shown at side of figure follow $(0.23)^n$ input.

A welcome surprise, discovered recently in another application of the camera, is its increased sensitivity. It has been found that under similar situations the planar streak camera is over 100X more sensitive than its conventional counterparts based upon the old streak tube design [10]. Several factors may account for this. The new camera uses a fiberoptics input window and the slit is placed in direct optical contact with the photocathode eliminating transfer optics. This results in an efficiency increase of about 16X. Other factors being equal, part of the additional sensitivity can be atrributed to the magnification problem in the old design. The planar 1:1 format results in roughly one order sensitivity increase over the old system.

The planar streak tube and proximity-focused follow-on image intensifier yield a small rugged camera system. When operated with the solid dielectric sweep module, the camera is capable of true sub-picosedond operation, while maintaining its other desirable features such as range and statistical qualtiy. This camera together with an OMA system makes a camera team of unparalleled advantages and has already unlocked many new fields of picosecond research.

REFERENCES

1. J. S. Courtney-Pratt, "A New Method for the Photographic Study of Fast Transient Phenomena," Research, London 2, 287 (1949).

2. R. Stoudenheimer and J. C. Moor, "An Image Converter Tube for High-Speed Photographic Shutter Service," RCA Review 18, 322, (1957).

3. M. Ya Schelev, M. C. Richardson, and A. J. Alcock, "Operation of a Grid-Shuttered Image Converter Tube in the Picosecond Region," Review of Scientific Instruments 43 #]2, 1819, (1972).

4. D. J. Bradley, B. Liddy, W. Sibbett, and W. E. Sleat, "Picosecond Electron Optical Chronography," Applied Physics Letters 20 #6, (1972).

5. S. W. Thomas, J. W. Hougton, G. R. Tripp, and L. W. Coleman, "The LLL Compact 10 ps Streak Camera-1974 Update," Proceedings of the 11th International Congress on High-Speed Photography, 101, London (1972).

6. R. Rolaert, J. Rodiere, "Improvement of the Performance of High-Speed Cinematography Through the Use of A Proximity-Focused Microchannel Image Intensifier," ibid pg 170.

7. See Proceedings 12th International Congress on High-Speed Photography SPIE Vol. 97, Toronto, Canada (1976).

8. C. F. McConaghy and L. W. Coleman, "Picosecond X-ray Streak Camera," Applied Physics Letters 25 #5, 268 (1974).

9. A. J. Lieber, R. F. Benjamin, H. D. Sutphin, and C. B. Webb, "Investigating of Micro-Channel Plates as Parallel-Bore Electron Collimators for use in a Proximity-Focused Ultra-Fast Streak Tube," Nuclear Instruments and Methods, 27, 87 (1975).

10. L. R. Veeser and J. C. Solem, "Exploratory Laser Driven Shock Wave Studies," Bull. APS 23, #1, Jan. 1978).

Design for Ultrafast Electrical Switching

R. Castagne, R. Laval, S. Laval, and A. Merriaux

Institut d'Electronique Fondamentale, Laboratoire Associé au C.N.R.S.
Université Paris XI, Bâtiment 220
F-91405 Orsay Cédex, France

The realisation of ultrafast electronic devices implies the development of technics in which the parasitic capacitances must be suppressed. Thus, to observe picosecond switching, the switching device must be inserted in a propagation line with proper dimensions, which can be either a microstrip line or a coplanar structure line. The switch enables a quasi-instantaneous short circuit between the two parts of the transmission line. A microstrip structure has been initially used by LAWTON and SCAVENNEC [1] to develop a fast photoconductive detector, and by AUSTON [2] who first demonstrated picosecond switching and gating. We have obtained picosecond gating using a coplanar propagation line [3].

In order to characterize the operation of such switching devices, and to forecast their ultimate performances we have derived a theoretical model which has been checked by comparison with experimental results in the nanosecond range, and applied to obtain the picosecond operation characteristics.

As the switching element appears as a gap in the microstrip line, the voltage is initially present across it, and the short circuit can be obtained by a variation of the conductibility. Using a semiconducting substrate, this can be achieved by modulation of either the charge carrier density or the thickness of a channel. We are mainly interested here by the first case in which an external cause induces the fast apparition of an excess carrier density. A typical example is the illumination by a picosecond laser pulse of the microstrip gap realized on a high resistivity semiconducting material. In the following we assume that the carrier injection is uniform over the whole active region, and that the two parts of the propagation line end by two ohmic contacts. We have shown [4] that a quite good approximation to describe the conduction process inside the microstrip gap is given by an ohmic law. The conduction current density in the gap is then written as a function of space and time :

$$j(x, t) = q(\mu_n + \mu_p)\, n(t)\, \mathcal{E}(x, t)$$

where $n(t)$ is the density of injected carriers ($n = p$), μ_n and μ_p the electron and hole mobilities, and $\mathcal{E}(x, t)$ the electric field at a given point x in the gap and at time t.

A small portion dx of the active region can be characterized by an elementary resistance

$$dR = \frac{1}{q(\mu_n + \mu_p)\, n(t)} \, \frac{dx}{eL}$$

where e is the mean depth over which the injection is settled, L is the width of the region under injection, which is assumed to be equal to the line strip width.

The output voltage reaches its maximum value when the residual resistance R_s of the whole gap may be neglected with respect to the characteristic impedance R_o of the adapted propagation line. Let us define the saturation time t_s as the time needed to get $R_s = 0.1\ R_o$.

The electric field $\mathcal{E}(x, t)$ depends on the speed of the injection process and on the electrical properties of the semiconducting material, but also on the geometrical dimensions of the device. If the transit time t_t for an electromagnetic wave to propagate across the gap is much smaller than t_s, then the propagation phenomena can be ignored and the electric field reduces to its electrostatic component. Furthermore, if the carrier density risetime is smaller than the transit time t_c for the carriers across the gap, no space charge appears inside the active region and the electric field can be considered as equal to its Laplace's component, i.e. :

$$\mathcal{E}(x, t) = V(t)/L = \mathcal{E}_o(t)$$

where V(t) is the voltage across the gap at time t.

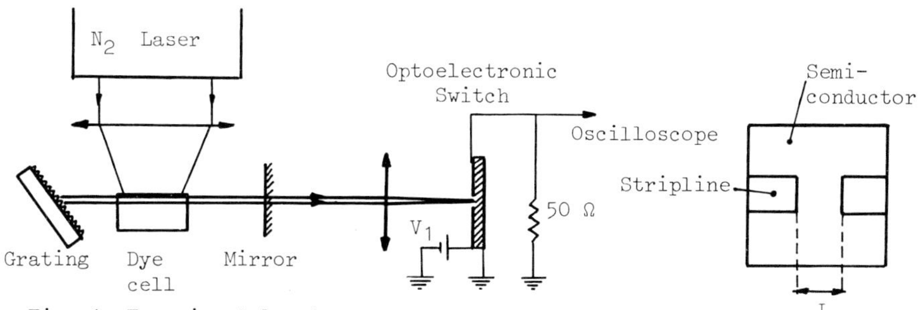

Fig. 1 Experimental set-up

Experiments have been done under these conditions, using the set-up represented in Fig. 1. The carrier injection is produced by a laser pulse coming from a nitrogen-laser pumped-dye laser. The pulse duration is about 4 ns and its energy is high enough to insure the saturation of the output voltage. The propagation line has a characteristic impedance of 50 Ω. The output voltage $V_s(t)$ is recorded on a fast oscilloscope and compared to the computed value Fig. 2 . Such a device can be used to determine the laser pulse shape, from the experimental knowledge of $V_s(t)$ Fig. 3.

However, in order to describe the operating mode of such devices in the general case, the propagation effects must be taken into account. Before a more detailed discussion, a qualitative description can be given.

Let us assume that the high density carrier plasma is created in a time much shorter than the EM propagation time t_t. Before the excitation, the electric field pre-exists in the gap. When the carrier plasma is quasi-instantaneously created, after a time of the order of the momentum relaxation time τ_m ($\simeq 10^{-13}$ s in Si or AsGa), the carriers have got a speed proportional to the local electric field $\mathcal{E}(x, t)$. When the high density carrier plasma is established, the voltage across the gap begins to decrease and the phenomena can be described by two linearly increasing voltage waves which propagate in opposite direction from the initial voltage distribution. Thus, the output voltage increases linearly with time.

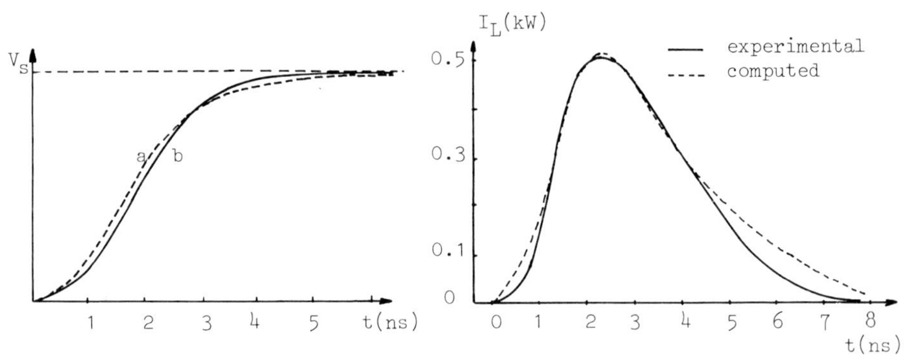

Fig. 2 Output voltage versus time a = experiment ; b = theory (assuming a gaussian laser pulse)

Fig. 3 Experimental and computed laser pulse versus time

We have worked out a theoretical model in which the gap is represented by a portion of a propagation line in which a variable resistance R(x, t) is included in series with the inductive part to take into account the resistance of the injected region, and whose characteristic impedance is adapted to 50 Ω when R(x, t) cancels.

The maximum output voltage and the risetime of the output electrical pulse are calculated as functions of the material characteristics (mainly the mobility) and of the injection risetime. Some results are given in Fig. 4. Two distinct regions appear :

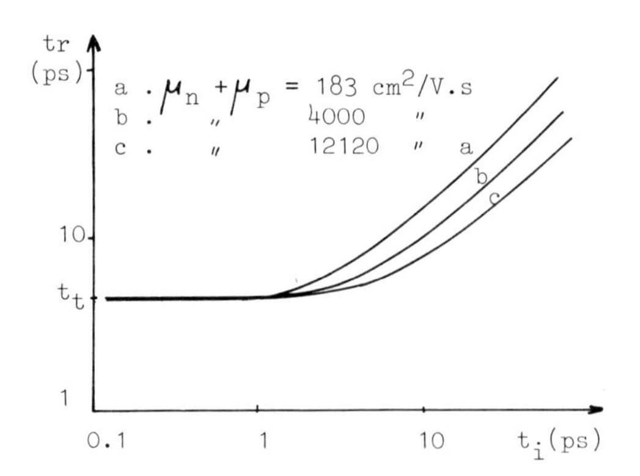

Fig. 4 Electrical pulse risetime versus injection risetime

- When $t_s > t_t$, the electrical pulse risetime is proportional to the carrier creation one (or to the laser pulse risetime in the case of an illumination) and the electrical pulse shape is related to the laser pulse shape.
- When $t_s < t_t$, the electrical pulse has a risetime limited by t_t and it grows linearly with time.

So far, the carrier velocity has been assumed proportional to the electric field. However, when the electric field exceeds a critical value $\mathcal{E}_c$, (4 kV/cm in GaAs, 10 kV/cm in Si), the carrier velocity saturates at about 2.10^7 cm/s. The mobility is no longer a constant and the previous analysis must be revised.

The velocity saturation appears for carriers accelerated by a field higher than the critical one after a time τ_E of the order of the energy relaxation time, i.e. $\simeq 10^{-12}$ s. For times shorter than τ_E, the mobility can be supposed to have still its low-field value and then the electrons can reach velocities far larger than the saturation value. This phenomenon will influence the electrical pulse shape, on condition that the latter is not limited by the propagation time t_t of the EM wave, in the following way :
- If the gap length is adjusted in such a way that the carrier propagation time t_c across the gap is longer than τ_E, the velocity overshoot exists during a time τ_E. The output pulse will present a breakage corresponding to the fast decrease of the carrier velocity with thermalization.
- If, according to the gap length, $t_c < \tau_E$, the carriers remain out of thermal equilibrium, and the electrical pulse risetime will correspond to an apparent mobility equal to the low-field value.

This potential possibility of characterizing the dynamical transport phenomena of semiconducting materials is important for picosecond electronics.

References

1 A.R. Lawton, A. Scavennec, Electronics Letters, 11, 74 (1975).
2 D.H. Auston, Appl. Phys. Lett., 26, 101 (1975).
3 R. Castagné, S. Laval, R. Laval, Electronics Letters, 12, 438 (1976).
4 A. Merriaux, R. Castagné, R. Laval, S. Laval, Electronics Letters, 13, 245 (1977).

Small Area Metal-Oxide-Metal Junctions as Picosecond Photoemissive Detectors[1]

C. Slayman, M. Guedes, T.K. Yee, and T.K. Gustafson

Electronics Research Laboratory, University of California
Berkeley, CA 94270, USA

Internal photoemission [1 - 5] in MOM (metal-oxide-metal) junctions can be used as a means for fast photodetection at visible and ultraviolet wavelengths. Figure 1 shows the geometry and energy band diagram of such a photodetector. Incident photons of energy $h\nu$ excite electrons from the Fermi sea of one metal. If $h\nu$ is greater than the oxide barrier height ϕ (typically around 2 eV), "hot" electrons can be emitted over the barrier into the second metal, creating a photoemissive current in the MOM.

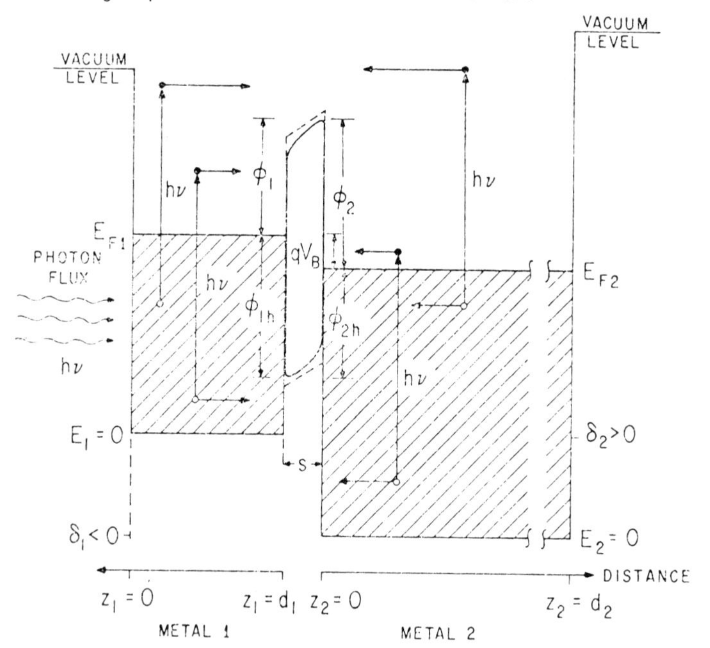

Fig.1 Energy band diagram of the MOM. Energies are measured from the conduction band bottoms. V_B is the bias voltage across the detector.

The front electrode (Metal 1) is chosen to be absorbing, while the back electrode (Metal 2) is reflecting, so that essentially all of the photoexcita-

[1]Work supported by AFOSR Grant F44620-76-C-0100 and NASA Grant NSG-2151.

tion will occur in Metal 1. Making the thickness d_1 of Metal 1 small (~ 300 Å) insures that a large fraction of electrons excited throughout the metal will reach the oxide before scattering. Such thin films can actually absorb more radiation than bulk metals due to the reflections at the interfaces, further increasing the efficiency of the device [6].

If the junction is illuminated uniformly across an area A, the photocurrent due to unscattered electrons is [7]

$$I_{ph} = qA \int_0^{d_1} dz_1 \int_{E_{F_1}}^{E_{F_1}+h\nu} dE_1 \int_0^{\pi/2} d\theta_1 \; G(E_1,\theta_1,z_1)\, P(E_1,\theta_1,z_1)\, D(E_1,\theta_1) \qquad (1)$$

where E_1 is the electron kinetic energy (measured from the conduction band bottom), and θ_1 is the angle of the electron trajectory from the z_1 axis. $G(E_1,\theta_1,z_1)$ is the generation rate of electrons at z_1 to energy E_1 and angle θ_1. $P(E_1,\theta_1,z_1)$ is the probability that the electron will reach the barrier unscattered, and $D(E_1,\theta_1)$ is the transmission probability into Metal 2.

In an isotropic metal, if every absorbed photon generates one hot electron, then

$$G(E_1,\theta_1,z_1) = -\vec{\nabla}\cdot\vec{S}\; g(E_1)\, \sin\theta_1 / 2h\nu \qquad (2)$$

where $\vec{S}$ is the time average Poynting vector. The generation rate of electrons to energy E_1, $g(E_1)$, will depend on the band structure of the metal and whether a direct or indirect transition model is used [8]. If ℓ_1 is the energy dependent mean free path in Metal 1, then

$$P(E_1,\theta_1,z_1) = e^{(z_1-d_1)/\ell_1\cos\theta_1} . \qquad (3)$$

For a step-function transmission probability (i.e., the only transmitted electrons will be those whose "normal" component of energy $E_1\cos^2\theta_1$ is greater than the total barrier energy $E_{\phi_1} = E_{F_1}+\phi_1$), (1) becomes

$$I_{ph} = -\frac{qA}{2h\nu} \int_0^{d_1} dz_1 \int_{E_{\phi_1}}^{E_{F_1}+h\nu} dE_1 \int_0^{\cos^{-1}\sqrt{E_{\phi_1}/E_1}} d\theta_1\, \vec{\nabla}\cdot\vec{S}\, g(E_1)\, e^{(z_1-d_1)/\ell_1\cos\theta_1} \sin\theta_1 . \qquad (4)$$

Numerical integration of (4) has given good agreement with experimentally observed photocurrents in large area (300 μm × 300 μm) $Ag\text{-}Al_2O_3\text{-}Al$ junctions around $h\nu = 4$ eV [9].

The quantum efficiency (transmitted electrons/incident photon) increases with larger $h\nu$, making the device more sensitive in the UV. Zero bias quantum efficiencies on the order of 10^{-3} have been observed in the $Ag\text{-}Al_2O_3\text{-}Al$ junctions. At higher photon energies, the current due to the emission of scattered electrons will become significant, further improving the detector's sensitivity.

Hole emission from Metal 1 through the oxide valence band (see Fig.1) into Metal 2 is the opposite polarity of the electron photocurrent. Thus, it is desirable to have the hole barrier ϕ_{h_1} (measured down from E_{F_1}) larger than ϕ_1. This condition is determined by the material properties of the metal-oxide interfaces as well as the junction bias. A positive bias on Metal 2 will lower the electron barrier and raise the hole barrier.

The transit time of photoexcited electrons is extremely short. Typical Fermi velocities in metals range between 1× to 2× 10^6 m/sec, so electrons above E_F will have larger velocities. Thus, the transit time of a hot electron through a 300 Å electrode is well into the sub-picosecond range.

Two dominant energy loss mechanisms for hot electrons are electron-electron and electron-phonon scattering. In general, mean free paths are less than 1000 Å, so hot electron mean free times are less than 10^{-13} sec. Any electrons not collected after several mean free times will have lost too much energy to be transmitted over the barrier. Thus, there will be no slow tail to the photocurrent.

When a hot electron is emitted into Metal 2, the free charges redistribute themselves rapidly. The speed at which this process occurs governs how soon the current will appear at the electrode. This is the conductivity relaxation time and is roughly ε/σ, where ε is the dielectric constant and σ is the conductivity. For metals, the relaxation time is on the order of 10^{-18} sec.

Therefore, the limiting speed of the MOM will be determined by its RC time constant. Figure 2 shows the lumped equivalent circuit of the detector. R_j is the tunneling resistance, C_j is the junction capacitance, and R_L is the load impedence. If a perfectly terminated 50Ω transmission line is used

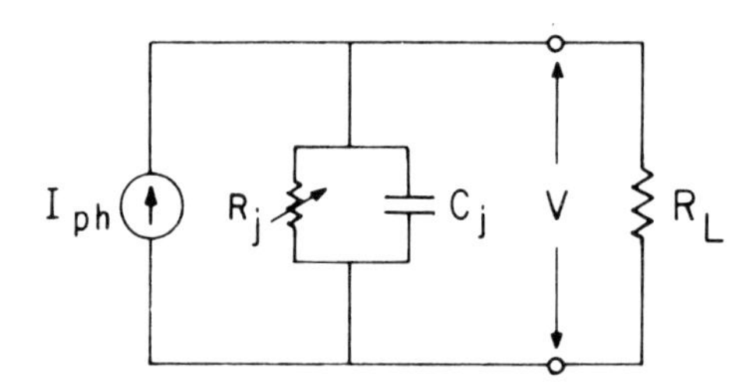

Fig.2 Lumped equivalent circuit of an MOM photodetector.

as the load, R_L will be 50Ω. For a 4 μm × 4 μm area junction (as shown in Fig.3), with ∿30 Å oxide thickness of relative dielectric constant $\varepsilon_r \sim 3$, the junction capacitance is $C_j \sim 0.1$ pF. The time constant of the photodetector will be

$$\tau = R_L C_j < 10 \text{ psec} . \quad (5)$$

It should be noted that for small area junctions with thick (> 20 Å) oxides, R_j is large and must be shunted by R_L in order to obtain fast RC decay times. The response time can be further reduced by decreasing the junction area and increasing the oxide thickness. However, decreasing the area of the junction beyond the focusing limit of the light beam will decrease the signal, and thicker oxides (i.e., on the order of hundreds of Angstroms) might introduce transit time problems through the oxide conduction band. Nevertheless, response times less than 10 psec do not appear difficult to obtain under properly optimized conditions.

The MOM photodetector displays two types of nonlinearities in photocurrent vs. incident power that might prove useful in correlation experiments. The tunneling resistance R_j is a nonlinear function of the voltage V developed across the junction. The photocurrent also depends nonlinearly on V due to the change in ϕ. If the photocurrent becomes large enough (for a short high

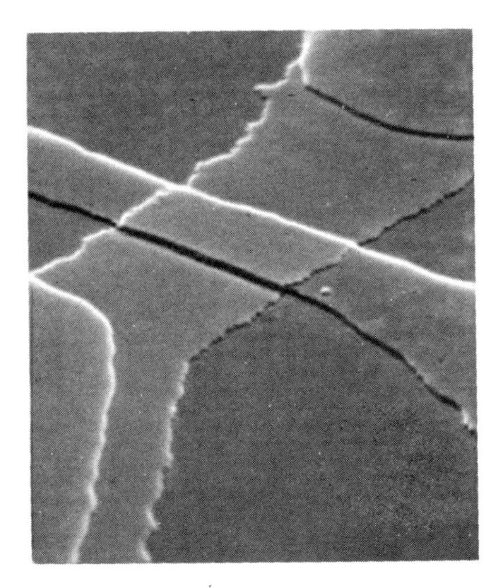

Fig.3 Scanning electron microscope photograph from a 60° angle of a 4 μm × 4 μm Ni-Al_2O_3-Al junction. The top electrode is a thin (∿400 Å) Ni film. The bottom electrode is Al. A thick Al electrode was evaporated on top of the Ni for good electrical contact. Magnification × 8000.

power laser pulse), the self-bias will introduce nonlinearity in the detector response, and twice the incident power will not necessarily give twice the signal.

A second mode of operation for the MOM is electron-hole generation in the barrier. This process will occur when $h\nu > E_g$, the gap energy between the barrier valence and conduction bands. Semiconductors or insulators could be used for the barrier. The $\vec{E}$ field in the barrier (due to bias or surface charges) would then sweep out electrons and holes in opposite directions similar to a pin photodetector. Metal 1 should be thin and semitransparent, and Metal 2 should be reflecting. Carrier generation can then be localized to very small dimensions (on the order of hundreds of Angstroms).

The MOM photodetector offers a promising potential for fast pulse detection. Work is now in progress to determine the photoemissive response time of the 4 μm × 4 μm Ni-Al_2O_3-Al junctions shown in Fig.3.

REFERENCES

1 R. K. Jain et al., Appl. Phys. Lett. 48, 1543 (1977).
2 G. M. Elchinger et al. J. Appl. Phys. 47, 591 (1976).
3 K. H. Gundlach, J. Kadlec, Thin Solid Films 28, 107 (1975).
4 Z. Burshtein, J. Levinson, Phys. Rev. B 12, 3453 (1974).
5 A. Braunstein et al., Phys. Rev. Lett. 14, 219 (1965).
6 M. Guedes et al., AIAA Progress in Astronautics and Aeronautics 61, (1978).
7 C. N. Berglund, W. E. Spicer, Phys. Rev. 136, A1030 (1964).
8 J. F. Janak et al., Solid State Comm. 8, 271 (1970).
9 M. Guedes et al., to be published.

VI. Coherent Techniques and Molecules

Picosecond (Coherent) Transients in Molecular Mixed Crystals

W.H. Hesselink and D.A. Wiersma

Laboratory for Physical Chemistry, University of Groningen, Nijenborgh 16
NL-9747 AG Gronigen, The Netherlands

Abstract

Results of picosecond photon echo and pump-probe experiments on pentacene in naphthalene and p-terphenyl are reported. The photon echoes are detected by optical mixing of the echo with a synchronized probe pulse in an ADP crystal. From the temperature dependence of the photon echo it is concluded that the pure electronic transition of pentacene in both naphthalene and p-terphenyl is relaxed by phonon scattering off a resonant phonon. Time-resolved absorption measurements on some selected excited state vibrations of pentacene in naphthalene show that drastic differences exist in vibrational relaxation times.

Narrative

As of today very few picosecond experiments have been performed on well-resolved electronic and vibronic transitions of molecules in solids. With the recent development of tunable picosecond dye-lasers [1] such experiments have become feasible,and considerable progress in this field of spectroscopy is therefore expected. In the branch of coherent optical spectroscopy, picosecond excitation is also expected to rapidly gain importance as optical dephasing processes, that occur on a picosecond time scale, become accessible. Picosecond relaxation processes of current interest are vibrational dephasing [2], exciton - [3] and polariton - [4] scattering and pure electronic dephasing at elevated temperature [5]. Among the optical coherent transients the photon echo [6] is most suitable to study these phenomena as its intensity is a direct measure for the occurring dephasing processes. We have recently shown [5] that picosecond photon echoes are easily generated by amplified pulses of a synchronously pumped dye-laser and detected by optical mixing with an amplified pulse of a second synchronously pumped dye-laser system.

In this paper we will report new results of picosecond experiments on pentacene in naphthalene at low temperature. Pentacene, a large organic molecule, has recently received considerable attention [7,8]. From optical free induction decay experiments [9] the transition dipole of the lowest singlet-singlet transition is known(1.0 $\pm$ .3 D), which enables us to calculate the necessary laser power density to create a substantial coherent excitation. The origin of this transition is the most prominent line in the spectrum and in naphthalene at 1.5 K occurs at 6028 Å [10].

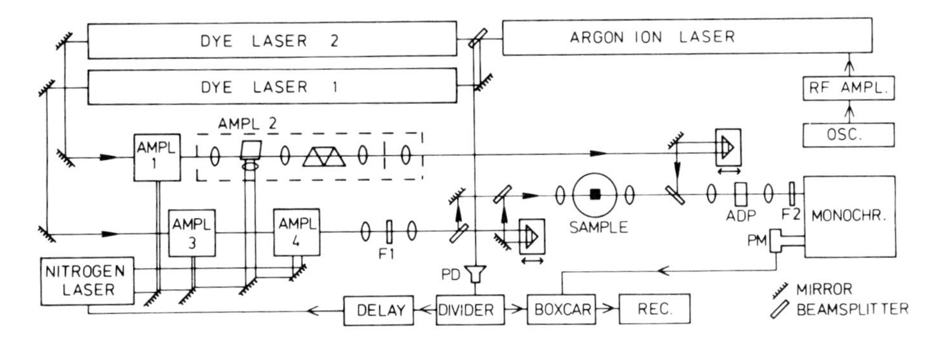

Fig. 1. Experimental arrangement for picosecond photon echo studies

Vibronic transitions selected for pump-probe experiments are found at 5934 and 5768 Å. As a point of practical interest we note here that pentacene in naphthalene mixed crystals of excellent optical quality may be grown.

The experimental set-up used for the photon-echo study is shown in Fig. 1. As a detailed description of the system is given in Ref. 5, we will only recall those features that are of interest for the present presentation. Both dye-lasers, when equipped with a three-element birefringent filter, produce transform limited pulses of 8 ps, as determined in auto-correlation measurements [11]. One of the dye-lasers (DL-1) uses Rhodamine-6G and the other (DL-2) Rhodamine-B as lasing medium. After amplification (using a Molectron UV-22 nitrogen laser) pulses of 200 KW peak power at 10 Hz are available. We note here that during the experiments auto- and cross-correlation measurements are used to check the pulse width of and the jitter between the dye-lasers. As the jitter between the dye-lasers ultimately limits the time resolution in the photon echo experiment it is important to minimize it. We have however not been successful in tracking down the cause of the remaining jitter and found it to vary from day to day between 8 and 25 ps. At the time of the experiments reported here the jitter between the dye-lasers was ca 13 ps as shown in Fig. 2. We also note that amplification of the dye-laser pulses does not lead to any measurable temporal or spectral broadening of the pulses.

We now proceed by discussing the resultsof picosecond photon echo experiments on pentacene in naphthalene. In this experiment DL-1 was tuned to 6028 Å while DL-2 was tuned for maximum output (6290 Å). The photon echo of pentacene was studied using both 8 and 18 ps excitation pulses. With 8 ps excitation pulses we find a very distinct difference between the width of the echo and the excitation pulses as shown in Fig. 2. The interpretation of this effect is quite straightforward if one realizes that the excitation pulse width is shorter than the dephasing time (T_2*) of the *inhomogeneous* optical line(width πT_2^*). Indeed from the 0.85 cm^{-1} optical linewidth of the origin of pentacene in naphthalene, T_2^* is calculated to be 12.5 ps. Takeuchi [12] has shown that, as long as the excitation pulse width does not exceed 0.7 T_2^*, the photon echo width is $2(2\ln 2)^{\frac{1}{2}}T_2^*$ which in this case amounts to 29.5 ps. Taking into account the pulse width of the probing laser (8 ps) and the jitter between the lasers (13 ps) we calculate for the experimental photon echo width 33 ps. This is in fair agreement with the observed width of 36 ps. We note that this is the first time that photon echoes have been generated such that the full inhomogeneous width contributes to the echo formation.

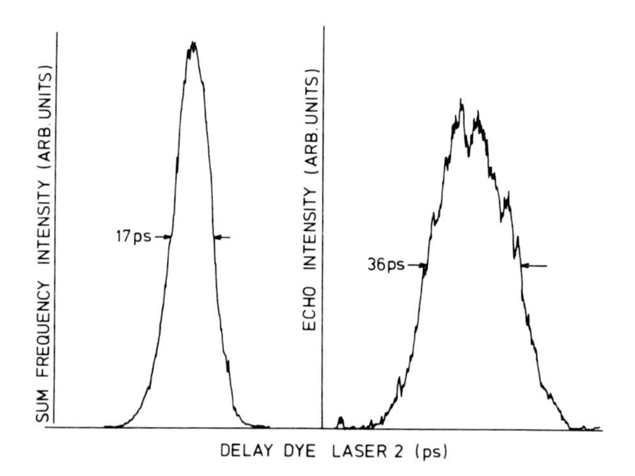

Fig. 2. Cross-correlation measurements between the probe pulse and excitation pulse (left) and between the probe pulse and photon echo of pentacene in naphthalene at 1.6 K (right). The photon echo was excited with a 116 ps separation between the exciting pulses.

We have also tried to measure the temperature dependence of the photon echo of pentacene in naphthalene using 8 ps excitation pulses. Very irreproducible results however were obtained, especially when the crystal was cycled between 1.5 and 4 K. A possible explanation is that at ca 4 K crystal surface damage occurs under excitation with 8 ps pulses focused to power densities of ca 300 MW/cm^2. With longer ($\sim$ 18 ps) excitation pulses this problem does not occur and the temperature dependence of the photon echo was therefore studied using 18 ps excitation pulses. We have measured the temperature dependence of the echo for pulse separations of 100, 150 and 300 ps. Fig. 3 shows the result obtained for a 100 ps separation between the excitation pulses. In the figure is also included a similar measurement for pentacene in p-terphenyl. The dotted curves in the figure are fittings whereby $T_2(T)$ is assumed to be of the following form $T_2(T) = k_\infty e^{\Delta E/kT}$. For pentacene in naphthalene we find: $k_\infty = 4 \pm 2$ ps and $\Delta E = 16 \pm 2$ cm^{-1}, while for pentacene in p-terphenyl [5] the values $k_\infty = 3 \pm 1.5$ ps and $\Delta E = 28 \pm 2$ cm^{-1} are obtained. The observed exponential activation of T_2 with temperature indicates that optical dephasing of pentacene in these crystals is predominantly caused by scattering processes that involve a resonant phonon mode of the crystal. In the mixed crystals of porphyrin in n-octane [13] and tetracene in p-terphenyl [14] T_2 was also found to be exponentially activated.

It therefore seems a rather general phenomenon that at low temperature specific phonons in the ground and/or excited state relax the optical transition. There are several theories that predict such an exponential

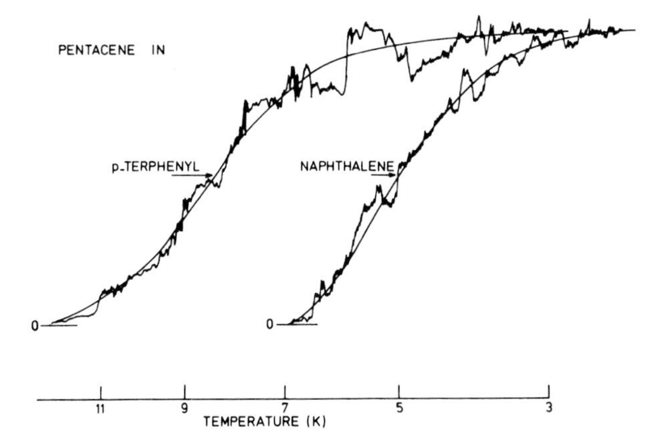

Fig. 3. Temperature dependence of the photon echo intensity of pentacene in p-terphenyl and naphthalene for a 100 ps excitation pulse separation. The dotted lines are curves assuming $T_2(T) = K_\infty e^{\Delta E/kT}$ in
$k_\infty = 3 \pm 1.5$ ps, $\Delta E = 28 \pm 2$ cm^{-1} (p-terphenyl)
$k_\infty = 4 \pm 2$ ps, $\Delta E = 16 \pm 2$ cm^{-1} (naphthalene)

activation of T_2 in case the pseudo-local phonons are treated as local phonons [15,16]. Especially the exchange theory [16] based on a stochastic treatment of the optical lineshape [17] is physically very appealing and has been applied to the interpretation of the results obtained for porphyrin in n-octane [13]. We have recently shown [18] that the exchange theory in the optical domain can only be applied if the excited state resonant phonon coupling is identical to the ground state resonant phonon coupling. As this is generally not the case, the exchange theory may not always be applicable. We therefore propose [18] that a more general treatment of the resonant phonon-induced relaxation should be based on Redfield relaxation equations [19]. It can be shown that the exchange theory is a particular solution of the Redfield equations. In the context of the present paper we will close this discussion by noting that it is still very peculiar, in our opinion, that up to ∿ 10 K the effect of the bulk phonons on optical dephasing is negligible.

Let us now turn to discussion of the second topic of this paper namely vibrational relaxation of pentacene in naphthalene. Recently it has been shown that dephasing of groundstate vibrations can be measured by time-resolved coherent anti-Stokes Raman scattering experiments [20]. This technique however is not applicable to a study of dephasing of vibronic states and we therefore intend to study these states by photon-echo measurements. As the echo relaxation time is always equal or shorter than the vibrational relaxation time we have used a pump-probe absorption technique [21] to try and measure directly the vibrational relaxation times. The experiments were conducted on the electronic origin of pentacene in naphthalene (6028 Å) and two vibronic transitions at 5934 and 5768 Å.

The set-up used employs only a single modelocked cw dye laser (DL-1), producing 8 ps pulses, and is further very similar to the one described in in Ref. 21. The main problem in performing these experiments is that high

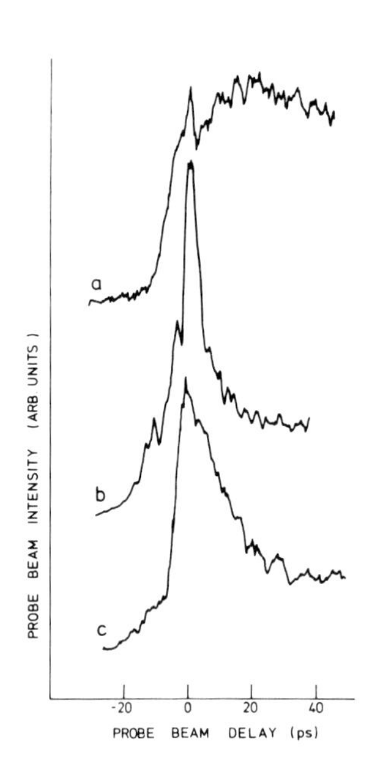

Fig. 4. Results of pump-probe experiments at 1.6 K on pentacene in naphthalene.
a. origin at 6028 Å
b. vibronic transition at 5934 Å
c. vibronic transition at 5768 Å

optical quality crystals are needed, as otherwise, beam wandering of the probe pulse obscures the measurements. (At the time of the experiments reported here, the optical quality of the crystals was not superior!) The preliminary results obtained so far are shown in Fig. 4. The upper curve (a) is due to the induced absorption change on the zero phonon line at 6028 Å. As expected, after temporal overlap of the pump and probe pulse, which leads to a coherent coupling artifact [22] the absorption stays constant at a lower level and no further change on a picosecond time scale occurs. The middle curve (b) is obtained when the pump-probe experiment is performed on the vibronic transition at 262 cm^{-1} from the origin. In this case the observed probe beam intensity may be interpreted as the sum of a very fast transient absorption and a coherent coupling artifact. This indicates that vibrational relaxation is faster than the excitation pulse width. This conclusion is in agreement with the fact that the 5934 Å vibronic transition in the optical spectrum exhibits a Lorentzian-like absorption line, with a halfwidth of 2.4 cm^{-1}. If this width would only be caused by vibrational relaxation, the corresponding vibrational relaxation time would be 2.2 ps. The lower curve (c) in Fig. 4 (obtained when the vibronic transition at 747 cm^{-1} from the origin is excited) however shows the expected tail due to transient absorption by the probe pulse. Preliminary analysis of these curve and other ones shows that the relaxation time of this vibronic level is ca. 13 ps. This is not inconsistent with the observed linewidth of 0.85 cm^{-1} of the Gaussian-like broadened vibronic transition at 5768 Å. This transition therefore seems to be a potential candidate for a photon-echo measurement.

We once again emphasize that the results obtained should be considered preliminary. Two conclusions however may be drawn.
First, pump-probe experiments on single crystals of organic material are feasible and with high optical quality crystals should give dependable

results. Second, drastic variations occur in the relaxation times of different vibrations which points at the danger of measuring relaxation rates of molecules in liquids at room temperature, whereby averages are taken over several vibrational states. It seems appropriate to note here that the difference in linewidth of the 262 cm^{-1} and 747 cm^{-1} vibrational absorption also persists when pentacene is dissolved in p-terphenyl. This indicates that the intrinsic (molecular) contribution to vibrational dephasing is quite important.

In conclusion we dare state that "high-resolution" picosecond (coherent) spectroscopy of molecular crystals, as a branch, has a bright future.

References

1 S.L.Shapiro (ed): *Ultrashort Light Pulses*, Topics in Appl.Phys., Vol.18 (Springer, Berlin, Heidelberg, New York 1977)
2 A. Laubereau, G. Wochner, W. Kaiser: Phys. Rev. A13, 2212 (1976)
3 D.M. Burland, D.E. Cooper, M.D. Fayer, C.R. Gochanour: Chem. Phys. Lett. 52, 279 (1977)
4 S.L. Robinette, G.J. Small: J. Chem. Phys. 65, 837 (1976)
5 W.H. Hesslink, D.A. Wiersma: Chem. Phys. Lett. June 1 issue (1978): Picosecond Phenomena meeting, Hilton Head, South Carolina, May 1978.
6 N.A. Kurnit, I.D. Abella, S.R. Hartmann: Phys. Rev. Lett. 13, 567 (1964), Phys. Rev. 141, 391 (1966)
7 T.J. Aartsma, D.A. Wiersma: Opt. Commun. 18, 201 (1976), Chem. Phys. Lett. 42, 520 (1976)
8 T.E. Orlowski, K.E. Jones, A.H. Zewail: Chem. Phys. Lett. 54, 197 (1978)
9 H. de Vries, D.A. Wiersma: J. Chem. Phys. to be published
10 A. Brillante, D.P. Craig: J. Chem. Soc., Faraday Trans 2, 71, 1457 (1975)
11 H.P. Weber, H.G. Danielmeyer: Phys. Rev. A2, 2074 (1970)
12 N. Takeuchi, IEEE J.QE-11 236, (1975)
13 S. Voelker, R.M. Macfarlane, J.H. v.d. Waals: Chem. Phys. Lett. 53, 8 (1978)
14 T.J. Aartsma, J.B.W. Morsink, D.A. Wiersma: Chem. Phys. Lett. 47, 425 (1977)
15 M.A. Krivoglaz: Sov. Phys. Jetp. 21, 204 (1965)
16 C.B. Harris: J. Chem. Phys. 12, 5607 (1977)
17 R. Kubo: Adv. Chem. Phys. XV, 101 (1969) and References therein
18 P. de Bree, D.A. Wiersma: to be published
19 A.G. Redfield: Adv. Magn. Res. vol I (ed. J.S. Waugh, Acad.Press, 1965)
20 D. von der Linde, A. Laubereau, W. Kaiser: Phys. Rev. Lett. 26, 954 (1971)
21 E.P. Ippen, C.V. Shank, A. Bergman: Chem. Phys. Lett 38, 611 (1976)
22 C.V. Shank, D.H. Auston: Phys. Rev. Lett 34, 479 (1975)

Nonlinear Optical Techniques for Probing Dephasing of Molecules and Excitons: Nano and Picosecond Dynamics

R. Shah, D. Millar, and A.H. Zewail

Noyes Laboratory of Chemical Physics, California Institute of Technology
Pasadena, CA 91125, USA

A brief account of some of our recent work at CalTech on the time resolved spectra of gases and solids in the nano and sub-picosecond time regimes was given. First, we discussed the concepts involved in nonlinear optical techniques and their relationships to optical T_1 and T_2 measured in large (pentacene) and small (iodine) molecules. Second, a description for nanosecond techniques was given. Third, we presented a picosecond reflection technique for probing exciton dynamics and surface states in molecular crystals. Finally, we mentioned a curious effect which is in analogy with the optical Ramsey fringes. This effect is observed in molecules excited to one-photon states in contrast to previous observations made in two-photon transitions. Since there was necessarily some overlap with the material presented at the conference on *Advances in Laser Chemistry* organized by one of the authors (A.H.Z.), the reader is referred to [1].

1. A.H. Zewail (ed): *Advances in Laser Chemistry*, Springer Series in Chemical Physics, volume 3 (Springer, Heidelberg, Berlin, New York 1978)

Optical Coherent Transients by Laser Frequency Switching: Subnanosecond Studies[1]

R.G. De Voe and R.G. Brewer

IBM Research Laboratory, San Jose, CA 95193, USA

ABSTRACT: By extending the laser frequency switching technique to a 100 psec time scale, we have observed for the Na D_1 line the first order free induction decay (FID), its inhomogeneous dephasing time T_2^*, its interference with the nonlinear FID, and a 1.8 GHz interference beat of the ground state hfs. Detailed theoretical predictions of these new coherence effects are faithfully observed.

The method of *laser frequency switching* [1], utilized recently in generating coherent optical transients, has provided new ways for examining the dynamic interactions occurring in molecules [1], solids [2], and atoms [3]. In this work, the frequency of a cw dye laser is abruptly switched by means of voltage pulses applied to an intracavity electro-optic phase modulator. A resonant sample in the path of this light exhibits coherent transients which are detected in the forward beam, allowing dephasing and population decay times to be measured thus far in the range 1 microsecond to 5 nanoseconds.

In the present study, laser frequency switching is extended to a time scale of 100 picoseconds. This fifty-fold increase in time resolution is achieved without sacrificing the previous advantages of heterodyne detection, high sensitivity, and the ability to monitor the entire class of coherent optical transients by preselecting the voltage pulse sequence. Hence, quantitative studies of coherent optical transients in this time domain are now feasible.

New coherence effects may also arise at these short times as illustrated here for the optical free induction decay [4] of an inhomogeneously broadened transition. This transient displays a polarization containing both a first order and a nonlinear laser field dependence having different decay times, heterodyne beat frequencies, and laser tuning characteristics. The first order FID, which was predicted [5], decays rapidly in the time of an inverse inhomogeneous linewidth T_2^* and is observed in the time domain for the first time by laser frequency switching. The well-known nonlinear FID [4] may be long-lived with a decay time determined by the power-broadened homogeneous

---

[1] Work supported in part by the U. S. Office of Naval Research
Published in Physical Review Letters 40, 862 (1978); permission granted for publication in this volume

linewidth. We view these two forms of FID as the transient analogs of steady-state linear and nonlinear (hole-burning) laser spectroscopy of an inhomogeneously broadened transition. Furthermore, the increased time resolution permits the first observations of very high frequency interference beats, for example due to the 1.8 GHz hfs splitting of the sodium ground state.

The above properties of FID follow from a density matrix solution [5] of the Schrödinger wave equation where we assume that a Doppler-broadened atomic transition $1 \rightarrow 2$ with center frequency ω_{21} is coherently prepared in steady-state by a cw laser beam. The system then radiates the FID transient when the laser frequency is suddenly switched from $\Omega \rightarrow \Omega'$. We make use of an earlier derivation [5] of the Doppler-averaged polarization and express the FID heterodyne beat signal E_b^2 in terms of an error function of complex argument $w(x+iy)$, [6,7]

$$E_b^2(L,t) = KE_o^2 e^{-t/T_2} e^{-(\sigma t/2)^2} e^{-i(\omega_{21}-\Omega')t}$$

$$\times \left\{ (1+\eta) w\left(\frac{i\sigma t}{2} + \frac{i\Gamma-\delta}{\sigma}\right) + (\eta-1) w\left(-\frac{i\sigma t}{2} + \frac{i\Gamma+\delta}{\sigma}\right) \right\} + \text{c.c.} \qquad (1)$$

Here, all frequencies are in angular units; the dipole dephasing time is T_2; the population decay time T_1; the Doppler linewidth is σ; we define the inhomogeneous dephasing time $T_2^* \equiv 2/\sigma$; the quantity $\delta \equiv \omega_{21}-\Omega$; the power-broadened linewidth is $\Gamma \equiv \sqrt{1/T_2{}^2+\chi^2 T_1/T_2}$; the Rabi frequency is $\chi \equiv \mu_{12}E_0/\hbar$ where E_0 is the laser field amplitude and μ_{12} the transition matrix element; and the saturation parameter $\eta \equiv 1/(T_2\Gamma)$. This expression is nearly exact within the rotating wave approximation and the slowly varying envelope approximation for an optically thin sample of length L.

In the asymptotic limit [7] as $t \rightarrow \infty$, Eq. (1) yields

$$E_b^2(L,t) = 4(\eta-1)E_o^2 K\, e^{-t/T_2} e^{-\Gamma t} e^{-\delta^2/\sigma^2} \cos(\Omega-\Omega')t, \qquad (2)$$

the anticipated nonlinear FID transient [4,5] where the emission occurs at the initial laser frequency Ω and produces with the laser beam a heterodyne beat of frequency $\Omega-\Omega'$. Note that the signal is nonlinear in the laser intensity due to the factor $E_0^2(\eta-1)$, the signal is absorptive, and the decay is an exponential with decay rate $1/T_2+\Gamma$.

In the short time limit $t<2\Gamma/\sigma^2$, we expand the error function [7]

$$w(z) = 1 + \frac{2iz}{\sqrt{\pi}} + \ldots \qquad \text{for} \quad z<1,$$

retaining only the first two terms to obtain the initial behavior of the first order FID

$$E_b^2(L,t) \cong 4E_0^2 K\, e^{-t/T_2}\, e^{-(\sigma t/2)^2}$$
$$\times \left\{ \frac{1}{T_2}\left(\frac{1}{\Gamma} - \frac{2}{\sqrt{\pi}\sigma}\right) \cos(\omega_{21}-\Omega')t - \frac{2}{\sqrt{\pi}} \frac{(\omega_{21}-\Omega)}{\sigma} \sin(\omega_{21}-\Omega')t \right\}. \quad (3)$$

This radiation peaks at the Doppler line center at frequency ω_{21} and produces with the laser beam a beat of frequency $\omega_{21}-\Omega'$. The signal amplitude is linear in the laser intensity and contains both an absorptive and a dispersive part, which depends on the initial laser frequency through the factor $(\omega_{21}-\Omega)$. Here, the decay is a Gaussian, $e^{-(\sigma t/2)^2}$. Consequently, at short times the first and higher order FID terms interfere in a variety of interesting ways, dependent on the particular conditions of light intensity and laser tuning. Numerical solutions of the general expression (1) are obtained using an error function subroutine, and the results of computer plots are given in Fig. (1) for the Na D_1 line where $(\Omega-\Omega')/2\pi=5$ GHz, $\sigma=5.8$ GHz, $\Gamma=1.6$ GHz, $\chi=2.3$ GHz, $T_2=32$ nsec and $\eta=0.019$. At lower light intensity when $\chi=0.8$ GHz, the interferences are less striking and more closely resemble the observations of Fig. 2, which we now consider.

In the present optics, light from a Coherent 599 cw dye laser is frequency switched by a travelling wave electro-optic phase modulator [8] that is external to the laser cavity. This beam excites a sample resonantly and with the forward FID emission strikes a fast GaAs photodiode. The photo-detector of 2 mil diameter has a 30 psec response time or less and is mounted directly on an S-4 Sampling Head (25 psec response) of a Tektronix 7904 oscilloscope with a 7S11 sampling unit. The time-averaged transient signals are then sampled and stored digitally using a local computer.

The optical phase modulator is a lithium tantalate crystal ($0.5\times0.65\times50$ mm^3) mounted in a stripline microwave transmission line. A square wave dc voltage pulse (Tektronix 109 pulser with a 700 Hz repetition rate) applied to this configuration propagates with the light wave down the length of the crystal in a time of 1.5 nsec, changing the optical refractive index from n_0 to n_1. Since the propagation velocities of the optical (c) and dc (v) waves differ, the light will experience a time-varying refractive index and hence a frequency shift

$$\Omega-\Omega' = \Omega \frac{n_1-n_0}{n_0} \cdot \frac{v}{c\pm v} \quad (4)$$

where the waves may travel either in the same direction (minus sign) or opposite direction (plus sign). Frequency shifts in the range 0 to 10 GHz have been achieved with a duration of 1.5 nsec, a rise time of ~100 psec, and at a rate of 17 MHz/volt.

In testing the above FID theory, we examined the behavior of the sodium D_1 line at 16,956.16 cm^{-1} using a 10 cm path length of Na vapor at $\sim 2\times10^{-6}$ Torr with 5% of the laser beam being absorbed. Figure 2 shows the observed FID heterodyne beat signals for different values of $\omega_{21}-\Omega$ where the frequency jump $(\Omega-\Omega')/2\pi=5$ GHz and the laser power density is 3 W/cm^2 $(\eta=0.055)$. Note that the large frequency switch makes it possible for the first time to switch completely outside the Doppler linewidth, which for Na is 0.77 GHz. We see that as the initial laser frequency Ω is tuned through the Doppler lineshape, the FID changes remarkably, displaying a rich assortment of new effects, in agreement with the results of Eqs. (1)-(3) and Fig. 1. Thus, in Fig. 2(a) where $(\omega_{21}-\Omega)/2\pi$= 1.5 GHz the detuning is large enough so that only the first order FID is evident and only the *off resonant velocity groups* contribute. The decay, which obviously is nonexponential, is complete in 500 psec, as expected since T_2^*=340 psec. Due to the high laser intensity, the signal is dispersive and thus reverses sign when the laser is tuned to the other side of the Doppler peak, as in Fig. 2(d) where $(\omega_{21}-\Omega)/2\pi$= -1 GHz, in agreement with (3) and Fig. 1(d). Furthermore, the first order FID beat frequency of 6.5 GHz in Fig. 2(a) is given by $\omega_{21}-\Omega'$, and is confirmed in other experiments by varying Ω'.

On the other hand, the nonlinear FID becomes more prominent when $\omega_{21}-\Omega$ is small, as it is in Figures 2(b)-(d), due to the large amplitude preparation of the *resonant velocity group*. Compare with Figs. 1(b)-(d). The decay, which persists beyond the first order FID, is exponential in agreement with Eq. (2), being given by $\sim e^{-\chi t}$ in this power-broadened regime with χ=0.8 GHz. Note that the decay is faster in Fig. 1 as χ=2.3 GHz. The observed beat frequency of 5 GHz is given by the laser frequency jump $(\Omega-\Omega')/2\pi$ and is verified in other experiments where this quantity is varied. These signals are also absorptive since their phase is invariant to laser tuning. In Fig. 2(c) where $\omega_{21}-\Omega$=0, a dramatic interference between the linear and nonlinear FID occurs near the time origin, as in Fig. 1(c). Thus, all of the predictions of Eq. (1) seem to be obeyed.

In addition, in Fig. 2(d) where $\omega_{21}-\Omega$=-1.0 GHz, the laser is tuned midway between the two hyperfine components producing a modulated pattern, not present in Fig. (1), due to a 1.8 GHz interference beat of the ground state hyperfine splitting. The origin of this beat is being studied further through its Fourier transform spectrum. Quantum beats have been observed previously in the upper state by spontaneous emission, in the lower or upper state by modulated photon echoes, and in superfluorescence but at frequencies which are at least one order of magnitude smaller [9].

In the case of extreme homogeneous broadening when $\Gamma/\sigma >> \sigma t$, Eq. (1) simplifies to give the NMR result

$$E_b^2(L,t) = KE_0^2\, e^{-t/T_2}\, e^{-i(\omega_{21}-\Omega')t}\, \frac{1/T_2-i\delta}{\Gamma^2+\delta^2} + \text{c.c.}, \tag{5}$$

and then T_2 may be obtained in the absence of the above interference phenomena.

In extending this subnanosecond technique, it is evident that other coherent transients involving pulse preparation such as adiabatic rapid passage and photon echoes can be applied in quantitative optical dephasing studies where the time scale may be reduced even further.

The technical assistance of D. E. Horne and K. L. Foster proved most valuable and is acknowledged with pleasure. Conversations with A. Z. Genack are appreciated also.

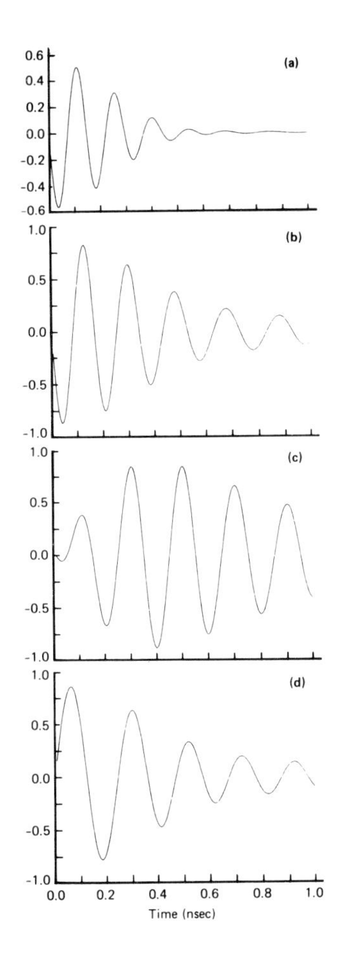

Figure 1. Numerical solutions of the FID heterodyne beat signal $E_b^2(L,t)$ of Eq. (1) where χ = 2.3 GHz, $(\Omega-\Omega')/2\pi$ = 5 GHz, and $(\omega_{21}-\Omega)/2\pi$ in GHz equals (a) 1.5, (b) 1.0, (c) 0.0, and (d) -1.0. See text for other parameters.

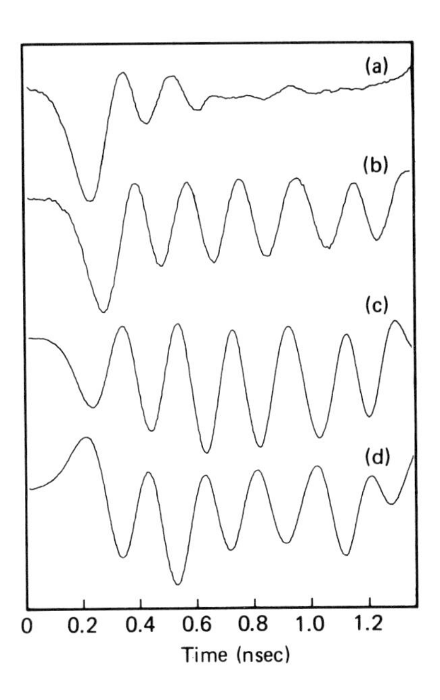

Figure 2. Experimental FID heterodyne beat signals where χ = 0.8 GHz and $(\omega_{21}-\Omega/2\pi$ in GHz equals (a) 1.5, (b) 0.7, (c) 0.0, and (d) -1.0. The laser frequency shift is $(\Omega-\Omega')/2\pi$ = 5 GHz, corresponding to a 335 volt dc square wave pulse. The gain in (a) is 2x.

REFERENCES

1. R. G. Brewer and A. Z. Genack, Phys. Rev. Lett. 36, 959 (1976); ibid, Phys. Rev. A, April 1978.

2. A. Z. Genack, R. M. Macfarlane and R. G. Brewer, Phys. Rev. Lett. 37, 1078 (1976); A. H. Zewail et al., Proc. of the Society of Photo-Optical Instrumentation Engineers 113, 42 (1977); H. de Vries, P. de Bree and D. A. Wiersma, Chem. Phys. Lett. 52, 399 (1977).

3. R. M. Macfarlane, A. Z. Genack and R. G. Brewer, Phys. Rev. B, April 1978.

4. R. G. Brewer and R. L. Shoemaker, Phys. Rev. A6, 2001 (1972).

5. K. L. Foster, S. Stenholm and R. G. Brewer, Phys. Rev. A10, 2318 (1974).

6. P. F. Liao, J. E. Bjorkholm and J. P. Gordon, Phys. Rev. Lett. 39, 15 (1977).

7. M. Abramowitz and I. A. Stegun, Handbook of Mathematical Functions (Dover, N.Y.), p. 297.

8. I. P. Kaminow, An Introduction to Electrooptic Devices, (Academic Press, N.Y., 1974), p. 213-237.

9. For a review see S. Haroche: In *High-Resolution Laser Spectroscopy*, ed. by K. Shimoda, Topics in Applied Physics, Vol. 13 (Springer, Berlin, Heidelberg, New York 1977) p. 253

Rise in Time of dc Electric-Field Induced Optical Rectification and Second Harmonic Generation in Liquids

W. Alexiewicz, J. Buchert, and S. Kielich

Nonlinear Optics Division, Institute of Physics
A. Mickiewicz University, Poznań, Poland

Picosecond pulse technique is well adapted to the study of nonlinear electro-optical effects [1,2]. When measuring the nonlinear electric polarizabilities of molecules it can occur that measurements, made at different moments of time can yield strongly differing results owing to the competitive nature of the various molecular contributions to electric third-order susceptibility.

The theory proposed by us [3] shows that the rise in time in dc electric-field induced optical rectification /DCIOR/ effect, lately studied by WARD and GUHA [4], subsequent on the application of external electric fields applied along the laboratory z-axis of coordinates, is given by the following formula:

$$\chi^{(3)}_{zzzz}(0;\omega,-\omega,0;t) \sim \frac{c}{6} + \frac{1}{5}\frac{mb}{kT}\left(1-e^{-t/\tau_1}\right) + \frac{2}{135}\frac{\gamma^2}{kT}\left(1-e^{-t/\tau_2}\right)$$

$$+\frac{2}{135}\frac{m^2\gamma}{(kT)^2}\left(1-\frac{3}{4}e^{-t/\tau_1}-\frac{1}{4}e^{-t/\tau_2}-\frac{3}{2}\frac{t}{\tau_1}e^{-t/\tau_1}\right)$$

where τ_1, τ_2 denote the relaxation times of the dipole moment and birefringence, respectively [3]. In this formula the parameters m, γ and b were, for simplicity, assumed to be independent of the electronic dispersion and absorption. The nonlinear susceptibility $\chi^{(3)}_{zzzz}(0;0,\omega,-\omega;t)$ related with DCIOR comprises contributions from:

- nonlinear mean second-order hyperpolarizability c ; in the approximation of third-order polarizability, this term does not affect the rise in time since, here, the primary role belongs to electron rather than orientational processes;
- a term, composed of two parts, and proportional to $(kT)^{-1}$, the part dependent on the product of the dipole moment m and mean first-order hyperpolarizability b of the molecule introduced a contribution, positive or negative according to the sign of b. The other part, which is positive, depends on the squared anisotropy of polarizability γ^2 ;
- a term containing the product $m^2\gamma$ with sign depending on that of γ .

Figure 1 shows the different time-evolutions of these terms vs the parameter t/τ_2.

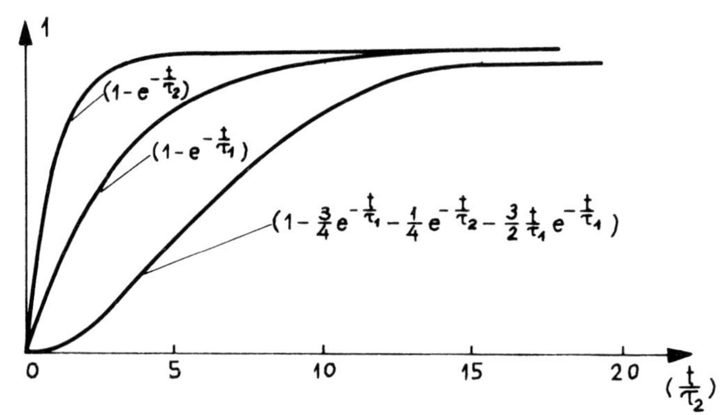

Fig.1. Normalized time-rise functions for different contributions in DCIOR effect vs t/τ_2.

In the general case of an ac field the susceptibility tensor moreover contains Debye relaxation factors which, at optical fields / difference between the various laser modes comparable with τ_1^{-1} / can modulate the contributions describing the phenomena [3].

In optical work i.e. if even one of external electric field frequencies is of the order 10^{14} - 10^{15} Hz, no term with m^4 and proportional to $(kT)^{-3}$ occurs, since the orientation of the dipole moment is unable to keep pace with the field frequency / the vanishing of this term is due to the relaxation factors [3] /. The term plays an important role in the phenomenon of electric saturation [3].

Figure 2 shows the rise in time process of the DCIOR effect in nitrobenzene. Since the values of the mean hyperpolarizability b from various authors diverge considerably, we plotted the curves for various available b-values [5].

Measurements of the time-characteristics of the DCIOR process in a time scale admitting of observation of the orientational molecular motions will permit the determination of the first-order hyperpolarizability of the molecule. Similar information can be derived from measurements of the various in electric permittivity tensor $\Delta\varepsilon_{ij}(0,0,\omega,-\omega)$ induced by intense light [3,6].

Figure 2. shows that reliable values of the nonlinear polarizability can be obtained for the stationary process i.e. for $t > 20\ \tau_2$. By performing simultaneous measurements of the time-characteristic of the second harmonic, producted in the presence of a dc field (DCSHG), we gain additional information concerning the values of b. The susceptibility related with DCSHG is given by the expression:

$$\chi^{(3)}_{zzzz}(-2\omega;0,\omega,\omega;t) \sim \frac{c}{6} + \frac{1}{10}\,\frac{m\,b^{2\omega}}{kT}\left(1-e^{-t/\tau_1}\right)$$

where the only nonzero term varying in a time scale comparable with the relaxation time τ_1 of the dipole moment is that containing the product mb.

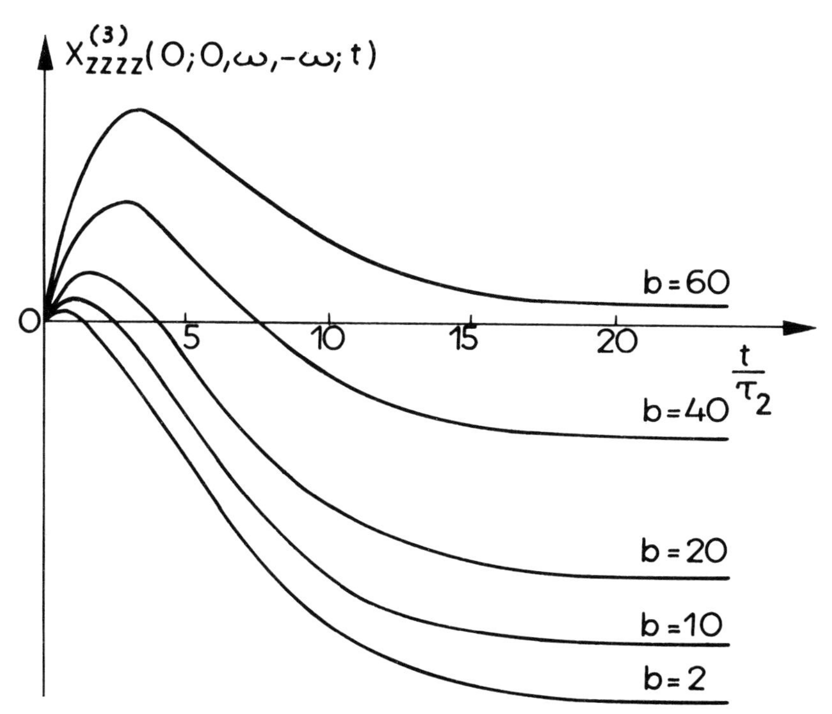

Fig.2. Rise in time process of the DCIOR effect in nitrobenzene for various values of mean hyperpolarizability b.

The preceding considerations are more of a qualitative nature inasmuch as the theory [3] is based on the rather simple equation of Debye rotarional diffusion for axially-symmetric molecules. If nonetheless enables us to gain insight into the time-evolution of important dynamical processes governing the molecular systems presently studied by the methods of laser pulse technique.

References

1. S.L.Shapiro(ed): *Ultrashort Light Pulses*, Topics in Applied Physics, Vol. 18 (Springer,Berlin,Heidelberg,New York 1977).
2. K.Sala, M.C.Richardson, Phys. Rev. A12,1036 (1975).
3. B.Kasprowicz-Kielich, S.Kielich, Adv. Mol. Relax. Processes, 7,275 (1975); W.Alexiewicz, J.Buchert, S.Kielich, Acta. Phys. Polonica A52, 445 (1977); J.Buchert, B.Kasprowicz-Kielich, Adv. Mol. Relax. and Interaction Processes, 11, 115 (1977).
4. J.F.Ward, J.K.Guha, Appl. Phys. Lett. 30, 276 (1977).
5. B.F.Levine, C.G.Bethea, J. Chem. Phys., 63, 2666 (1975); P.Bordewijk, Chem. Phys., 39, 342 (1976).
6. S.Kielich, Physica 34, 365 (1967).

Picosecond Molecular Dynamics in Liquids and Electron Localization

G.A. Kenney-Wallace[1]

Department of Chemistry, University of Toronto
Toronto, Canada M5S 1A1

The confluence of picosecond spectroscopy and molecular dynamics in liquids is a natural outcome of the fact that the duration of the probe pulse and the phenomena share the same molecular space and time-scale. A 2 psec pulse is far less than, or comparable to, molecular reorientational times, while in 10 psec most molecules in translational diffusion move but a few angstroms. At subpicosecond times the fluid becomes frozen to the observer, who only senses the vibrations of the otherwise stationary molecules. Picosecond spectroscopy (1) thus allows us a direct view of this microscopic world for the first time and we can begin to test models of dynamical behaviour in liquids with a quantitative perspective that previously had been feasible only in the gas and crystalline phases. And yet, reducing the myriad forms of picosecond data into the correct molecular correlation functions, which describe the equilibrium and dynamical interactions of molecules in dense fluids, is not simple and offers a major challenge to contemporary theory of molecular motions in liquids (2,3).

In this paper we describe some aspects of molecular dynamics of liquids as revealed through our picosecond spectroscopy studies of electron localization in organic liquids, the alcohols, and electronic optical Kerr effects, in which subpicosecond optical gating in liquid CS_2 is reported.

1. Molecular Motion in Liquids: Theory and Experiment

Exactly how does a molecule diffuse, reorient, collide, exchange energy and perhaps undergo chemical reaction, in a molecular milieu which displays continual energy, density and configurational fluctuations on a time scale comparable to nuclear motion? These are fundamental questions. Nor are the answers restricted to understanding molecular motions in liquids; they are also of pivotal interest to theories of energy transfer and relaxation, and to the development of microscopic kinetic theories of chemical reactions. For example, cross-sections for reactive scattering of two isolated molecules in the gas phase are strongly influenced by the internal energy of the two reacting partners, and thus the internal and recoil energy and angular momentum of the newly formed molecules carry the history of the collisional encounter.

---

[1]Alfred P. Sloan Research Fellow 1977-79.

However, in the liquid phase, the fastest bimolecular reactions that occur (with the obvious exception of tunnelling) involve diffusion-controlled processes, and the molecular dynamics of both the reacting particles and the liquid in which they diffuse dominate events. The omnipresent intermolecular forces give a dynamical molecular structure to the liquid and introduce collective or many-particle effects to the simpler notion of binary collisions. Thus molecular dynamic theory can yield information on the collisional or microscopic origin of the motion, while hydrodynamics focuses on the macroscopic or collective features.

Current theory (4) seeks to evaluate these individual and collective effects to the overall molecular motion. Likewise different experimental techniques can probe the single-particle (property P_j) or many-particle (ΣP_j) interactions. In many instances, it is not clear that (for example) vibrational and rotational processes can be separated, nor molecular reorientation from translational motion. Thus the molecular dynamics of liquids must be examined experimentally from several different techniques in order to establish which time correlation functions may be retrieved, and whether they are auto or cross-correlation functions (between different variables).

A remarkable observation in recent years has been the fact that hydrodynamic behaviour is adequately predictive at the molecular level, in many cases (2). A natural corollary would be that collective effects are dominant in these liquids. Recent molecular dynamics theory (4) has shown that, in motion of small spherical top molecules, collective effects contribute about 25% to the diffusion constant, D, which is written as the sum of a collisional (Enskog) constant D_e and the hydrodynamic (Debye-Stokes-Einstein) value D_s, for a liquid of viscosity η:

$$D = D_s + D_e = D_s(1+3\eta/\beta) \qquad (1)$$

$$D_s = kT/8\pi\eta R^3 \qquad (2)$$

It is β, a parameter describing the boundary condition, which takes us from a molecular dynamics description of a molecule rotating in a bath to a hydrodynamic and thus collective description of the total system. The details of the collisional angular momentum transfer in the liquid, and the coupling of rotational to translational motion, determine whether or not the stick (hydrodynamic, $\beta=\infty$) or slip (molecular dynamic, $\beta\sim$ small) boundary limits apply. Another parameter has been introduced (2) to describe the coupling of the different motion and non-diffusive contributions to D:

$$D = D_s\kappa^{-1} \qquad 0\leq\kappa\leq 1 \qquad (3)$$

when $\beta=\infty$, $\kappa=1$ and small β values are thus consistent with small κ. The detail and application of these and other theories are currently the focus of much discussion, which we cannot review here. However, picosecond spectroscopy can obtain data relevant to β and κ (although at the present time the theory has not been extended to include other than quasispherical molecules). If $\beta=\infty$ (stick) then the angular velocity of the rotating molecule matches the collective tangential velocity of its neighbours, and D will be much smaller than under slip conditions, where β is small compared to η. By observing molecular reorientation as a function of the size and shape of the molecule in comparison to the solvent molecule, we gain information on the boundary conditions which can bridge the two major theoretical approaches to molecular motion.

As we stated earlier, dynamical information must be gleaned from several different techniques. Picosecond experiments involving induced-dichroism (5) and fluorescence depolarization (6) have explored the applicability of the hydrodynamic theory to the rotation of large molecules in hydrogen-bonded solvents. Coherent raman scattering (7) from small molecules yields T_2 and T_2^* vibrational dephasing times, which are a measure of the intermolecular potentials and the torques experienced by the molecules during collisions in the liquid. Optical Kerr (8) studies have examined the rotational relaxation behavior of molecules by observing the decay of the induced birefringence. Finally collisional and resonance energy transfer are both influenced by the dynamics of the molecular environment hosting the participating molecules (9). From all of these studies will emerge dynamical details that can be integrated with the growing body of data from frequency-domain measurements, such as dielectric relaxation, four-wave mixing and transient grating techniques, and laser light scattering.

We now turn to quite a different system, one involving the dynamics of electron solvation and the orientational relaxation of a liquid as described by dielectric and Kerr-relaxation effects.

2. Electrons in Fluids

Following photoionization or ionization of a molecule, or photodetachment of an ion in a liquid, the ejected electron scatters as e^-_{qf} through the conduction band of the fluid until strong local interactions block the quantum transport and induce localization to an electronically bound state of the system (10). As e^-_s, it has a kinetic lifetime $\lesssim 10^{-6}$s. Figure 1 illustrates the energy level scheme for a disordered system including the quasilocalized states, (located below the mobility edge V_o separating the extended from localized states) from which thermal fluctuations can promote the excess electron back into the extended state. In the ground electronic state, e^-_s

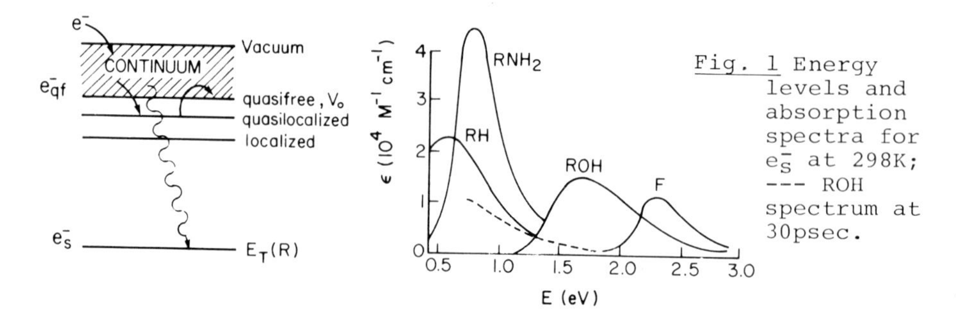

Fig. 1 Energy levels and absorption spectra for e^-_s at 298K; --- ROH spectrum at 30psec.

exhibits an intense and broad optical absorption spectrum, whose features lead us to deduce that the binding and transition energies are dominated by the short-range local structure of the molecular trapping site. For example, alkane liquids (RH) support localized electron states with infrared absorptions, ethers and amines (RNH_2) appear at slightly higher energies while electrons in n-alcohols (ROH) display strong visible bands. F-centres (F) such as in KBr exhibit as intense absorptions but the spectral linewidths are typically 0.4eV (fwhm) in comparison to ∿1eV for e_s^-. The sequence of alcohols from methanol to undecanol show absorption maxima which are remarkably similar, $E_{max} = 1.82 \pm 0.06$eV. Perturbation of the local liquid -OH structure as in the case of branched alcohols, leads to a more weakly bound electron, and the red-shift in E_{max}. High dipole-density fluids, such as ethylene glycol $(CH_2OH)_2$, form efficient chelating structures about $\bar{e}$ and provide a deep trapping potential.

The e_s^- absorption is homogeneously broadened (11,12), and following optical excitation rapid nonradiative relaxation occurs to the ground state (11), reminiscent of the ultrafast relaxation in several dye molecules (13). In competition with this relaxation is a very efficient chemical channel for e_s^- in ROH originating in the excited state e_s^{-*}. This reaction involves the motion of a OH bond of one of the ROH molecules comprising the molecular trap ultimately to give RO^-. The branching ratio between e_s^{-*} photochemistry and vibronic relaxation in CH_3CH_2OH is 0.65 but in CH_3CH_2OD it is ∿0.2. This fact, coupled with our other evidence that the loss of e_s^{-*} occurs within vibrational timescales, points unequivocally towards the -OH mechanism, and we are presently working out further details. The dynamics, however, are subpicosecond and as yet have not been resolved.

It was the purpose of the following experiments to establish the mechanism for the quasifree-localized transition of electrons in fluids, by observing the appearance of population in the localized state e_s^- via picosecond absorption spectroscopy. Polar liquids were chosen in which the probability of field-induced molecular reorientation to form the most stable and configurationally relaxed trapping site would offer potential correlations with the dynamics of the liquid itself. To this end we have also carried out independent studies on molecular motions in the liquids under study, via ^{13}C nmr (14) and fluorescence depolarization techniques (9). Finally, because strong local field effects must play a role in any mechanism trapping a free charged particle, we are undertaking experiments to look at the rotational motion of molecules in strong (optical) fields via the optical Kerr effect, with a view to observing any difference between the rise and decay of the transient birefringence as a function of field and molecular shape.

3. Picosecond Spectroscopy of Electron Solvation

Two pump-probe techniques (12,15) have been used to study electron localization in liquids at 300K; both employ picosecond absorption spectroscopy with a picosecond continuum but differ in the generation of electrons. Repetitively pulsed electron beams facilitate the study of very short-lived transients in low dipole-density or dilute fluids and gases, since the LINAC beam can be operated $\leq 60Hz$. This permits us to record very small absorption signals with signal averaging and sampling techniques. The target molecule is directly ionized by a single, picosecond, electron pulse and the kinetics and spectra of the transients are interrogated with synchronously pulsed lasers or a picosecond Cerenkov continuum (15a). Electrons may also be readily generated at observable levels via the laser photoionization of impurity molecules in the pure liquid under study. Mode-locked Nd or ruby lasers are used to generate UV radiation via harmonic generation, which in turn pumps the molecule into the continuum via a two or three photon transition. The actual mechanism and threshold of photoionization is now amenable to study too (16), as is the geminate ion-recombination, by varying the incident photon energy. The sequence of non-linear processes involved in generating e_s^- require the high peak-powers of these single-shot lasers. Thus, picosecond laser spectroscopy and e^--beam techniques are complementary in the picosecond time domain for studies of electrons in fluids. For the future, the latest developments in the subpicosecond dye laser (17) promise high laser peak-power, improved time resolution and a 10Hz repetition rate, a combination that will be the most versatile for spectroscopic studies of many picosecond phenomena.

Figure 2 illustrates our single pulse, neodynium phosphate system for picosecond absorption spectroscopy (12) and nonlinear optical effects in liquids. Recent developments in the phosphate glass promise future operation at 10Hz. A single pulse, from 2-6psec duration, is extracted from the mode-locked train via a polarization gate; full details of this system will appear elsewhere (18).

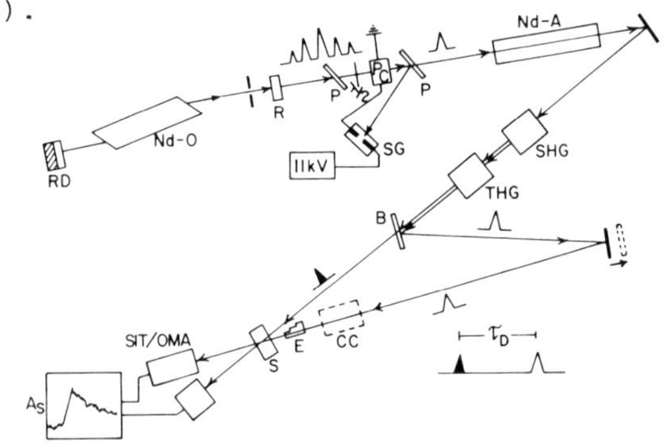

Fig. 2 Picosecond laser apparatus: oscillator amplifier (Nd-O/A); pulse extractor, P,P; harmonic generation (SHG,THG); echelon (E); continuum (CC); vidicon/OMA II detection. Ionizing ▲, probe ⋀ pulses interact with sample S at delay τ_D.

4. Dynamics of Electron Solvation

Since e_s^- in linear alcohols (ROH) show similar optical spectra we presume the electrons are located in similar trapping sites and that by following the dynamics of the quasifree-bound transition any systematic variation in the appearance of a stable e_s^- population can be correlated to any molecular rearrangement necessary to provide this site. Furthermore, by observing the same transition in inert fluids supporting a very low density of dipolar molecules it may be possible to observe the synthesis of the molecular trap, not just local rearrangements, and ultimately to distinguish collective from single particlar motions.

Pure Alcohols. Immediately (10ps) after electron injection into pure alcohols at 298K a featureless, infrared spectrum is present which over 10^{-11} to 10^{-9} evolves into the characteristic visible e_s^- spectrum, peaking at about 625nm in the case of decanol (C_{10}), illustrated in Figure 3(a). The time-dependent spectral changes are rapid in the case of methanol and become systematically slower as the chain length (n) of the alcohol increases. Analyses of the kinetics at several wavelengths on the high energy side of E_{max} yield exponential plots from which rate constants k_s for the red to blue spectral shift can be extracted (15). Table 1 lists k_s^{-1} or τ_s for several alcohols (±10%), and compares the values to τ_{rot}, data extracted from dielectric dispersion measurements, and assigned (19) to the rotation of single alcohol molecules.

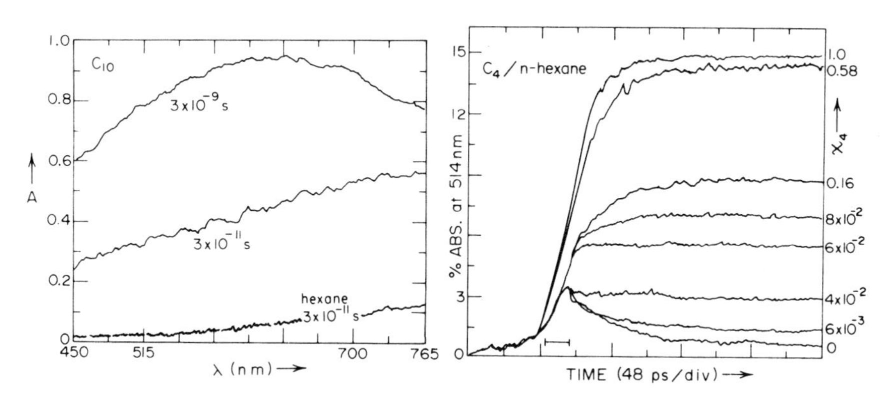

Fig. 3(a) and (b)

Table 1. Electron solvation times in $CH_3(CH_2)_nOH$ at 298K

n	0	1	2	3	4	7	9
τ_s(ps)	10	18	25	30	34	45	50
τ_{rot}(ps)	12	20	22	27	28	39	48

The agreement is quite persuasive, and, given the correct assignment of τ_{rot}, suggests that we are observing the rotational diffusion of alcohol molecules about the electron as it acquires a configurationally relaxed, ground electronic state within its molecular trap. However, since the electron presents a strong local field, we are surprised at such agreement, which implies that the polar molecules cannot respond any faster to this situation than to the presence of a weak, external field. But what if the dipoles were screened from the local charge by an intervening layer of molecules? We will pursue this point and return to possible non-linear field effects later.

The alcohols are, to various degrees, associated fluids, but if we temporarily neglect that fact and calculate the energy of the charge-dipole interaction when screened by media of different static dielectric constant, ε, it appears that the interaction energy reaches thermal energies at $5A<r<20A$ over the range of possible ε_{ROH}. Clearly these are crude limits, but the statistical structure of the fluid and the notion of screening play seminal roles in the cluster model of electron solvation to be outlined later.

Dilute alcohols. If we next examine the effect of reducing the density of ROH molecules on the dynamics of electron trapping, then three distinct regions of behaviour emerge as a function of χ_4, the mole fraction of alcohol in alkane. The kinetic profiles are shown in Fig. 3(b) for the appearance of $e_{\bar{s}}$ in the butanol (C_4)/n-hexane system. In neat n-hexane (χ_4=0) the initial IR spectrum is present, but at a considerably lower intensity (the signal in Fig. 3(a) for n-hexane at 30 psec has been multiplied by 10 for clarification). The kinetics at 514 nm show a rapid absorption (within the pulse) and decay. Not until $\chi_4=6\times10^{-3}$ is there any change, when suddenly the decay appears to level off and by $\chi_4=4\times10^{-2}$ the absorption signal rises to a plateau. No spectral changes are discernible over this range. By $\chi_4=8\times10^{-2}$ we observe amplitude changes in the IR spectrum and begin to see signs of the evolution of a visible $e_{\bar{s}}$ spectrum. By χ_4=0.16 the system appears to behave almost like a pure fluid and the kinetic profile, when normalized to the χ_4=0.58 data, is superimposable with that curve. The solvation time is about 51 psec in comparison to 30 in the pure fluid (20).

5. Cluster Model of Electron Solvation

We will now attempt to integrate these observations into a general mechanism for electron trapping and solvation in hydrogen-bonded fluids.

The cluster model envisages the quasifree electron, scattering through the fluid at subpicosecond times, inducing only electronic polarization in its wake - until it encounters a small "frozen" density or configurational fluctuation, which promotes localization to a quasilocalized or weakly bound state in the fluid. In alcohols, these initial trapping sites are the molecular clusters, namely dimers or trimers that comprise the statistical geometry of the fluid prior to electron injection. Localization begins in the potential well formed by the OH dipoles. It is this localized, weakly bound state that we assign to the infra-red spectrum, which from recent laser data has still been seen at <2 psec. In pure n-hexane, the much weaker IR spectrum is evidence of a quasilocalized state, anticipated for an electron of mobility $0.09cm^2V^{-1}s^{-1}$ (21). In a polar fluid, neighbouring molecules now begin to realign in the screened field about the newly formed cluster ion, moving against the intermolecular forces and building up the size of the cluster, whose internal structure relaxes to minimize repulsive hydrogen atom and dipole-dipole interactions. The dynamics of solvation in these polar liquids are thus a balance of time-dependent forces and torques. If we examine the functional dependence of k on such molecular properties as the orientational polarizability (α_O) and viscosity (η) of the fluid, it appears that the clustering step can be modelled in the Debye-Stokes-Einstein framework for D_S, as Figure 4 shows (20b).

If we presume that the linearity of the plot is sufficient to say that we are seeing a hydrodynamic-like response of the fluid to the perturbation presented by the electron, then the slope of the plot yields a hydrodynamic volume, from which an effective radius R can be calculated. However, the data give R=1.1A, clearly far too small but consistent with the much underestimated values for R usually found when $\beta=\infty$, or stick boundary conditions are assumed (4-6). Can we evaluate β? And exactly how slippery are the ROH molecules? Although the ROH are not spherical molecules, we can estimate the rotational-translational coupling and degree of slip. Taking 5<R<10A for the clustering molecules, then $0.01<\kappa<0.1$. This implies extensive slip. Furthermore, the intercept of τ_S v. η is non-zero, as $\eta\to0$, but gives a τ_S of 6±1 psec, which implies that there is some non-diffusive motion involved. But further interpretation, already complicated by the associated nature of the liquids, awaits more data.

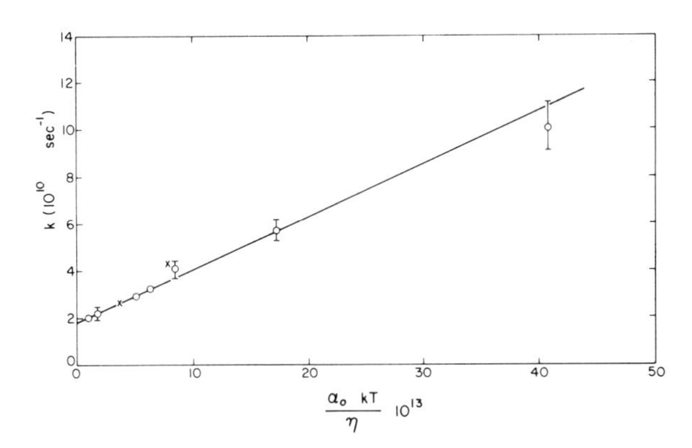

Fig. 4 Electron solvation dynamics for linear O, and branched X, ROH in terms of matrix properties. See text for details.

What mechanisms dominate the low dipole-density fluids? Figure 3(b) presents a profile of the solvation sequence at 514 nm. Here picosecond spectroscopy directly reveals the clustering phenomena for the first time. We can turn solvation on and off! The rapid decay of e^-_{ql} in pure n-hexane is consistent with ion recombination and other fast chemical reactions of the quasilocalized electron (21). Not until a mole fraction of butanol $\chi_4=4\times10^{-3}$ can we observe any difference in the kinetics, and the significance of this threshold in χ is that it coincides with the onset of clustering observed in C^{13} nmr of the molecular dynamics of the same systems (14). Small clusters are indeed a prerequisite for trapping electrons in an alcohol environment, and their appearance heralds a competition with the positive ions for e^-_{ql}. Translational diffusion of distant ROH is too slow to be effective at times $<10^{-10}$sec. As χ_4 increases, so do the size and distribution of clusters until by $\chi_4=4\times10^{-2}$ the fast chemical channel appears to be eliminated.

By $\chi_4=8\times10^{-2}$ we observe the small "cluster-ions" begin to build up in size due to rotational diffusion of neighbouring molecules, which co-exist in the pools of alcohol molecules that initially trapped the electron. Clearly, at low χ_4 k no longer bears the same proportionality to $\alpha_0\eta^{-1}$, and $\beta<<\infty$; but we are approaching experimental conditions that are much more amenable to theoretical treatment.

Future studies will continue to explore the dynamics of the electron clustering phenomenon in non-hydrogen bonded fluids and in previously characterized clusters on the surface of low temperature matrices, through a range of picosecond spectroscopy techniques.

6. Non-linear Optical Phenomena

We posed earlier the problem of how to describe the dynamical response of a polar molecule to a strong local field. An analogous question would be whether or not the rise and decay transients of the optical Kerr effect are symmetrical in time. This will depend on the reorientational motion of the dipole moment vector and induced moments in the molecule. In the associated alcohols, the cross-correlation terms will be particularly important. We are therefore investigating the dynamics of the optical Kerr effect, where the time-dependence of the nonlinear index of refraction $\delta n(t)$ is given by (8):

$$\delta n(t) = \frac{\eta_2}{\tau} \int_{-\infty}^{t} \bar{E}^2 (t')[\exp-(t-t')/\tau]dt' \qquad (4)$$

The nonlinear index n_2 is the sum of an electronic n_2^e and orientational polarizability n_2^o. By considering the circularly polarized analogue of the optical Kerr effect, it can be shown (18,22) that the orientation response terms cancel, and:

$$\delta_{n_+}-\delta_{n_-} = \frac{\pi}{2n}[\sigma_{1122}-\sigma_{1212})+(B_W-2A_W)]E^2 \qquad (5)$$

where σ are the electronic, and A_W, B_W the nuclear, response functions associated with a Raman resonance at the difference frequency W. Near an electronic resonance, this induces a polarization rotation $\Delta\Theta$ due to the applied field w_2, of w_1

$$\Delta\Theta = \frac{\pi w_1}{4nc}[\sigma_{1122}-\sigma_{1212}]|E(w_2,r)|^2 \qquad (6)$$

while in the vicinity of a vibrational Raman resonance,

$$\Delta\Theta = \frac{\pi w_1}{4nc}[B_W-2A_W]|E(w_2,r)|^2 \qquad (7)$$

We are presently investigating the former effect in CS_2, where our initial results indicate that the electronic response is about 20% of the normal Kerr effect, in agreement with known η_2 values. The second effect has been observed in C_6H_{12} and C_6H_6 in the context of the Raman-induced Kerr effect (23). It is pertinent to note that both of these effects, in conjunction with crossed polarizers, can be used as subpicosecond optical gates for subpicosecond laser pulses, since the σ response is of the order of electronic transitions or $\sim10^{-15}$sec and A_W, B_W respond approximately within a vibrational period, namely 10^{-13} to 10^{-14}sec. Thus unlike the Kerr cell, this effect is not Debye-limited in response or decay times. And so it is, a study of possible nonlinear dynamical responses of molecules to applied local fields has both fundamental significance to motion in liquids and a direct application to the development of subpicosecond optical gating techniques - a confluence of ideas that is reminescent of our opening remarks.

Acknowledgements are due to Dr. B.A. Garetz and to Dr. C. D. Jonah for their contributions to these studies.

1. S.L.Shapiro(ed): *Ultrashort Light Pulses*, Topics in Applied Physics, Vol.18 (Springer,Berlin,Heidelberg,New York 1977)
2. D. Kivelson in New Aspects of Molecular Relaxation Processes, Faraday Symposium 11 (Chemical Society, 1977).
3. B.J. Berne and G.D. Harp, Adv. Chem. Phys. 17, 63 (1970); G. Williams in Chem. Reviews, 89 (1978).
4. J. Hynes, R. Kapral, M. Weinberg, J. Chem. Phys. in press (1978).
5. T. Chuang and K. Eisenthal, Chem. Phys. Lett. 11, 368 (1971).
6. K. Spears, L. Cramer, Chemical Physics, 30, 1 (1978); G. Fleming, J. Morris, G.W. Robinson, Chem. Phys. 17, 91 (1976).
7. A. Laubereau and W. Kaiser, Ann. Rev. Phys. Chem. 26, 83 (1975).
8. K. Sala and M. Richardson, Phys. Rev. A12, 1036 (1975); P. P. Ho and R. Alfano, J. Chem. Phys. 68, 4551 (1978).
9. This laboratory, unpublished data.
10. G.A. Kenney-Wallace, Acct. of Chem. Res., in press (1978).
11. G.A. Kenney-Wallace and K. Sarantidis, Chem. Phys. Lett. 53, 495 (1978), and published data.
12. D. Huppert and P.M. Rentzepis, J. Chem. Phys. 64, 181 (1976).
13. E. Ippen, C. Shank, O. Teschke, Chem. Phys. Lett. 45, 291 (1977).
14. R. Ling, G.A. Kenney-Wallace and W.F. Reynolds, Chem. Phys. Lett. 54, 81 (1978).
15. a G.A. Kenney-Wallace, C.D. Jonah, Chem. Phys. Lett. 39, 596 (1976); b W. Chase, J.W. Hunt, J. Phys. Chem. 79, 2835 (1975).
16. G.E. Hall and G.A. Kenney-Wallace, Chem. Phys. 28, 205 (1978).
17. E.P. Ippen and C.V. Shank in ref. (1), and this Volume.
18. B.A. Garetz and G.A. Kenney-Wallace, submitted for publication.
19. S.K. Garg and C.P. Smyth, J. Phys. Chem. 69, 1294 (1965).
20. a G.A. Kenney-Wallace, C.D. Jonah, Chem. Phys. Lett. 47, 362 (1977) and to be published. b G.A. Kenney-Wallace, Can. J. Chem., 55, 2009 (1977).
21. A.O. Allen and R.A. Holroyd, J. Phys. Chem. 78, 796 (1974).
22. R.W. Hellwarth, Prog. Quant. Elec. 5, 1 (1977).
23. D. Heiman, R.W. Hellwarth, M.D. Levenson and G. Martin, Phys. Rev. Lett. 36, 189 (1976).

Collisionless Intramolecular Vibrational Relaxation[1] in SF_6

H.S. Kwok and E. Yablonovitch

Gordon McKay Laboratory, Harvard University
Cambridge, MA 02138, USA

1. Introduction

There has always been great interest in the limits of applicability of statistical mechanics. For example the point is often made that statistical concepts describe macroscopic systems very well. On the other hand, a tiny dynamical system such as a single isolated molecule may behave in a non-ergodic manner.

While the question of ergodicity may seem highly theoretical, it has rather direct experimental consequences. Consider, for concreteness, the resonant excitation [1] of the ν_3 vibrational mode in SF_6 by an ultra-short infrared laser pulse. As a result of the energy deposition, the fundamental absorption spectrum will experience an anharmonic shift. There are two possibilities for the spectral shift:

(1) For energy localized within the ν_3 mode the anharmonic shift will be $\sim$5 cm^{-1} per 1000 cm^{-1}. This *self*-anharmonicity is known from the overtone spectrum [2] of the ν_3 mode.

(2) For energy randomly distributed in all the modes the anharmonic shift will be 2.6 cm^{-1} per 1000 cm^{-1}. This *cross*-anharmonicity is determined from the temperature shift of the fundamental ν_3 spectrum.

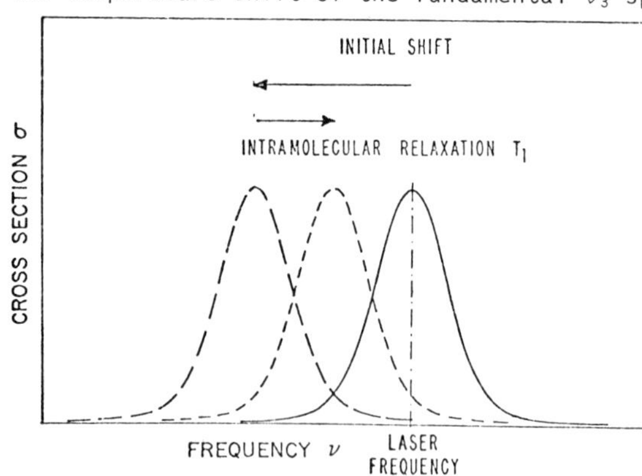

Fig.1 Behavior of the ν_3 spectrum under infrared pumping by a powerful ultrashort CO_2 pulse

[1] Research supported by Joint Services Electronics Program and Office of Naval Research.

Therefore, we may anticipate the behavior shown in Fig.1. An ultrashort infrared pulse will initially deposit energy in the ν_3 mode, causing a large anharmonic shift. Subsequently, the energy may randomize among all the vibrational modes, causing the ir spectrum to shift part way back to the original position. Time resolved vibrational spectroscopy can monitor the amount of energy deposited, its distribution among the modes and can provide an operational monitor for ergodicity. The anharmonic shifts in effect measure the probability distribution of the molecule in its phase space, thereby testing for ergodicity and the rate at which it sets in.

The redistribution of energy from a driven mode should be regarded as a T_1-type relaxation process [4], where the rest of the molecule acts as a heat bath.

In the experiments described below, SF_6 gas is pumped by a powerful picosecond CO_2 laser pulse. The drop in absorption cross-section which results should not be regarded as a true 2-level saturation effect. Instead the absorption saturation is best interpreted as due to a shift off resonance of the absorption spectrum as indicated in Fig.1. We may hope to see a strong absorption saturation effect when the ν_3 mode is initially excited, followed by a partial recovery of absorption as the vibrational energy is redistributed among the modes.

2. Experiment

In this series of experiments the time resolved saturation and partial recovery of absorption is probed at the same wavelength as the pump beam. The full saturation spectrum may be obtained by a simple extension of the techniques employed here.

The experimental apparatus is illustrated in Fig.2. A single mode Tachisto TEA CO_2 laser is followed by a plasma shutter [5] which truncates the laser pulse. The beam then passes through an optical free induction decay [6] cell which generates the 30 psec pulses. This light source is employed in a standard [7] picosecond pump and probe arrangement, the time delay being supplied by a translation stage. The probe pulse were observed with a Ge:Au detector and averaged with a boxcar integrator.

In one of the experiments a comparison was made between the behavior of heated and unheated SF_6. Rather than using an oven, the pre-heating was accomplished by means of the truncated optical pulse from the plasma shutter. An optical delay of 32 nsec after the heating pulse allowed collisional equilibration before the picosecond pump and probe pulses arrived. The irradiation sequence experienced by the molecules is shown in the insert of Fig.2. However, most of the experiments were performed at room temperature with the pre-heating beam not in use.

3. Collisional Effects

In this type of experiment it is important to separate out collisional effects and to operate in a pressure regime where they are negligible. Figure 3 shows a series of pump probe saturation scans which were performed at three different SF_6 gas pressures. The graphs show a strong absorption saturation followed, as anticipated, by a partial recovery on a rapid collisionless time scale and then a slow pressure dependent recovery. The pressure dependent recovery time constant is 13.5 ± 3 nsec-torr. This

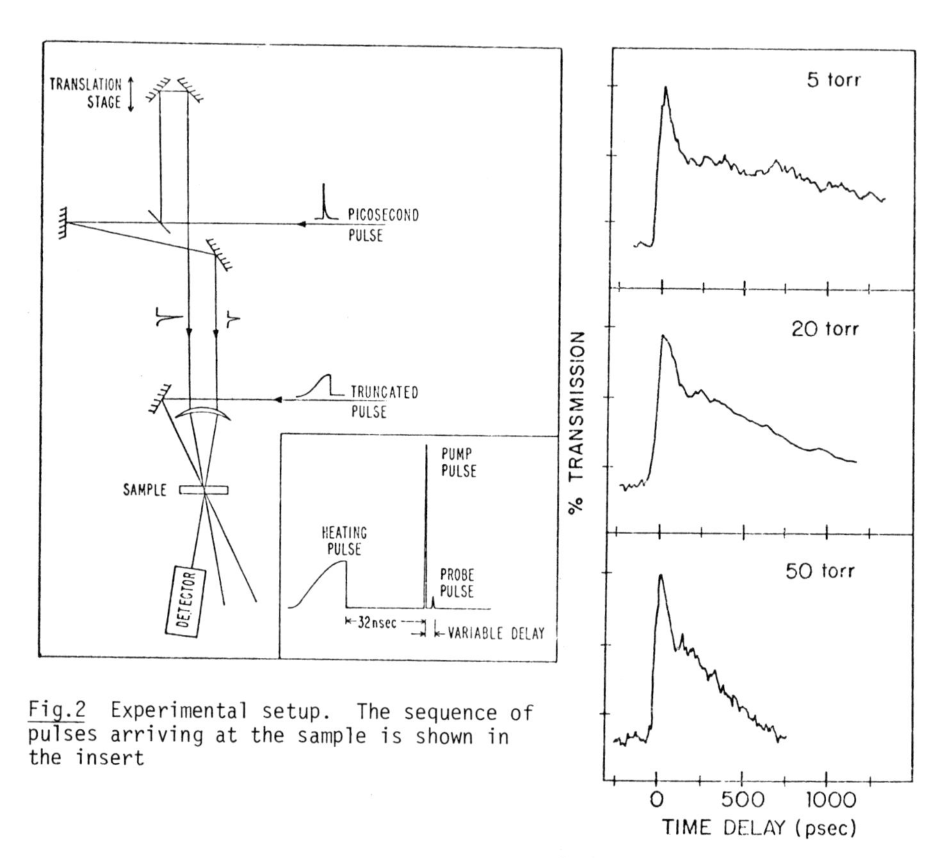

Fig.2 Experimental setup. The sequence of pulses arriving at the sample is shown in the insert

Fig.3 Recovery of saturation is shown for three gas cell pressures. 50 psec pulses at 66 MW/cm^2 were used

corresponds to a very large gas kinetic cross-section and sets the limit for true collisionless operation of our experiments.

Previous workers [8,9], who observed a saturation recovery in the nanosecond regime, speculated that they might be seeing collisionless effects. The improved time resolution in our experiments shows that those effects were most likely collisional in nature. Deutsch and Brueck [10] have seen some interesting effects even on a microsecond time scale. It is clear from our measured pressure-time constant that their observations were firmly in the collision dominated time scale.

4. Results and Discussion

Our main results are shown in Figs.4 and 5. They were taken at a sufficiently low pressure so that collisional effects can be neglected. Figure 4 compares the saturation recovery in SF_6 with that observed in the saturable

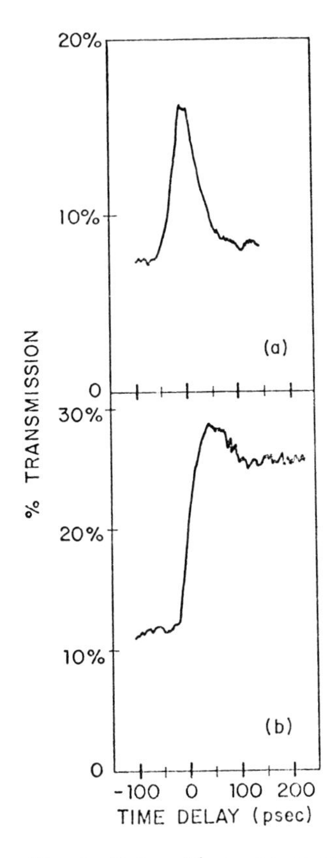

Fig.4 Saturation recovery scans for (a) p-Ge at an intensity of 150 MW/cm^2, (b) 20 torr of SF_6 at 0.3 GW/cm^2. 30 psec pulses were employed in these scans

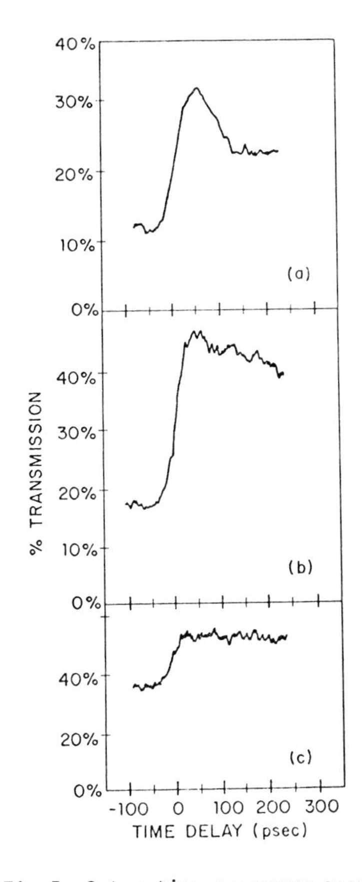

Fig.5 Saturation recovery scans to show the disappearance of the initial fast component as the internal vibrational energy of the SF_6 molecules is increased. See text for explanations of each curve

absorber p-Ge [11]. The saturation recovery time in p-Ge had been shown to be $\lesssim$1 psec which is essentially instantaneous on the time scale of our experiments. Therefore, Fig.4(a) is a type of pulse autocorrelation scan and it enables us to fix an accurate zero of time delay.

The saturation recovery of SF_6 under three different conditions of excitation is shown in Fig.5. The scan in Fig.5(a), which resembles those of Figs.3 and 4(b), was taken under conditions where on the average, $\lesssim$1000 cm^{-1} of energy was deposited per molecule by the pump pulse [12]. In Fig.5(b), the pump power was increased so that the average energy deposited per molecule was $\sim$3000 cm^{-1}. It can be seen that a steady state was reached within the 50 psec duration of the laser pulse, and the fast recovery component is absent. In Fig.5(c) the experiment was repeated [13] in SF_6 preheated to 650°K, a temperature at which the average internal vibrational energy is equivalent to $\sim$3500 cm^{-1}. Once more, a steady state is reached within 50 psec and the fast recovery component is absent.

The spectral shifts depicted in Fig.1 may result not only from the internal redistribution of energy, but also from the loss of molecular energy due to adiabatic decay of level populations which follow the decaying laser pulse. This loss of internal energy simply reduces the anharmonic shift. We interpret the fast recovery component which is seen in Figs.3,4(b) and 5(a) as due to the adiabatic decay of level populations in an energy region where the vibrational level density is relatively sparce [14]. It is readily seen from Fig.4 that the fast recovery component roughly follows the autocorrelation function.

On the other hand, Figs.5(b) and (c) are typical of experiments on molecules with >3000 cm^{-1} excitation. This is the so-called quasi-continuum regime [15] where the vibrational level density is very high. In these cases, there is no fast recovery component and the saturation approaches steady state during the laser pulse duration.

In either of the above cases, the saturation level achieves a final steady state during or immediately following the laser pulse. The conclusion seems inescapable therefore that the intramolecular damping occurs within 30 psec. The proper interpretation of this point requires a clear conceptual picture of intramolecular damping.

Ignoring spontaneous emission, dissociation, and collisional effects, the energy levels of the exact vibrational Hamiltonian, though they may be very dense, are rigorously discrete. A molecule with energy localized in a given mode is inherently nonstationary due to anharmonicity and must be described as a coherent superposition of discrete stationary states of differing energies. Therefore the lineshape of intramolecular damping is best regarded as the *envelope* function of those discrete levels which contribute to the nonstationary superposition. The reciprocal of the width of that envelope function is the intramolecular damping time.

5. Conclusion

The experimental conclusion is that the linewidth of the envelope function of such a nonstationary state must be greater than the reciprocal of 30 psec. If the envelope function were any narrower, then the decay of the coherent superposition would produce temporal structure on a time scale longer than the pulse duration. Thus the redistribution of energy from the driven normal mode must occur within an upper limit of 30 psec.

A lower limit of 1 psec may be determined from the absorption linewidth [3] of heated SF_6 which is never much greater than $\sim$20 cm^{-1}. Therefore, the intramolecular relaxation time must fall within the limits 1 psec $< T_1 <$ 30 psec.

Although the final steady state is achieved very rapidly, this experiment does not establish that the final state is necessarily exactly ergodic. Indeed, no one individual measurement [16] is sufficient for this purpose. The accessible vibrational phase space contains thousands of cells and therefore thousands of different experiments would be required to show that each cell is occupied with equal probability.

References

1. For a review, see N. Bloembergen and E. Yablonovitch, Phys. Today 31, No. 5, 23 (May 1978).
2. R.S. McDowell, J.P. Aldridge and R.F. Holland, J. Phys. Chem. 80, 1203 (1976); H. Kildal, J. Chem. Phys. 67, 1287 (1977).
3. A.V. Nowak and J.L. Lyman, J. Quant. Spectr. Rad. Transf., 15, 1945 (1975); J.F. Bott, Appl. Phys. Lett. 32, 624 (1978).
4. E. Yablonovitch, Opt. Lett. 1, 87 (1977).
5. H.S. Kwok and E. Yablonovitch, Appl. Phys. Lett. 30, 158 (1977).
6. E. Yablonovitch and J. Goldhar, Appl. Phys. Lett. 25, 580 (1974).
7. S.L.Shapiro (ed.): *Ultrashort Light Pulses*, Topics in Applied Physics, Vol.18 (Springer, Berlin, Heidelberg, New York 1977).
8. D.S. Frankel, Jr., J. Chem. Phys. 65, 1696 (1976).
9. R.S. Taylor, T.A. Znotin, E.A. Ballik and B.K. Garside, J. Appl. Phys. 48, 4435 (1977).
10. T.F. Deutsch and S.R.J. Brueck, to be published.
11. F. Keilmann, Appl. Phys. 14, 29 (1977).
12. The measured energy deposited with pump pulses of varying duration in the picosecond range will be discussed in a forthcoming paper. H.S. Kwok and E. Yablonovitch, to be published.
13. The scan in Fig.5(c) was actually taken with the P(28) line of the CO_2 laser since the ν_3 resonance was temperature shifted by preheating. All other scans were made at P(20).
14. The usual condition for adiabatic decay of a given level population in $\Delta\omega/\Delta\omega_L >> 1$ where $\Delta\omega$ is the frequency mismatch and $\Delta\omega_L$ is the laser linewidth.
15. J.G. Black, E. Yablonovitch, N. Bloembergen and S. Mukamel, Phys. Rev. Lett. 38, 1131 (1977).
16. The anharmonic shift observed by W. Fuss, J. Hartmann, and W.E. Schmid, Appl. Phys. 15, 297 (1978), is consistent with the deposited energy being distributed among all the modes rather than localized in ν_3.

Resonant Rayleigh-Type Mixing Spectroscopy Using a Picosecond Dye Laser System

H. Souma, Y. Taira, and T. Yajima

Institute for Solid State Physics, University of Tokyo
Roppongi, Minato-ku, Tokyo 106, Japan

Resonant Rayleigh-type optical mixing [1-5] has recently been used as a useful method to determine the ultrashort relaxation times of far below 1 psec associated, for example, with electronic transitions of dye solutions. In this frequency-domain method both the longitudinal and transverse relaxation times can be determined from the dispersive characteristics of the third-order nonlinear susceptibility, $\chi^{(3)}(\omega_3 = 2\omega_1 - \omega_2)$, as a function of $|\omega_1 - \omega_2|$ under the conditions of Rayleigh-type resonance ($|\omega_1 - \omega_2| \simeq \gamma$) and near one-photon resonance (ω_1, $\omega_2 \simeq \omega_0$). Here, ω_1 and ω_2 are the frequencies of two incident light waves, γ is the relaxation rate of various kind, and ω_0 is the frequency of a material transition which is usually inhomogeneously broadened.

In this paper we will report the result of the extension of this method to solid materials, solid sheets containing dye molecules. A newly constructed picosecond dye laser system was used instead of the previous nanosecond dye laser system pumped by a nitrogen laser [4] to overcome a serious limitation set by the optical damage of the solid sample due to strong absorption of incident light beams. Even with the picosecond pulses the experimental situation is still quasi-steady as far as the pulse width is much longer than the relaxation times under study. We will also discuss briefly the relaxation mechanisms through the comparison between the results for dye molecules in solid and liquid.

An experimental system of resonant Rayleigh-type mixing for solid samples is schematically shown in Fig. 1. The light sources consist of two independently tunable picosecond dye lasers with rhodamine 6G laser dye synchronously pumped by the second harmonic of a mode-locked Nd:YAG laser accompanied by several amplifier stages. The output pulses typically have the repetition rate of 10 Hz, the duration of 30 psec, the spectral width of 1 Å and the peak power of more than 150 kW. The frequency selection and tuning of the dye lasers were accomplished by a diffraction grating with 1800 ℓ/mm. The two laser beams were noncollinearly incident on the sample with the angle of about 1° to make the compromise between the spatial filtering and the output efficiency [4]. A streak camera and a microscope were used to realize the exact overlapping of the two picosecond pulses at the sample temporally and spatially, respectively. The detection system was the same as in the previous experiment with nanosecond lasers.

As the solid samples to test we prepared sheets of high polymers containing dye molecules for the reasons of their good optical homogeneity and flexibility in selecting sample conditions and, in addition, for the convenience to consider the ultrafast relaxation mechanisms through the comparison with the

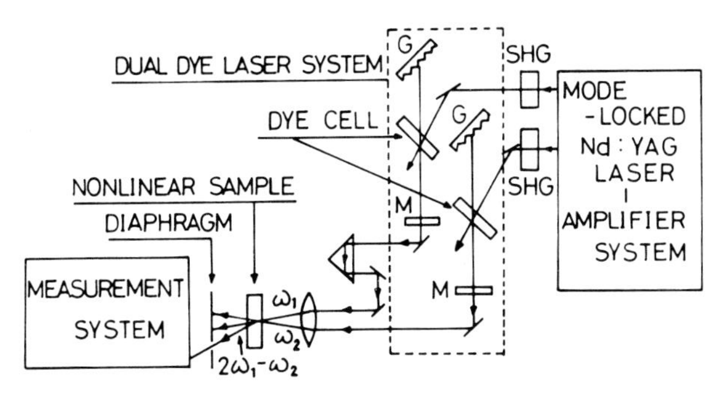

Fig.1 Experimental system for resonant Rayleigh-type mixing with picosecond dye lasers. G: grating, M: output mirror

same dye in liquid solutions. The selection of dye compounds was made in consideration of their chemical stability in time to perform the systematic study of relaxation properties. Rhodamine B and erythrosine B, being selected in this experiment, belong to the class of xanthene dyes, and are known to be comparatively stable. Typical dye concentration in solid or liquid and sample thickness were 1×10^{-3} M and 0.2 mm, respectively.

Figure 2 shows the absorption spectra of rhodamine B in both solid polystyrene and ethanol and erythrosine B in both solid gelatin and water, respectively, which correspond to the $S_0 \rightarrow S_1$ electronic transition. There appear some differences in the absorption spectra between solid and liquid samples, but we will not discuss here about their origins. In the figures are simultaneously shown the typical wavelength positions of both the incident light waves within the tuning range of the picosecond rhodamine 6G dye laser and the output light wave. The corresponding frequency separation, $|\omega_1 - \omega_2|$, is near the value of T_2^{-1} for each dye solution determined later, where T_2 is the transverse relaxation time. For the measurement of nonlinear dispersion,

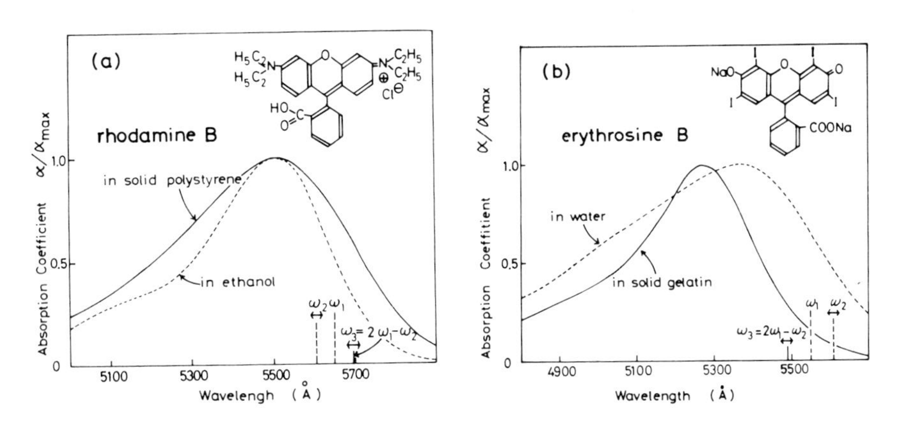

Fig.2 Absorption spectra corresponding to the $S_0 \rightarrow S_1$ electronic transition. (a) rhodamine B in solid polystyrene and ethanol, (b) erythrosine B in solid gelatin and water

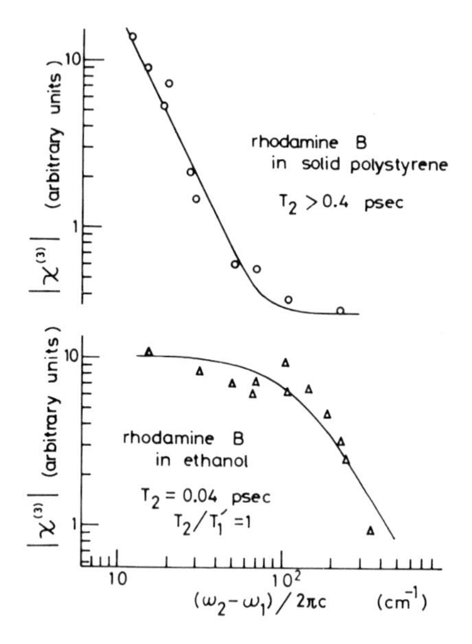

Fig.3 Measured frequency responses of $|\chi^{(3)}|$ for rhodamine B in both solid polystyrene and ethanol under the same experimental condition. The solid lines denote the theoretical fits of (1)

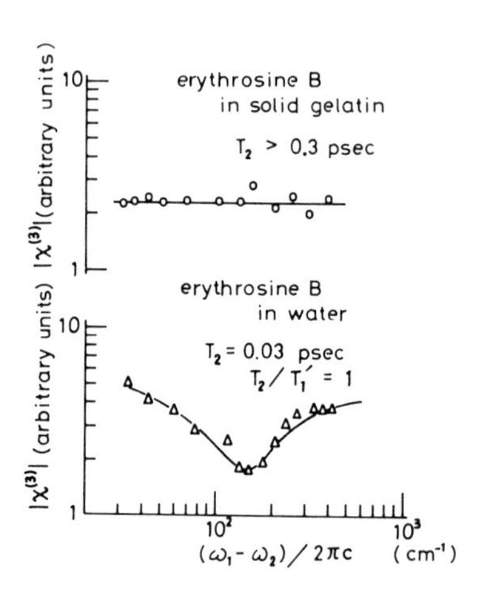

Fig.4 Measured frequency responses of $|\chi^{(3)}|$ for erythrosine B in both solid gelatin and water under the same experimental condition. The solid lines denote the theoretical fits of (1)

ω_2 was scanned with fixed ω_1 and the output intensity at ω_3 was measured as a function of $|\omega_1 - \omega_2|$.

The results of measured $|\chi^{(3)}|$ in solid samples are shown in Figs. 3 and 4 together with those for the same dyes in solutions. The correction for the wave propagation effect [3] has been performed in these results. The measured points can well be fitted to the theoretical formula [3,4]

$$|\chi^{(3)}| = \left|\frac{iK}{\{1 + i(\omega_2 - \omega_1)T_1'\}\{1 + i(\omega_2 - \omega_1)T_2\}} + \chi_{NR}^{(3)}\right|, \qquad (1)$$

where K is a constant nearly independent of $|\omega_1 - \omega_2|$, T_1' is the effective longitudinal relaxation time, and $\chi_{NR}^{(3)}$ is the non-resonant part of $\chi^{(3)}$. When the spectral cross relaxation is important, T_1' is given by $T_1' = (T_1^{-1} + T_3^{-1})^{-1}$ where T_1 and T_3 are the longitudinal and cross relaxation times, respectively. A variety of frequency characteristics of $|\chi^{(3)}|$ appear as seen in Figs. 3 and 4, depending on the relative magnitude between the Rayleigh-type resonant part (the first term) and $\chi_{NR}^{(3)}$, and also on the values of $(T_1')^{-1}$ and T_2^{-1} relative to the scanned range of $|\omega_1 - \omega_2|$. Even the cases where only the upper or lower limit of relaxation times can be determined are of value. The solid lines for rhodamine B are the theoretical fits with $T_2 > 0.4$ psec for the solid sample, and $T_2/T_1' = 1$, $T_2 = 0.04$ psec for the liquid sample. In the case of erythrosine B, the solid line for the liquid sample is the theoretical fit with $T_2/T_1' = 1$, $T_2 = 0.03$ psec. Due to the limited accuracy of the fitting, the values of T_2 thus determined have the error of ±30%, and those of T_2/T_1' involve the ambiguity of a factor of 2 or 3 in both solutions. The dip appearing in Fig. 4 reflects the interference effect between the resonant and non-resonant terms of $\chi^{(3)}$. The dispersionless behavior for the solid sample leaves ambiguities in its interpretation including the possibility that the resonant part is hindered by the larger $\chi_{NR}^{(3)}$. It is likely, however, that the resonant part will appear in the lower $\omega_1 - \omega_2$ region outside the measured range. Under this assumption we get the value of $T_2 > 0.3$ psec. For the more accurate determination of T_2 in solids, it is required to perform a higher spectral-resolution measurement which is presently limited by the spectral purity of light sources. Nevertheless, we can say from the experiment that the value of T_2 for a dye in solid is longer at least by one order of magnitude than that in liquid. The study of whether the same relation holds for other dye materials is planned.

There are a number of relaxation processes contributing to the dephasing of electronically excited dye molecules in liquid and solid caused by both the intermolecular and intramolecular perturbations. Energy changing processes such as internal conversion and intersystem crossing are the parts of these. There are also adiabatic dephasing prosesses such as phase-changing elastic collisions, energy exchange between similar molecules, and two-phonon processes with simultaneous creation and annihilation of vibrational quanta. Intermolecular perturbations in liquids are much more intense and rapid to increase the rate of a particular dephasing mechanism, because a molecule can migrate freely to interact with surrounding molecules. Furthermore, the migration process brings about additional relaxation processes. When the electrostatic perturbation of surrounding molecules is slow enough, it contributes to only the inhomogeneous spectral broadening. On the other hand, when the perturbation becomes rapid enough, it causes a homogeneous dephasing or rapid spectral cross relaxation. Therefore the dephasing rates in solids are naturally expected to be much smaller than those in liquids. Although details of individual relaxation processes are left to be examined, above considerations are consistent with our experimental observations.

Further quantitative studies will clarify the dephasing mechanisms of the electronically excited dye molecules. Our result also implies that this kind of solid dye materials will provide a sample having fairly long T_2 convenient for the study of coherent transient phenomena in the picosecond range. The resonant Rayleigh-type mixing with the present type of picosecond dual dye laser system will further be useful for the study of ultrafast relaxation processes in other types of solid such as semiconductors.

We wish to thank Dr. Hiroshi Ueba of Toyama University and Syunji Takada of Fuji Photo Film Co., Ltd. for the information of sample preparation.

References

1. T. Yajima: Optics Commun. 14, 378 (1975)
2. T. Yajima, H. Souma and Y. Ishida: IXth International Conference on Quantum Electronics (Amsterdam, June 1976) Q7; Optics Commun. 18, 150 (1976)
3. T. Yajima and H. Souma: Phys. Rev. A 17, 309 (1978)
4. T. Yajima, H. Souma and Y. Ishida: Phys. Rev. A 17, 324 (1978)
5. T. Yajima: J. Phys. Soc. Japan 44, 948 (1978)

VII. Solids

Picosecond Luminescence Spectroscopy of Highly Photo-Excited Semiconductors

H. Mahr, T. Daly, and N.J. Frigo

Materials Science Center, Cornell University
Ithaca, NY 14853, USA

Over the past few years we have used a picosecond time resolution luminescence spectrometer [1-4] for the study of time resolved luminescence spectra of highly photo-excited semiconductors. At He-temperatures in CdSe single crystals, we find a single, broad luminescence band, centered at about 6820Å (Fig. 1). The width of the band changes with time after excitation [4]. The total emission in this band decays with a nearly exponential, but distinctly non-exponential time dependence over a time span of about 500 psec. Considerable effort was spent in identifying the mentioned luminescence band as an intrinsic property of highly photo-excited CdSe. Samples of various types and origins were found to give rise to bands of similar location and width, particularly at early time delays. Considerable care was exercised to identify and stay clear of stimulated emission [4] which was clearly present in some samples. We are fairly confident, then, that the emission band shown in Fig. 1 is an intrinsic, spontaneous emission phenomena originating in highly photo-excited CdSe. The luminescent shape of Fig. 1 is, however, influenced by self-absorption. Luminescence spectra taken with the E-vector of the emitted light parallel to the c-axis of CdSe can be different from luminescent spectra taken for E⊥c. Another complication over earlier reports is the fact that the luminescent band of Fig. 1 begins to shift and change width again after considerable time delays approaching nanoseconds. Finally, 5 nanoseconds after excitation, very narrow, characteristic impurity exciton bands appear.

These studies of the behavior of photo-excited CdSe at He-temperatures are continuing. More recently, luminescence spectra of mixed crystals CdS/CdSe were obtained at He-temperatures and of CdSe at liquid Nitrogen temperatures.

This work was supported by the Materials Science Center at Cornell University, grant #DMR-76-81083, technical report #3054.

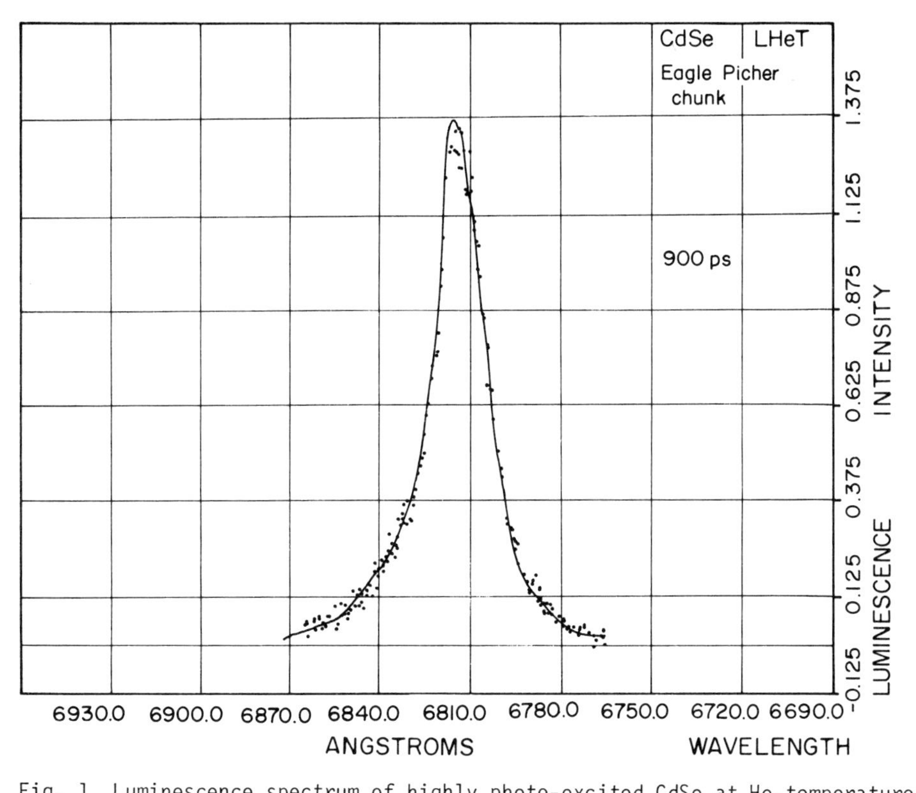

Fig. 1 Luminescence spectrum of highly photo-excited CdSe at He-temperatures 900 ps after excitation

References

1. Mahr, H. and Hirsch, M.D., Optics Commun. 13, 96 (1975)
2. Mahr, H., Hirsch, M.D., Marcus, M.A. Lewis, A. and Frigo, N.J., Biophys. J. 16, 399 (1976)
3. Frigo, N.J., Daly, T. and Mahr, H., IEEE J. Quant. Electron. QE-13, 101 (1977)
4. Daly, T. and Mahr, H., Solid State Commun. 25, 323 (1978)

Picosecond Absorption Spectroscopy of Direct Semiconductors

D. von der Linde and R. Lambrich

Max-Planck-Institut für Festkörperforschung
D-7000 Stuttgart, Fed. Rep. of Germany

Time-resolved measurements of optical spectra can provide detailed information on relaxation processes of optically excited semiconductors. We present measurements of the time evolution of the absorption spectra of GaAs and CdTe after excitation by intense picosecond light pulses. The time-resolved spectra are measured using picosecond probe pulses whose frequency is continuously tunable across the absorption edge of the semiconductors. Because significant variations of the spectra occur on an energy scale of typically 1 meV we choose our spectral resolution to be 0.5 meV. The uncertainty principle then limits the time resolution to approximately 10 ps.

Our experimental system is based on a passively mode-locked Nd-YAG laser generating bandwidth-limited pulses of 25 ps duration. A fraction of the laser output directly serves for excitation of the semiconductors via two-photon absorption. This method of excitation ensures uniform carrier density also for relatively thick samples. The remaining fraction of the laser output is amplified and frequency-doubled and drives a $LiNbO_3$ parametric oscillator (OPO). The synchronously mode-locked OPO generates probe pulses of ~20 ps duration synchronized to within ~1 ps with the excitation pulses at 1.064 μm. The signal wavelength of the OPO is tunable from 650 nm to about 900 nm by temperature control.

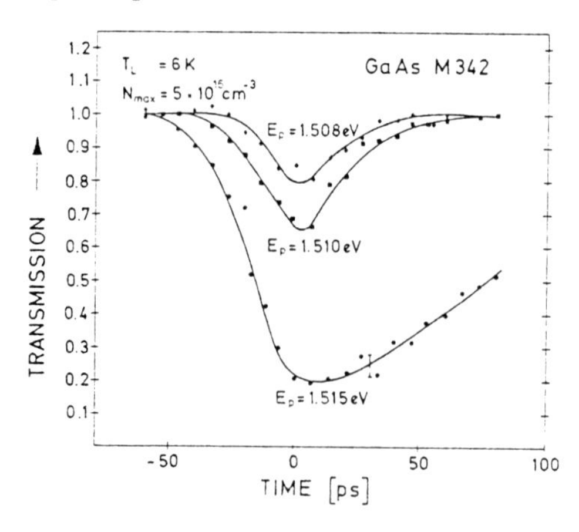

Fig.1 Transmission as a function of time for three different probe energies E_p. T_L: lattice temperature N_{max}: estimated maximum carrier density.

Measurements were done on high purity epitaxial layers of GaAs of approximately 30 μm thickness having a free carrier concentration of less than 10^{14} cm^{-3}. Experimentally we measure the sample transmission as a function of the delay time of the probe pulse with respect to the exciting pulse. Fig. 1 shows the transmission as a function of time of the GaAs layer for three probe wavelengths near the exciton ground state energy (E_{x1}= 1.515 eV). The density of optically excited carriers in these experiments is calculated from the pulse intensity and the known two-photon absorption coefficient to be about 5×10^{16} cm^{-3}. From Fig. 1 the following points should be noted: (i) a transient induced absorption $\Delta\alpha$ is observed at probe wavelengths corresponding to energies less than the lowest exciton energy; (ii) the build up of $\Delta\alpha$ follows the integrated pulse intensity; (iii) $\Delta\alpha$ is larger at shorter probe wave lengths; (iv) the decay of $\Delta\alpha$ is faster at longer wavelengths. At $h\omega_p$= 1.515 eV we estimate the maximum absorption change to correspond to $\Delta\alpha\simeq200$ cm^{-1}.

Quite different observations were made at higher carrier densities and longer probe wavelengths. Fig. 2 shows a transmission curve measured at $\hbar\omega_p$= 1.506 eV for an initial carrier density of about 2×10^{17} cm^{-3}. Here the build up of the induced absorption is followed by a very rapid decay in less than 20 ps. For delay times greater than 20 ps we observe gain. The gain reaches a maximum at 60 ps and then decays in about 100ps.

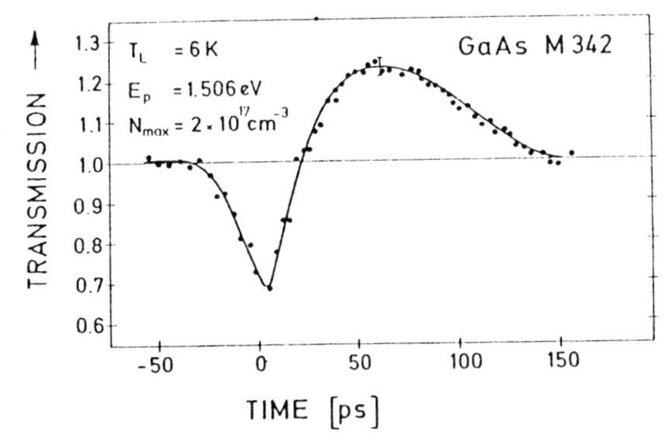

Fig.2 Transmission curve showing the transition from induced absorption to gain. The carrier density is four times larger than in Fig.1

Now we describe our experimental results on high purity (zone refined) bulk samples of CdTe. Fig. 3 depicts transmission curves of CdTe for three different probe wavelengths. It is interesting to point out the distinct differences between the observations on CdTe and GaAs: (i) absorption changes of the order of 100 cm^{-1} are observed at carrier densities of only about $10^{16}cm^{-3}$; (ii) the decay pattern does not change with probe wavelength; (iii) there appears to be a rather distinct cut-off of the induced absorption for probe wavelengths corresponding to energies below $E_{x1}-h\omega_{LO}$= 1.573 eV (788 nm).

We find that the experimental results on GaAs can be readily explained in terms of rapid cooling processes of a hot interacting electron-hole plasma (EHP). This interpretation is based on the following model. Initially electrons and holes are generated at a large excess energy E_e= 0.72 eV and

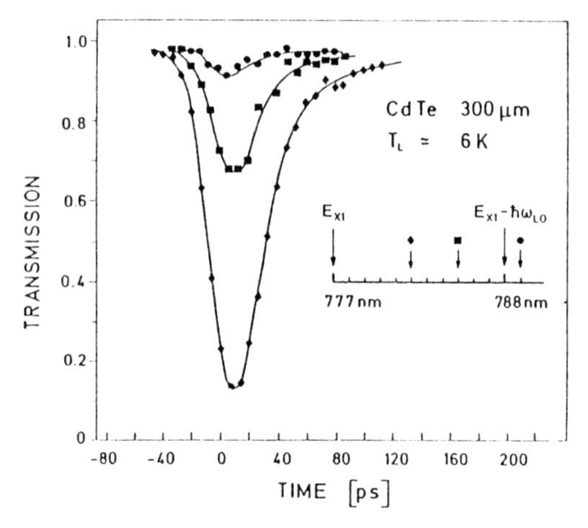

Fig.3 Transmission as a function of time for CdTe. The insert indicates the probe wavelengths corresponding to the three curves. Note also the position of the exciton ground state energy E_{X1} and the LO-phonon energy $\hbar\omega_{LO}$

E_h=0.1 eV, respectively. We estimate that a major fraction of ~90 % of the excess energy is very rapidly removed by LO-phonon emission. The remaining energy is randomized by fast carrier-carrier collision processes [1]. After the excitation pulse has passed the EHP is in a quasi-thermal equilibrium state characterized by a density n and a plasma temperature $T_p \gg T_L$. For our experimental conditions we calculate this temperature to be ~100 K. The subsequent evolution of the EHP is dominated by two processes: (i) further cooling of the plasma by LO-phonon emission; (ii) a decrease of the plasma density due to carrier recombination.

In order to relate this model to the observed time dependent optical properties of the excited semiconductor we calculate the reduced band gap $E_g'(n) < E_g$ and the chemical potential $\mu(n,T_c)$ [2]. Because both T_p and n are time-dependent variables due to the various relaxation processes, μ and E_g' also become functions of time. If μ and E_g' are known, the induced absorption $\alpha(\omega_p,t)$ can be calculated. Fig. 4 shows as an example the result of such a calculation which was aimed at qualitatively describing the results of Fig. 2. These calculations assume a relaxation of the k-selection rule [3] and a k independent matrix element. For $T_p(t)$ and n(t) we use exponential functions with adjustable time constants. It is seen from Fig. 4 that the theory indeed predicts induced absorption for short times, and a subsequent rapid transition to gain. We find that the latter behavior is essentially due to cooling of the EHP while the final decay of the gain can only be accounted for by carrier recombination. A fit of the theory to the experimental data leads to an energy relaxation rate of ~3×10^8eV/s. This value agrees very well with the theoretical energy relaxation rate of 7×10^8 eV/s for the LO-phonon process in GaAs at 70 K [4].

It turns out that the results on CdTe are very difficult to account for in the frame work of EHP theory. A necessary condition for the observation of induced absorption below the band gap is $\hbar\omega_p > E_g'$. The required E_g' to explain the data

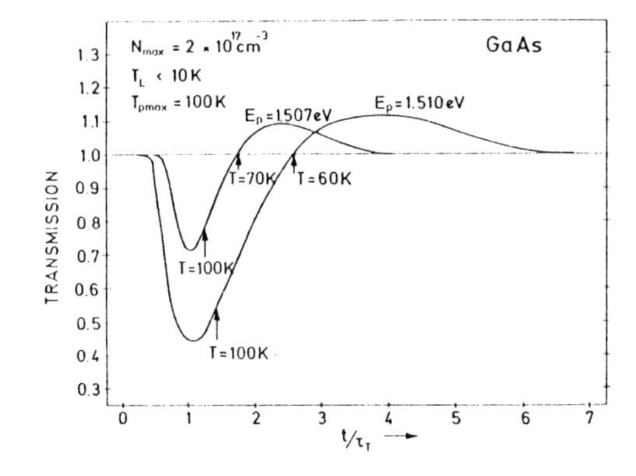

Fig.4 Transmission as a function of time calculated from EHP theory for two different probe energies. The calculation includes both a fast temperature decrease at a rate τ_T, and a slow carrier recombination. The arrows indicate the instantaneous plasma temperature.

would be achieved only at plasma densities in excess of $10^{17}cm^{-3}$.

For an alternative explanation we note that a very large non-equilibrium population of LO-phonons [5] is produced by the relaxing EHP. The phonon mode occupation number is estimated to be a few percent near the center of the Brillouin zone (at $\sim 10^6 cm^{-1}$). Since the Fröhlich constant of CdTe is about 4 times that of GaAs very strong phonon-assisted absorption processes occur for energies within $\hbar\omega_{LO}$ of the lowest excited states. An estimate based on Segall's theory [6] gives a value of $\Delta\alpha \sim 100$ cm^{-1} for our experimental carrier density of 10^{16} cm^{-3}, corresponding to a transmission decrease down to a few percent in our .3 mm thick CdTe crystal, in very good agreement with the experimental data. We therefore believe that a phonon-assisted process was observed and that the decay of $\Delta\alpha$ in CdTe reflects the relaxation of the non-equilibrium LO-phonons. From a semi-log plot of the data of Fig. 3 we find a well-defined exponential decay with a time constant (phonon lifetime) of (15 ± 3) ps.

In conclusion we have shown that in GaAs picosecond time-resolved absorption spectra can be directly related to important carrier relaxation process. Due to the more polar character of CdTe, phonon-assisted processes dominate in this material. We have utilized these processes to demonstrate a novel technique for measuring LO-phonon lifetimes in polar semiconductors.

The authors are indebted to K. Rother and H. Klann for expert technical contributions to this work. The loan of the excellent CdTe crystals by Dr. P. Siffert is gratefully acknowledged.

1. R. Stratton, Proc. R. Soc. A 246, 406 (1958)
2. W.F. Brinkman and T.M. Rice, Phys. Rev. B7, 1508 (1973)
3. E. Göbel, Appl. Phys. Lett. 24, 492 (1974)
4. E.M. Conwell, in Solid State Physics, Suppl.9, H. Ehrenreich, F. Seitz, and D. Turnbull eds., Acad. Press New York (1967)
5. J. Shah, R.C.C. Leite, and J.F. Scott, Sol. State Comm. 8, 1089 (1970)
6. B. Segall, Phys. Rev. 150, 734 (1966)

Fast Laser Excitations in VO_2 at the Semiconducting-Metallic Phase Transition

M.F. Becker, R.M. Walser, and R.W. Gunn

The University of Texas at Austin, Department of Electrical Engineering
Austin, TX 78712, USA

Modeling the 341K semiconductor-metal phase transition (SMT) in VO_2 has proved successful in the case of the electronic and magnetic properties of the high temperature (tetragonal-metallic, M) and low temperature (monoclinic-Semiconducting, SC) equilibrium phases. There are, however, several conflicting theoretical viewpoints as to the origin of the forces which drive the SMT [1]. Reduced to its most elementary form, the debate centers on the relative importance of electron and lattice effects.

We demonstrate for the first time the use of picosecond laser kinetic methods to study cooperative electron-lattice reactions that occur at coincident electronic and structural transformations in many transition metal oxides and chalcogenides. Our experiments show that the fast optical gap excitations induce an instability to a non-equilibrium state having the characteristics of a quasi-static bond distortion. The anomalously long decay of this state with increasing temperature and excitation are suggestive of an increasing decoupling of the electronic excitations from a softened phonon spectrum. This may be an observation of ANDERSON'S "infrared catastrophy" [2].

There is presently strong support for models which view this gap transition as a Mott-Hubbard correlation gap which is determined primarily by the Hubbard intra-atomic correlation energy [3]. Transitions across the correlation gap are assumed to occur between a ground state consisting of paired localized electrons occupying one set of orbital d states and another set occupied by excited itinerant electrons. In the absence of a dynamic theory for the SMT properly incorporating electron correlations and electron-phonon interactions, the physically appealing possibility that the SMT can be electronically initiated by gap excitations must be regarded as a viable concept. The main objective of our research was to put this question to a direct experimental test.

The pyrolysis-reduction of $(C_5H_7O_2)_4VO$ produced the VO_2 films used in our experiments. The films used were 0.1 to 0.4 μm thick on ½" diameter sapphire and quartz substrates. Their quality is comparable to the best VO_2 films produced by other techniques. They exhibit a 10^3-10^4 change in conductivity at 341K over a $\Delta T \simeq 2$-6K. The films are highly oriented polycrystalline deposits.

We used a passively mode-locked and Q-switched Nd:YAG laser to measure the picosecond rise time of the excited state. The laser and single pulse selecting switch produced 41 psec FWHM, 180 μjoule, 1.06 μm (1.18 ev) pulses at a 20 Hz rate. An optical delay line apparatus was used in the pulse probe mode to interrogate the sample transmission.

The rise time of the laser excited state, measured by monitoring the transmission of the film at 1.06 μ, is a prompt threshold effect to within the system resolution of <10 psec. This result can be clearly distinguished from the $\sqrt{2}$ longer rise time one obtains when the transition follows the integrated excitation energy. The laser-excited change in the 1.18 ev transmission was observed to have both threshold and saturation excitation intensities. The threshold pulse energy density, E_{th}, corresponded quite closely to that needed to heat the lattice from the ambient substrate temperature, T_s, to a temperature T_t' just below the SMT. Saturation occurred at a pulse energy density $E_{sat} \simeq E_{th} + L$, where $L = 230\ J/cm^3$, the latent heat of the SMT. The saturated change in transmission is ~25% of the change observed for the thermal equilibrium SMT, and no additional change was observed for excitations up to $2.5\,E_{sat}$.

We measured the longer relaxation times of the excited states using Q-switched laser excitation at 1.06 μm and a continuous probe laser at 633 nm (1.98 ev) to monitor the film reflectivity. The excitation laser was a Q-switched Nd:YAG laser which produced pulses 22 nsec FWHM up to 10 mj in energy at a 20 Hz rate. The cw probe detection system had a 100 nsec response time.

Figure 1 shows characteristic reflectivity transients for variable 1.18 ev excitation pulse energy. Threshold and saturation states were again observed at E_{th} and E_{sat}. For $E > E_{sat}$, the initial reflectivity change is pinned at the value ΔR_{sat} which decays initially at the rate $\tau_1 \simeq 70\ \mu sec$ independent of excitation. Subsequent decay is by a faster mode, with rate τ_2, which increases as a function of excitation and T_s. For values of T_s near T_T', τ_2 increases to greater than 10 msec and obscures τ_1. A minimum value measured for τ_2 was 1.3 μsec for $T_s = 296K$ and $E \leq E_{sat}$. This time is within 10% of the thermalization time for a 3600 Å film on a sapphire substrate.

Figure 2 shows reflectivity transients for different initial substrate temperatures through the SMT. For values of T_s near T_t', τ_2 increases to greater than 10 msec. For substrate temperatures above T_t', the static film reflectivity begins to increase. Since the laser excited state has a constant saturated reflectivity, the laser induced change in reflectivity is decreasing. When $T_s = T_t''$, the laser induced reflectivity change vanishes. The reflectivity is now independent of laser excitation for substrate temperatures from T_t'' to T_t where the SMT is complete. These relationships are illustrated as a function of laser energy in Fig.3. We note that ΔR_{sat} corresponds to a change in reflectivity ~70% of the change for the thermal equilibrium SMT.

We conclude that the laser excitation has produced a non-equilibrium state having the lifetime characteristics of a bond distortion. This atomic displacement at the SMT is similar to those at Martensitic phase transitions for which

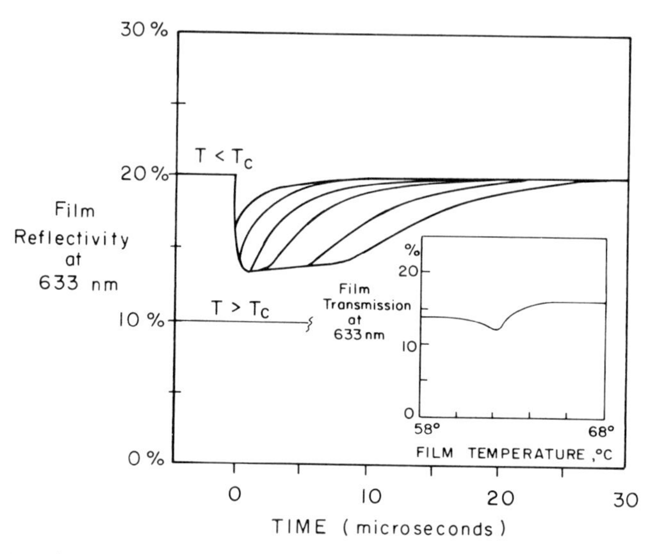

Fig.1 VO_2 film reflectivity as a function of time for several incident energy densities. Insert shows anomalous transmission for the thermally induced SMT

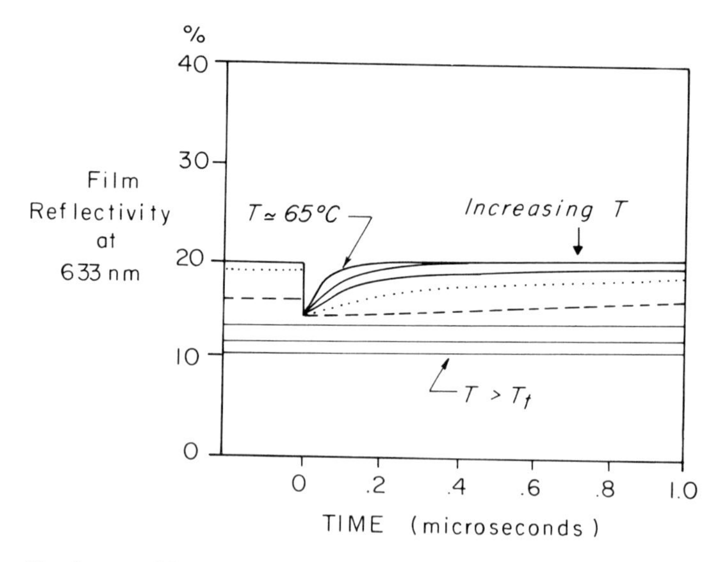

Fig.2 VO_2 film reflectivity as a function of time for several substrate initial temperatures through the phase transition

TONG and WAYMAN [4] estimated transition times of less than 10^{-12} sec. We cannot assume, therefore, that the observed change in 1.18 ev transmission is

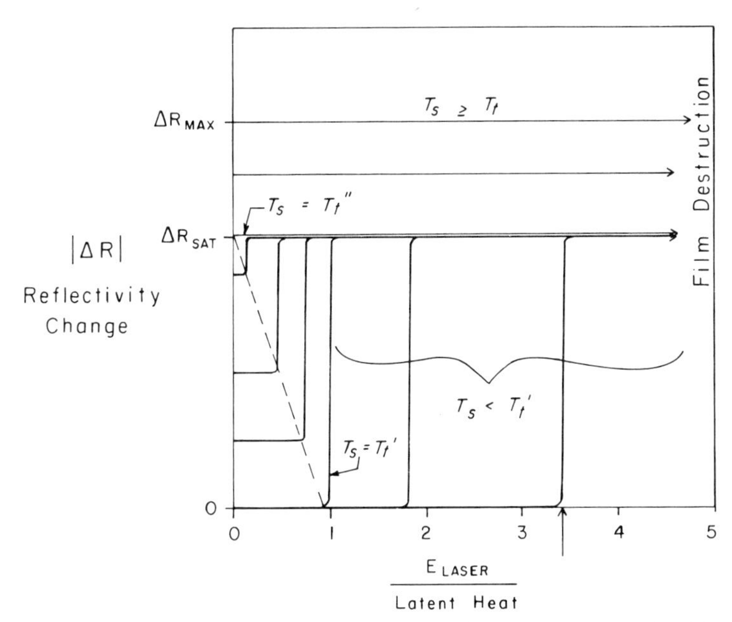

Fig.3 Interpretive plot of VO_2 film reflectivity versus incident energy density for several substrate initial tempertures through the phase transition

due solely to an electronic reorganization.

In terms of a Mott-Hubbard correlation gap [3] and recent molecular cluster calculations [5] for SC VO_2, we interpret the observed non-equilibrium state as a quasi-one dimensional conducting state along the V-V chains. The electrons excited into delocalized states become decoupled from the localized grounds states as the lattice distorts. The orthogonality of these two states closely resembles the infrared catastrophy [2]. In no case was a state characteristic of the static M phase of VO_2 observed with 1.18 ev excitation. This state has been reportedly [6] accessed with 1.78 ev laser excitation.

This research was supported by the DOD Joint Services Electronics Program through the Air Force Office of Scientific Research Contract F49620-77-C-0101.

References

1. J. M. Honig and L. L. van Zandt, in Ann. Rev. of Materials Sci., 225 (1975)
2. P. W. Anderson, Phys. Rev. Lett. 18, 1049 (1967)
3. A. Zylbersztyn and N. F. Mott, Phys. Rev. B 11, 4383 (1975)
4. H. C. Tong and C. M. Wayman, Phys. Rev. Lett. 32, 1185 (1974)
5. C. Sommers, R. deGroot, D. Kaplan, and A. Zylbersztyn, J. Physique Lett. 36, L-157 (1975)
6. W. R. Roach and I. Balberg, Solid State Comm. 9, 551-554 (1971)

Energy Transport in Molecular Solids: Application of the Picosecond Transient Grating Technique

D.D. Dlott and M.D. Fayer

Department of Chemistry, Stanford University
Stanford, CA 94305, USA

J. Salcedo and A.E. Siegman

Department of Electrical Engineering
Stanford University, Stanford, CA 94305, USA

The nature of energy transport in molecular solids is an area of great practical importance in addition to being one of the fundamental problems confronting both theoretical and experimental solid state physics. Considerable information about the nature of transport of electronic excitations has been obtained by indirect methods [1]. However, detailed understanding has been elusive due to the lack of an experimental observable which permits direct determination of the rate of exciton transport. Here we present a brief description of and the first results from a unique experimental method which provides the necessary direct observable. The technique involves the production of a picosecond transient diffraction grating [2] in the sample and permits electronic excitation migration to be studied at the microscopic level in molecular solids. Well-defined distance and time scales are provided and the method allows direct measurement of the rate of exciton transport along any crystallographic direction. By varying the sample temperature, the grating method can be employed to investigate the effect of exciton-phonon scattering on exciton transport.

The grating method involves the diffraction of a picosecond probe pulse from a grating which is optically produced in the sample by the interference of two coherent beams of picosecond excitation pulses. Optical absorption by the sample in the beam overlap region results in a spatially varying sinusoidal distribution of density of electronic excited states. The delayed probe pulse will encounter a sinusoidal spatial variation in the complex index of refraction, which can be substantial if the molecular ground state and excited state have different absorption cross sections at the probe wavelength. The intensity of the Bragg diffracted probe pulse is proportional to the square of the difference in excited state concentration at the grating peaks and nulls. Two dominant time dependent processes affect the decay of the diffracted probe pulse: the decay of excited state population, and the exciton transport which reduces the peaks and fills in the nulls. The signal as a function of probe pulse delay t, is the convolution $Ip(t)*[Ie(t)*h(t)]^2$, where $Ip(t)$ and $Ie(t)$ are gaussian probe and excitation pulse intensities respectively. $h(t)$ is the system impulse response function. For the experimental system described below, excitation transport is diffusive and therefore $h(t)$ can be obtained from the Green's function solution of the diffusion equation including decay. The resulting time-dependent signal is

$$S(t) = A \exp[-2(\Delta^2 D + 1/\tau)t] \quad (1)$$

with A a constant depending on experimental details, D the diffusion coefficient, τ the excited state lifetime, and Δ the grating wave vector, i.e. $\Delta = 2\pi/d = 2\pi\theta/\lambda$, d is the fringe spacing, θ is the angle between the excitation beams, and λ is the excitation wavelength with λ and θ measured in the same medium. Eq. (1) provides a direct method for the determination of the diffusion coefficient in the direction defined by the grating axis.

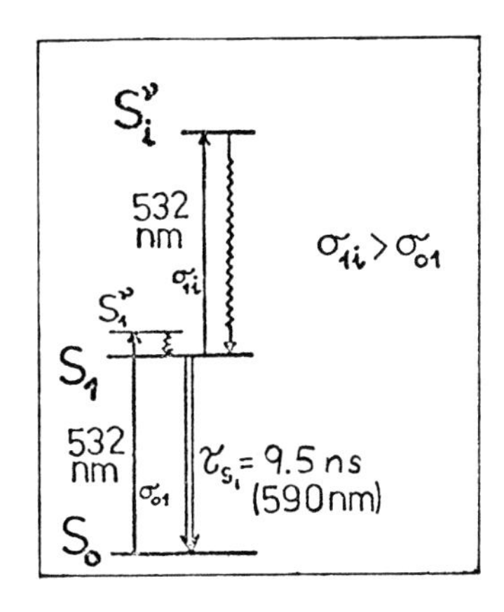

Fig. I

These experiments were performed using a high repetition rate continuously pumped, acousto-optically mode-locked and Q-switched Nd:YAG laser [3]. Single pulses are doubled, and split into two excitation beams and a delayed probe beam. The samples are 1 mm thick single crystals of p-terphenyl doped with $\sim 1 \times 10^{-3}$ M/M concentration of pentacene. The p-terphenyl acts as a transparent host crystal for the isolated pentacene molecules. An energy level diagram of pentacene is given in Fig.I. Excitation to $S1\nu$ is followed by rapid radiationless relaxation to vibrationally unexcited S1. The $S1\nu \rightarrow Si\nu$ transition was observed to be significantly stronger than the $S0 \rightarrow S1\nu$ transition, i.e. $\sigma 1i > \sigma 01$. Thus the sample becomes more absorbing after excitation. This difference in cross section gives rise to strong transient grating signals.

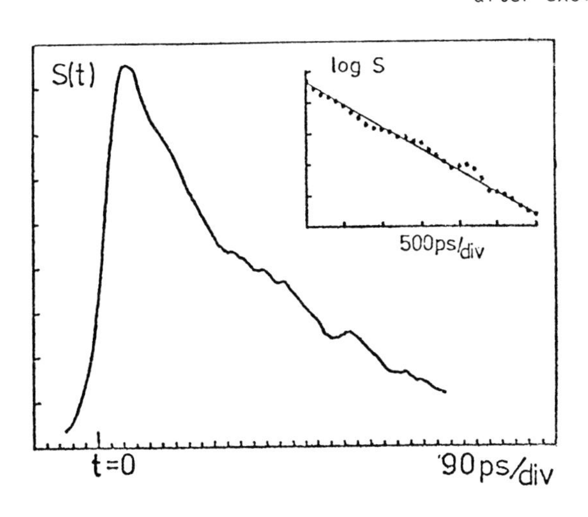

Fig. II

A typical result is shown in Fig.II. The inset shows a semi-log plot of the decay. The decay time, 1.5 ns, is significantly shorter than 4.75 ns, which would be produced by the excited state lifetime alone, since the independently measured lifetime at room temperature is τ = 9.5 ns. Fig.III displays a series of measurements on one crystal, but with different grating fringe spacings. The signal decay rate constants, K, are plotted against θ^2. Also, $2/\tau$ is plotted at $\theta^2 = 0$. From (1)

$$K = 2(\Delta^2 D + 1/\tau) = 8\pi^2 D\theta^2/\lambda^2 + 2/\tau \qquad (2)$$

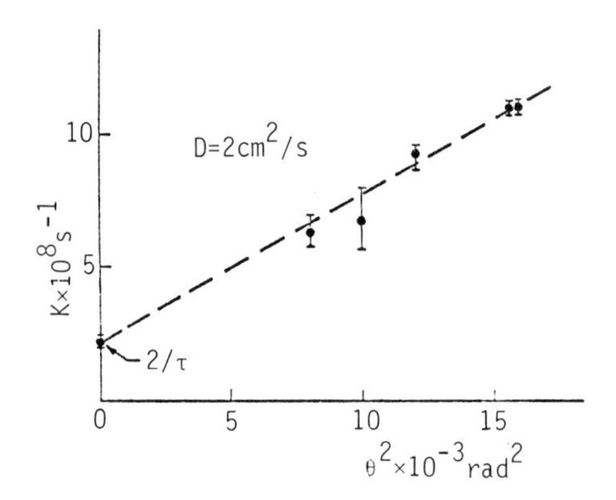

Fig. III

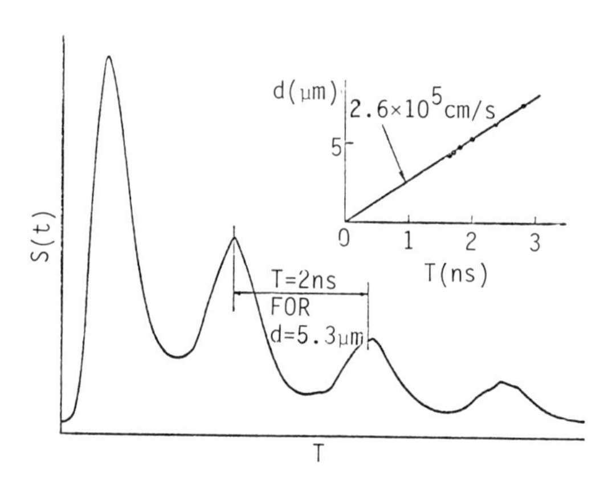

Fig. IV

The model involving both radiationless diffusion of electronic excitation from one pentacene to another and decay predicts that the K's should fall on a straight line and have $2/\tau$ as the intercept. The data is in good agreement with this prediction and yields a direct measurement of the diffusion coefficient, $D = 2.0 \pm 0.1$ cm^2/sec along the crystallographic b axis for a concentration 1.6×10^{-3} M/M. We are currently investigating the dependence on direction, concentration and temperature of the diffusion coefficient.

Another interesting aspect of this experiment is displayed in Fig.IV. For excitation intensities above 500 MW/cm^2 an oscillatory signal is observed. At these intensities, a significant population is built up in S1 (see Fig.I) Since $\sigma 1i > \sigma 01$, absorption to $Si\nu$ during the excitation pulse will occur, followed by immediate radiationless relaxation back to S1. (Another mechanism would involve S1-S1 annihilation. The net result will be the same.) The radiationless relaxation produces local heating of the sample, resulting in a density grating. The density grating can be visualized as a "frozen" sinusoidal stress pattern plus two counterpropagating acoustic waves superimposed on the excited state grating. The slope of the line on the inset of Fig.IV yields a velocity of 2.63×10^5 cm/sec along the grating direction. The speed of sound in this direction measured by the ultrasonic pulse echo overlap method is 2.65×10^5 cm/sec. An analysis of the beating effects shows that the observed pattern can occur only if the thermoacoustically generated grating is primarily an amplitude grating rather than a phase grating. We hypothesize that the acoustic waves modulate the optical absorption cross sections of the transitions involved, which is consistent with the reported pressure dependence of absorption of pentacene in p-terphenyl [4].

These preliminary experiments demonstrate that the picosecond transient grating method is a powerful tool for the investigation of energy transport in solids. This study involved electronic energy transfer between impurities, however, identical methods can be applied to exciton transport in pure crystals and other systems.

This work was supported by the National Science Foundation, Division of Materials Research, and the Joint Services Electronics Program of Stanford University. J. S. would like to acknowledge a NATO Fellowship and a Fulbright grant for partial financial support.

1. Y. Toyozawa, Prog. Theor. Phys. 20, 53 (1958); D. M. Burland, D. E. Cooper, M. D. Fayer, and C. R. Gochanour, Chem. Phys. Lett. 52, 279 (1977); A. H. Francis and C. B. Harris, Chem. Phys. Lett. 9, 181, 188 (1971); D. D. Dlott and M. D. Fayer, Chem. Phys. Lett. 41, 305 (1976); R. C. Powell and Z. G. Soos, J. Lumin. 11, 1 (1975); D. D. Dlott, M. D. Fayer, and R. D. Wieting, J. Chem. Phys. 67, 3803 (1977).

2. D. W. Phillion, D. J. Kuizenga, and A. E. Siegman, A. P. Phys. Lett. 27, 85 (1975).

3. D. J. Kuizenga, D. W. Phillion, T. Lund, and A. E. Siegman, Opt. Comm. 9, 221 (1973).

4. J. M. Vonnini, J. Chimie Phys. 71, 1543 (1974).

Ultrafast Processes in the Optical Response of the Electron-Hole Plasma in Germanium

B. Bosacchi, C.Y. Leung[1], and M.O. Scully

Optical Sciences Center, University of Arizona
Tucson, AZ 85721, USA

Abstract

We report an increase in the reflectivity of picosecond pulses in Ge at excitation energies immediately below the damage threshold, and discuss the results in terms of direct intraband transitions within the branches of the valence band.

****

Picosecond spectroscopy represents a valuable tool for the study of the ultrafast processes that accompany, or immediately follow, the creation of an electron-hole pair in a semiconductor by a photon of energy $h\nu$ greater than the forbidden gap E_g. In the case of Ge, such studies have been extensively carried out over the past few years by our group in Tucson [1-5] and by the group of AUSTON, SHANK, and coworkers at Bell Labs [6-8], using picosecond pulses from a Nd:glass laser. The basic physics of the interaction of intense picosecond pulses with semiconductor systems is becoming more clear, although many details remain to be clarified. In spite of the high intensity of the radiation, with photon densities well in excess of the density of states directly coupled by the radiation, complete "bleaching" of the sample is far from being achieved. This points to the existence of ultrafast processes that act to empty the coupled states, to enhance their density, or to introduce further absorption channels in times shorter or comparable to the duration of the exciting pulse. There is no shortage of candidates for mechanisms that can act in the above manner: for example, (1) intervalley scattering, (2) phonon relaxation, (3) Coulomb thermalization, (4) plasmon-assisted recombination, (5) Auger recombination, (6) free-carrier absorption (possibly aided also by Coulomb interaction), (7) direct intraband transitions between branches of the valence band, and (8) indirect intergap absorption, aided by Coulomb interaction. Since the spectral energy of the excitation is considerably larger than the energy gap, the carrier distribution will initially be very hot, i.e., characterized by a temperature T_c that greatly exceeds the lattice temperature T_L. Eventually, this energy will transfer from the carrier system to the lattice, through phonon emission processes, and the semiconductor system will ultimately regain its equilibrium by photon emission and transfer of heat to the thermal bath.

A theoretical model incorporating a subset of the above processes (intervalley scattering, phonon-assisted relaxation, Coulomb thermalization, phonon-aided free-carrier absorption, and plasmon-assisted recombination) in an attempt to account quantitatively for the experimental results, has been

[1]Present address: Department of Physics, National Central University, Ching-Li, Taiwan, Republic of China.

studied recently by ELCI et al. [9] (hereafter referred to as the ESSM model). The basic philosophy of this model is to develop a scheme for the analysis of the interaction of picosecond pulses with a semiconductor, in which the most important processes are contained. Obviously, in view of the complicated nature of the problem, the ESSM model has limitations that have been discussed elsewhere [10,11]. It represents, however, a first step toward an understanding of the excitation of a semiconductor with intense and ultrashort pulses, and, as such, it is open to revision and refinement. In particular, new experimental results that cannot be directly accommodated within its framework can provide useful insight. In the present paper we present some preliminary data on reflectivity, which argue for the inclusion of direct photon absorption within the valence band.

An example of a typical experimental result is reported in Fig. 1. This figure shows the reflectivity at room temperature of a Ge single crystal as a function of the intensity of the excitation with picosecond pulses from a Nd:glass laser. The reflectivity is seen to remain practically constant over the whole intensity range, but it shows a marked increase (∿20%) just before the damage threshold is reached. The enhancement occurs in the same excitation range at which, in the transmission vs excitation experiment [2], the increase in the transmission tends to saturate. This suggests that a common mechanism might lie at the root of both effects. Details of the experiment and of the procedure to check that the reflectivity rises before surface damage occurs are provided elsewhere [12].

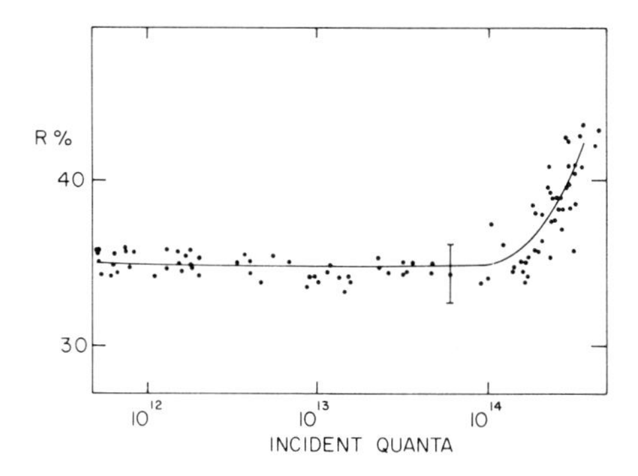

Fig. 1 Reflectivity of Ge at room temperature as a function of the energy of the pulses

Reflectivity enhancements under intense laser excitation in the nanosecond regime have already been reported several times [13-16]. These enhancements were first interpreted as plasma effects (occurrence of a plasma resonance at 1.17 eV) related to the high density of excited free carriers. It was soon realized, however, that this interpretation requires that carrier concentrations $N \gtrsim 2.0 \times 10^{21}$ cm^{-3} be reached, if an enhancement in reflectivity rather than a decrease, is to be seen [15]. On the other hand, it is doubtful that such high values of N can be achieved. In fact, both the band structure of Ge and the intervention of competing absorption and recombination processes seem to limit the highest concentration of carriers to $N_{max} \simeq 2 \times 10^{20}$ cm^{-3}. Probably because of these difficulties, the enhancement of reflectivity at high excitation intensities, using nanosecond pulses, is now commonly attributed to the formation of a molten layer with metallic properties at the incidence spot on the surface of the sample [17,18]. The damage that would be normally associated with melting can be avoided if annealing conditions prevail during the recrystallization of the molten spot. Indeed, the use of nanosecond pulses for annealing treatments on semiconductors is, by now, a well established process.

The molten-layer interpretation comes down to invoking a reversible change of the band structure of the sample at the incidence spot, so that constraints on the density of carriers, due to the intrinsic band structure of the crystal, can be circumvented. The applicability of this explanation to the case of picosecond excitation, however, is not straightforward. We have estimated [12] the temperature reached at the incidence spot, under the assumption that the energy of the pulse is entirely delivered as heat to the interaction volume in a time short enough to rule out processes like carrier diffusion and thermal conductivity, which would spread this energy over the whole sample. With an interaction volume $V \simeq 10^{-7}$ cm^3, and a pulse energy $E \simeq 3 \times 10^{14}$ photons, the increase in temperature at the surface is less than $100^{\circ}K$, so the final temperature would remain considerably lower than the melting point value ($T_M = 937^{\circ}C$). This estimate is only a rough evaluation, due to the severe indeterminacy in the values of both E and V, yet it suggests that the molten layer interpretation does not hold for our problem. Moreover, in order to use melting to explain our result, the melting temperature must be reached in a time shorter than the duration of the pulse; i.e., the excess energy of the excitation must be entirely transferred from the carrier to the lattice system in less than 10 psec. This also is doubtful, though the present knowledge of the energy relaxation process in Ge does not allow a clearcut assessment of this possibility. Using for the electron-optical phonon coupling constant $Q_0 = 9.6 \times 10^{-4}$ erg/cm, we have estimated for the energy relaxation time $\tau_E \simeq 15$ to 20 psec, but values of Q_0 ranging from 6 to 18×10^{-4} erg/cm have been reported in the literature [19]. It must be further noted that even if T_M is not reached during the duration of the pulse, there might still be sufficient temperature variation at the incidence spot to explain the observed enhancement as a thermal effect caused by temperature-induced changes in the optical constants rather than by melting.

On the other hand, as discussed later, a peculiar consequence of the use of picosecond pulses in the excitation of a semiconductor is the attainment of carrier distribution temperatures much higher than in the case of nanosecond excitation. This entails the possibility of highly nonequilibrium electronic processes, which are intrinsically fast and can affect the optical properties of a semiconductor over very short times. In this paper we will discuss one such mechanism, the direct intraband transitions (DIT) within the branches of the valence band.

The direct intraband transitions, as shown in the schematic band structure of Fig. 2, are a direct consequence of the multibranch nature of the valence band. DIT transitions have already been shown to be important in the interpretation of several experimental results [20,21], and their contribution to the dielectric constant in the far infrared has recently been considered [22]. With the excitation at 1.06 μm, these transitions occur relatively far from the center of the Brillouin zone, and, unless the excitation is sufficiently intense, they will usually be forbidden since both the initial and final states $|i\rangle$ and $|f\rangle$ are occupied. As shown in Fig. 2, even at the highest concentration of carriers that can be reached in the experiment ($\sim 2 \times 10^{20}$ cm^{-3}), the Fermi-level position is not sufficiently deep in the valence band to ensure, by itself, a significant depopulation of $|f\rangle$. The temperature of the carrier distribution is therefore the critical factor controlling the occupation of $|f\rangle$. This implies an important difference between nanosecond and picosecond excitation. Since the spectral energy of the

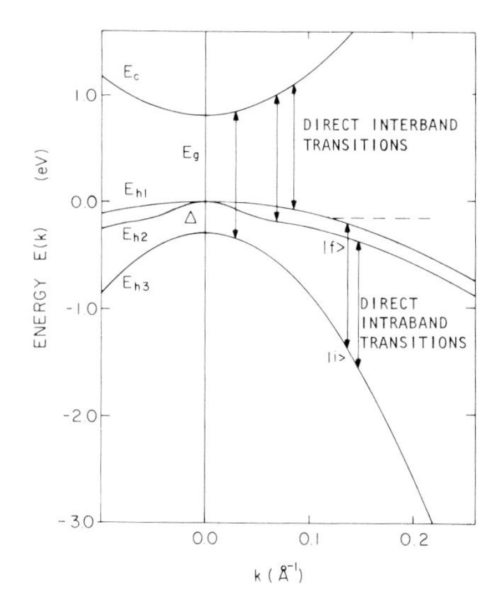

Fig. 2 Schematic band structure of Ge near the center of the Brillouin zone. (The dashed line indicates the Fermi-level position at $N = 2 \times 10^{20}$ cm^{-3}, $T_c = 300^{\circ}K$)

excitation considerably exceeds the energy gap, the carrier distribution is initially characterized by a temperature T_c that exceeds the lattice temperature T_L. This carrier heating effect, which takes place during the interaction of the pulse with the semiconductor, is opposed by phonon emission processes, which tend to transfer the excess energy from the carrier system to the lattice, thus increasing T_L at the expense of T_c. With nanosecond excitation, this transfer occurs effectively during the duration of the pulse, which is much longer than the time τ_E associated with the relaxation of the carrier distribution through optical phonon emission. We can, therefore, expect a high value of T_L at the incidence spot, but a relatively low value of T_c.

The reverse occurs in the highly nonequilibrium situation created by picosecond pulses, whose duration, particularly in a nonpolar compound like Ge, is comparable or shorter than τ_E. Distribution temperatures T_c of several thousand degrees Kelvin are indeed reached during the duration of the pulse, while T_L remains relatively low. Therefore, since the pulse interacts with a very hot distribution of carriers, not only the carrier density N, but also the carrier temperature T_c, must be taken into account in the discussion of the experimental results. Since the most direct and relevant effect of T_c appears to be on the statistical population of $|f\rangle$, direct intraband transitions are expected to play a relevant role in picosecond experiments, provided certain levels of excitation are reached.

The contribution to the dielectric constant of the crystal arising from the DIT transitions can be easily calculated, if we assume that the carrier distribution is described by a Fermi function. Actually, this assumption is justified only at sufficiently high carrier densities ($\sim 10^{20}$ cm^{-3}), such that the time τ_p (reciprocal of the plasma frequency) is much shorter than the duration of the pulse. These densities are indeed reached at the high excitation intensities ($\gtrsim 10^{14}$ photons/pulse) at which the reflectivity enhancement occurs.

The reflectivity as a function of the carrier density, calculated for various values of the carrier distribution temperature, is reported in Fig. 3. It must be noted that these curves do not carry individually a direct experimental significance, since, in a real experiment, we cannot vary the density while keeping the carrier temperature constant. Rather, the process is the opposite. We can think, in first approximation, that as we raise the intensity of the excitation, we first increase the density of carriers up to a limit, after which the energy we provide goes into raising the temperature through the intervention of processes (free-carrier absorption, DIT transitions, plasmon-assisted recombination, etc.) that keep the density constant. Thus, if the intensity of our excitation is such that we reach first a carrier density $\gtrsim 10^{20}$ cm^{-3} and then increase the temperature up to $\sim 4000^{\circ}K$, the curves of Fig. 3 show that a reflectivity enhancement from ~ 0.35 to $\gtrsim 0.40$ can be obtained, in good agreement with the experimental results.

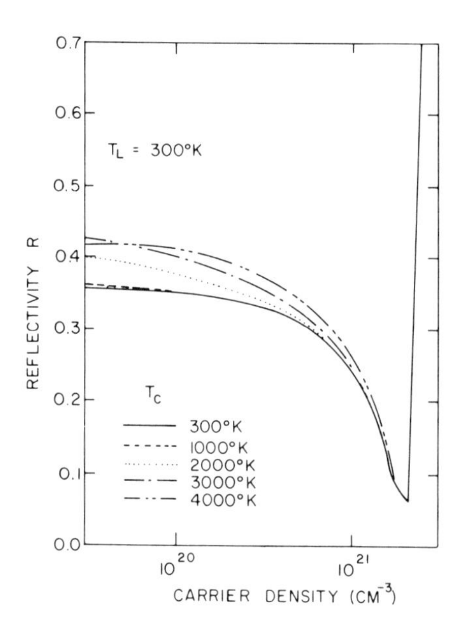

Fig. 3 Reflectivity vs carrier density, calculated for various carrier distribution temperatures T_c

As a check of the consistency of the above procedure, we report in Fig. 4 the temporal evolution, during the duration of the pulse, of the distribution temperature T_c, carrier density N, refraction index n, extinction coefficient k, and reflectivity R, for an excitation energy of 3×10^{14} photons/pulse [23]. The calculation has been performed in the framework of a revised ESSM model that includes the DIT transitions, assuming an electron-optical phonon coupling $Q_0 = 9.6 \times 10^{-4}$ erg/cm and the following parameters for the band structure of Fig. 2: $E_g = 0.805$ eV, $\Delta = 0.295$ eV, $m_c = 0.1\ m_e$, $m_{h1,2} = 0.34\ m_e$, $m_{h3} = 0.0665\ m_e$, where m_c, $m_{h1,2}$ and m_{h3} are the effective masses of conduction band and heavy, light, and split-off hole bands, respectively. Figures 4a and b show that temperatures close to $4000^{\circ}K$ and densities $\sim 10^{20}$ cm^{-3} are reached with a pulse of 3×10^{14} photons.

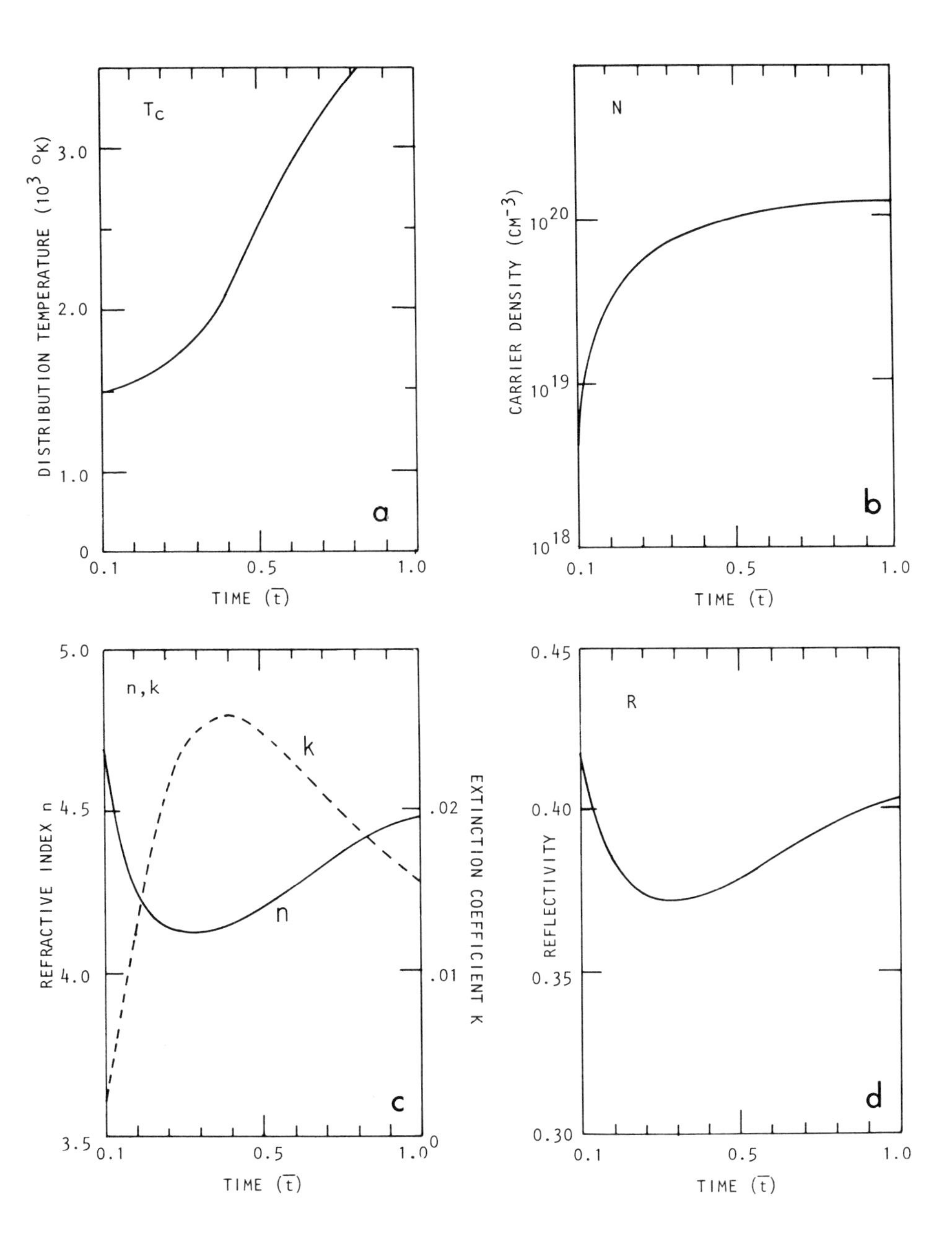

Fig. 4 Temporal evolution of T_c, N, n, k, and R during the duration $\bar{t}$ of the pulse (calculation for $\bar{t}$ = 11 psec).

In Figs. 4c and d, one can see that the average values of n and R seen by the pulse are indeed noticeably higher than the low intensity values ($n \simeq 4$, $R \simeq 0.35$). Obviously, the DIT process entails also a variation in k, which, though too weak to affect the reflectivity results, has an appreciable effect on the transmission experiment (transmission vs excitation intensity). Indeed, the inclusion of the DIT process in the ESSM model significantly improves the agreement between the results of the transmission experiment and the theoretical predictions [24]. Besides reproducing the saturation in the transmission enhancement observed at high excitation intensity, it leads to a much weaker role of the plasmon-assisted recombination process which, in agreement with the discussion of ref. 25, explains the lack of a strong pulse width dependence, as found in ref. 5.

An interesting feature of the DIT mechanism is its speed. Since the density of carriers is not sufficient to deplete the final states even at its maximum value, the process is actually controlled by the temperature of the distribution. The DIT process is activated at excitation values for which a critical temperature is reached, and is automatically switched off as the distribution cools down. It follows that the time during which the channel of the DIT process is open is controlled by the relaxation rate of the distribution temperature, rather than by the recombination rate of the carrier distribution. It is therefore an extremely fast process ($\sim$10 psec). The availability of a process that entails a substantial variation of the optical constants of the material over such short times obviously has interesting potential applications.

In conclusion, we have shown that the DIT mechanism can provide an explanation of the reflectivity enhancement we have observed, as an alternative to the thermal effects interpretation. The competition between the electronic and the thermal mechanisms is actually controlled by the electron-optical phonon coupling rate. This rate determines the highest carrier distribution temperature which can be reached, and the rate of transfer of the excitation energy to the lattice. A time resolved study of the reflectivity enhancement could provide a clear discrimination between the two alternatives. Unfortunately, owing to the limited range of intensities over which the enhancement occurs, its closeness to the damage threshold, and the irreproducibility of the Nd:glass laser pulses, we have been unable to obtain any conclusive result so far. Finally, we notice that, since the multi-branch structure of the valence band is a common feature of most semiconductors, the DIT process has a very general validity, particularly with the high intensities involved when picosecond pulses are used. Therefore, an indiscriminate analysis of experimental results involving free carrier effects simply in terms of the classical Drude model might lead to deceiving results.

Acknowledgments

We are pleased to thank Prof. Melvin Lax for a helpful discussion. This work was supported by the Air Force Office of Scientific Research (AFSC), United States Air Force, and the Army Research Office, United States Army.

References

[1] C. J. Kennedy, J. C. Matter, A. L. Smirl, H. Weichel, F. A. Hopf, S. V. Pappu, and M. O. Scully, Phys. Rev. Lett. 32, 419 (1974).

[2]A. L. Smirl, J. C. Matter, A. Elci, and M. O. Scully, Opt. Commun. 16, 118 (1976).
[3]H. M. van Driel, J. S. Bessey, and R. C. Hanson, Opt. Commun. 22, 346 (1977).
[4]H. M. van Driel, A. Elci, J. S. Bessey, and M. O. Scully, Solid State Commun. 20, 837 (1976).
[5]J. S. Bessey, B. Bosacchi, H. M. van Driel, and A. L. Smirl, Phys. Rev. B 17, 2782 (1978).
[6]D. H. Auston and C. V. Shank, Phys. Rev. Lett. 32, 1120 (1974).
[7]C. V. Shank and D. H. Auston, Phys. Rev. Lett. 34, 479 (1975).
[8]D. H. Auston, C. V. Shank, and P. LeFur, Phys. Rev. Lett. 35, 1022 (1975).
[9]A. Elci, M. O. Scully, A. L. Smirl, and J. C. Matter, Phys. Rev. B 16, 191 (1977).
[10]A. Elci, A. L. Smirl, C. Y. Leung, and M. O. Scully, Solid-State Electron. 21, 151 (1968).
[11]B. Bosacchi, in *Coherence in Spectroscopy and Modern Physics*, ed. by F. T. Arecchi (Plenum Press, New York, 1978).
[12]B. Bosacchi, C. Y. Leung, and M. O. Scully, submitted to Opt. Commun.
[13]W. R. Sooy, M. Geller, and D. P. Bortfeld, Appl. Phys. Lett. 7, 36 (1965).
[14]M. Birnbaum and T. L. Stocker, Brit. J. Appl. Phys. 17, 461 (1966).
[15]L. M. Blinov, V. S. Vavilov, and G. N. Galkin, Sov. Phys.-Solid State 9, 666 (1967).
[16]A. M. Bonch-Bruevich, V. P. Kovalev, G. S. Romanov, Ya. A. Imas, and M. N. Libenson, Sov. Phys.-Techn. Phys. 13, 507 (1968).
[17]G. N. Galkin, L. M. Blinov, V. S. Vavilov, and A. G. Golovashkin, JETP Lett. 3, 69 (1968).
[18]M. Birnbaum and T. L. Stocker, J. Appl. Phys. 39, 6032 (1968).
[19]See, for example, M. Costato, S. Fontanesi, and L. Reggiani, J. Phys. Chem. Solids 34, 547 (1973).
[20]R. Newman and W. W. Tyler, Phys. Rev. 105, 885 (1957).
[21]V. N. Murzin, V. A. Zayaks, and V. L. Kononenko, Sov. Phys.-Solid State 17, 1783 (1976).
[22]M. Combescot and P. Nozieres, Solid State Commun. 10, 301 (1972).
[23]Details of the calculation will be published elsewhere.
[24]C. Y. Leung, B. Bosacchi, and M. O. Scully, submitted to Phys. Rev. B. The relevance of the DIT mechanism in the transmission experiment has been independently suggested by S. McAfee and D. H. Auston (private communication).
[25]W. P. Latham, A. L. Smirl, A. Elci, and J. S. Bessey, Solid-State Electron. 21, 159 (1978).

Time-Resolved Induced Absorption Due to Exciton-Biexciton Transitions in CuC1

R. Levy, B. Hönerlage[1], and J.B. Grun

Laboratoire de Spectroscopie et d'Optique du Corps Solide Associé au C.N.R.S.
Université Louis Pasteur
F-67000 Strasbourg, France

Induced absorption due to exciton-biexciton transitions is observed in CuCl using an intense ultraviolet picosecond pulse excitation and a continuum as a probe beam.

Introduction

Light emission due to the radiative decay of biexcitons from their ground states to exciton states has been extensively studied in recent years. The reverse process where biexcitons are created from excitons by absorption of light (Induced Absorption), theoretically studied by GOGOLIN and RASHBA [1] , was observed in our laboratory in CuCl. A nitrogen laser pumped dye laser and a broad superradiant emission band were then used [2]. Since the induced absorption coefficient is proportional to the density of excitons, the mean distribution of the excitons in their bands were obtained. By similar experiments using a mode-locked laser excitation, the time dependence of this exciton distribution was then studied.

Experimental Set-up

A block diagram of the experimental set-up used is shown in Fig. 1. The picosecond laser system consists of a Yag (Nd^{3+}) oscillator, mode-locked by means of a saturable dye, producing a picosecond pulse train at 1060 nm. A pulse selector isolates from the middle of this train a single pulse of about 30 ps duration. This pulse is then fed into a singlepath amplifier, before being applied to two consecutive second harmonic generators. At the output of the second frequency doubler, the remaining green light (532 nm) and UV light (266 nm) are separated by a prism. The UV beam is suitably attenuated, and, after an adjustable optical delay, focussed onto the surface of the sample. The energy of the UV pulse, measured at the position of the sample with a calibrated calorimeter, is 50 μJ which corresponds to 7×10^{27} photons $s^{-1}cm^{-2}$. The green light, selected in amplitude, is focussed in a cell containing heavy water (D_2O). At the output of the cell, a broad emission of

[1] On leave from: Fachbereich Physik der Universität Regensburg, D-8400 Regensburg, Fed. Rep. of Germany

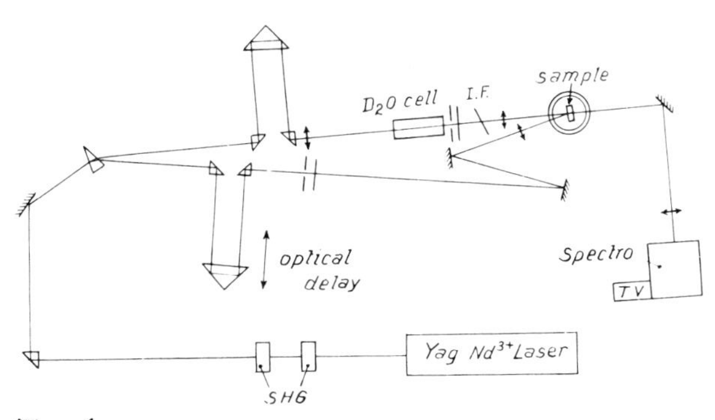

Fig.-1

weak intensity and of the same time duration as the laser, is obtained. A narrower band of this emission, called " continuum ", (21 meV at half-width) is selected by means of a tilted interference filter. The spatial coincidence of the UV beam and of the continuum on the sample is carefully checked. The continuum transmitted through the sample is focussed on the entrance slit of a 25 cm spectrograph. The spectra are recorded by an Optical Multichannel Analyser (P.A.R.), which is synchronized with the picosecond laser [3]. The samples used in these experiments are platelets of CuCl, grown from vapour phase transport, cooled down to liquid helium temperature.

Experimental Results

The origin of time taken is the onset of the induced absorption. $J_0(\omega)$ is the intensity of the continuum at frequency ω in the absence of UV light and $J(\omega)$ is the intensity of the continuum when the crystal is also excited by the UV pulse. In Fig. 2, $\ln[J_0(\omega)/J(\omega)]$ is given for different delays (R) between the continuum and the UV pulse. This quantity is proportional to the induced absorption coefficient, which is itself proportional to the density of excitons $n_{ex}(\omega)$. For small optical delays, a broad induced absorption band is observed. After about 200 ps, the shape of this absorption band changes drastically to a narrower structure. With increasing delay, its half-width decreases and its maximum shifts to higher energies.

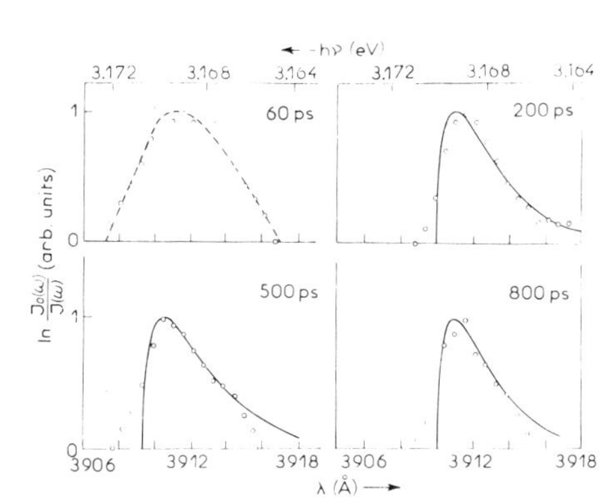

Fig.-2

In Fig. 3, the overall absorption $\int \ln[J_0(\omega)/J(\omega)]d\omega$ is reported in a logarithmic scale as a function of the optical delay After a sharp increase, a decrease is observed with a longer time constant of about 2 ns.

Similar results are obtained for different UV excitation intensities.

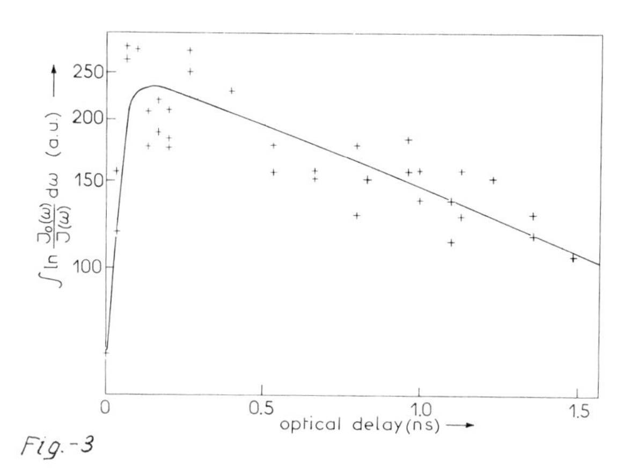

Fig.-3

Discussion of the Experimental Results

The induced absorption observed can be well explained by the reverse process of the radiative decay of biexcitons. The high energy side of the induced absorption band in Fig. 2 corresponds to the energy difference between the Γ_{5T} exciton and the Γ_1 biexciton states in CuCl, known from luminescence measurements.

Our time resolved measurements in Fig. 2 give us information on the evolution of the exciton distribution with time. The first spectra could correspond to an exciton packet which has only been thermalized by LO phonon emission but has not yet reached its internal equilibrium by mutual collisions. A quasi-equilibrium of the exciton gas seems to be achieved in the spectra obtained for longer delays. Then the absorption line has an inverse Boltzmann shape. The half-width of the absorption band decreases which corresponds to a decrease of the temperature of the exciton gas with a time constant of about 3 ns. This may be due to an energy transfer from the exciton and biexciton system to the lattice by emission of acoustical phonons or a heat diffusion from the excited region to the bulk of the crystal.

These two different relaxation processes appear also in Fig. 3 when we recall that $\int \ln[J_0(\omega)/J(\omega)]d\omega$ is proportional to the total number of excitons. The region of sharp increase of this curve corresponds to the creation of excitons. The processes involved are the relaxation of free carriers initially created by LO phonon emission, the creation of excitons either directly or from the biexciton population. Since the rise time is of the order of the pulse duration, it is difficult to conclude about the time constant of the processes involved. To deduce from the slope of the slow decay curve a value of the radiative lifetime of the excitons and biexcitons, we have studied the dynamic equations of the system. The best fit of our experimental results was obtained with the following radiative lifetimes τ_{ex} and τ_{Bi} of excitons and biexcitons :

$$\tau_{ex} \simeq 300 \text{ ps} \qquad \tau_{Bi} \simeq 700 \text{ ps}$$

The value for the biexciton lifetime is in good agreement with the one obtained from the biexciton oscillator strength, measured by two-photon absorption. It is also similar to the one deduced from the time-resolved study of the biexciton luminescence [3].

In conclusion, we have observed for the first time the formation of a quasi-equilibrium of the exciton population and also the further relaxation of this exciton distribution with the phonon system. We have also studied the time behaviour of the overall exciton population.

References

[1] A.A. Gogolin, Soviet Phys. Solid State, 15, 1824 (1974) ; A.A. Gogolin, E.I. Rashba, Zh. eksper. teor. Fiz. Pisma, 17, 478 (1973)

[2] A. Bivas, V.D. Phach, B. Hönerlage, J.B. Grun, Phys. Stat. Sol. (b), 84, 235 (1977)

[3] E. Ostertag, J.B. Grun, Phys. Stat. Sol. (b), 82, 335 (1977)

Broadly Tunable, Repetitive Picosecond Laser System and Its Application to Spectroscopy

Y. Tanaka, Y. Masumoto, S. Tanaka[1], and S. Shionoya

The Institute for Solid State Physics, The University of Tokyo
Roppongi, Minato-ku, Tokyo 106, Japan

A broadly tunable, repetitive picosecond laser system utilizing the optical parametric effect has been developed for the purpose of studying picosecond spectroscopy. It is possible to produce intense, monochromatic light pulses having 30-40ps duration and a tunable range from 240nm to 3.6μm with a repetition rate as high as 20pps. In a previous paper [1], the construction of an optical parametric oscillator was reported, in which a temperature-controlled $LiNbO_3$ crystal is pumped by the second harmonic of a mode-locked glass:Nd laser with one-shot operation. The same parametric oscillator is used in this paper changing the pumping source to a mode-locked YAG:Nd laser with repetitive operation. To demonstrate the usefulness of this tunable picosecond laser system, results of experiments on the two-photon resonant Raman scattering and luminescence processes associated with excitonic molecules in CuCl are also presented in this paper.

1. Laser System

Figure 1 shows a block diagram of the system. The 1.064μm output pulses of a repetitively mode-locked YAG laser oscillator are amplified by three YAG amplifiers, and pulses with ~1GW peak power and 18ps duration are obtained. The laser oscillator and amplifiers are decoupled by inserting saturable dye cells to increase the stability of the output power. The insertion of the cells was found effective to reduce the duration of output pulses, because the duration was 30ps without the cells.

The $LiNbO_3$ crystal ($4 \times 4 \times 50mm^3$) in the parametric oscillator, whose cavity length is adjusted to be equal to that of the YAG oscillator, is pumped by second harmonic 532nm pulses with ~100MW power. The crystal temperature is changed in a 50-450°C range with a 0.03°C stability. Thus parametric signal pulses of 625-900nm with ~3MW power and 30-40ps duration are obtained, which are highly directional ($\Delta\theta < 1$mrad). The idler pulses are expected to be extended to 3.6μm. The spectral width of signal pulses is less than 0.2nm below 760nm, while 0.5nm at 900nm. This width can be reduced to one third by inserting an etalon. Second harmonics of signal pulses obtained by using an RDP crystal have ~300kW power.

---

[1]Permanent address : Department of Electronics, Tottori University, Tottori 680, Japan.

The tunable range is widely extended by means of the sum or difference frequency mixing of signal pulses (ω_s) or their second harmonics ($2\omega_s$) with the 1.064μm fundamental pulses (ω_ℓ) by the use of a KDP crystal; namely, $\omega_s + \omega_\ell$: 394-488nm, 300kW power, $2\omega_s + \omega_\ell$: 241-316nm, 30kW, and $2\omega_s - \omega_\ell$: 443-780nm, 30kW. In this way the whole spectral region from 240nm to 3.6μm can be covered continuously without any missing regions. The spectral width less than 0.1nm can be obtained in the region below 760nm.

2. System of Spectral Measurements

To measure spectra of picosecond light pulses emitted or scattered by materials with sufficient sensitivity and accuracy, a microcomputer-controlled multichannel analyzer system using a television camera has been constructed. The block diagram is shown in Fig.1. Spectra are intensified by the image intensifier and detected by the television camera equipped with a high sensitivity image orthicon. Each scanning line of the camera is made to correspond to each wavelength component, so that the camera operates as a 500-channel analyzer. Repetitive output signals are, after digitalized, stored and summed by the microcomputer. The laser, the television camera, and the microcomputer are operated synchronously. It was shown that in data stored by hundred shots of the laser the signal-to-noise ratio is improved ten times.

3. Raman Scattering and Luminescence Associated with Excitonic Molecules in CuCl

Excitonic molecules in materials such as CuCl are created directly by resonant two-photon absorption. In CuCl this absorption line exists at 389.0nm. Excitonic molecules thus created emit luminescence leaving single excitons. When the wavelength of incident light is changed in the vicinity of this absorption line, two-photon resonant Raman scattering is observed [2]. In this sense the excitonic molecule in CuCl is an ideal system to investigate nature of secondary radiation including luminescence and Raman scattering by means of picosecond spectroscopy. We have performed time-resolved measurements for this luminescence and Raman scattering. This work is an extension of the early work by Ojima et at.[3] in our laboratory.

Figure 2 shows time-integrated emission spectra of CuCl oberved in the configuration of backscattering changing the wavelength of incident laser pulses around 389nm. The star indicates the position of the two-photon absorption line due to excitonic molecules, i.e., the position of just-resonance. It is seen that the shifts of the R_T and R_L Raman lines leaving transverse and longitudinal single excitons, respectively, are twice as much as the shift of the laser light. Further the spectral width of Raman lines is twice as broad as that of the laser light. These facts are reasonable, since one is looking at two-photon Raman scattering. In the case that the incident light wavelength is slightly off the resonance position, both the Raman lines and the M_T and M_L luminescence lines showing no shifts are observed. This may be due to the spectral tail part of the laser light which coincides with the two-photon resonance. In the just-resonance case, it is difficult to discriminate whether the observed secondary radiation lines, denoted by R_T^* and R_L^*, are Raman lines or luminescence lines.

Figure 3 shows time-resolved spectra in the case of just-resonance and also the time dependence curve of the R_T^* line. These spectra were measured by using a CS_2 optical Kerr shutter (see Fig.1). It is seen that the R_T^* line rises slightly delayed from the laser pulse and decays much more slowly than the laser pulse lasting until ~200ps. It is noticed in time-resolved spectra that the spectral width becomes narrower when the delayed time is more than 40ps. This indicates that the delayed component is dominated by luminescence whose line width is narrower than that of Raman scattering as seen in Fig.2. Time-resolved spectra were also measured in the off-resonance case where the luminescence and Raman lines are observed separately. It was found that the R_T Raman line grows and decays as fast as the laser pulse. This indicates that the Raman process is quite fast and its time response cannot be resolved in the present time resolution of ~20ps. The M_T luminescence line, on the other hand, rises a little delayed from the laser pulse, and decays much more slowly than the laser pulse.

The above-mentioned facts indicate that the R_T^* line consists of the Raman and luminescence components. Using knowledge of time dependence of the R_T Raman and M_T luminescence lines observed separately, it was attempted to decompose the observed time dependence of the R_T^* line into the Raman and luminescence components as shown in Fig.3. It seems that the two compents are nearly comparable in intensity in the present case. However, it is pointed out that the relative intensity of the two components depends very much on both the intensity of incident laser pulse and its spectral width.

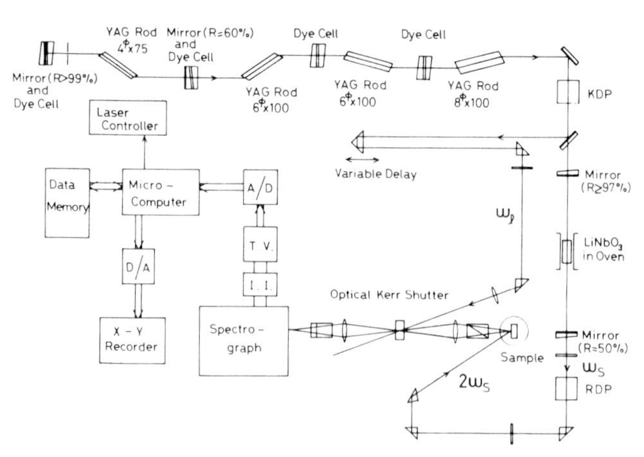

Fig.1 Blockdiagram of a tunable picosecond laser system and a system for measurements of time-resolved spectra

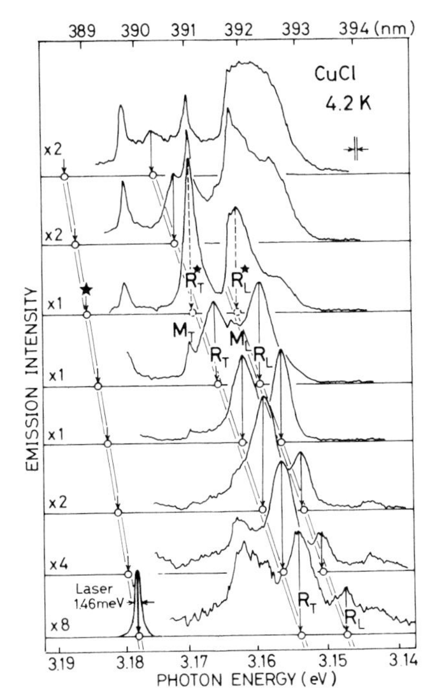

Fig.2 Time-integrated emission spectra of a CuCl crystal at 4.2K observed with changing the wavelength of incident laser pulses

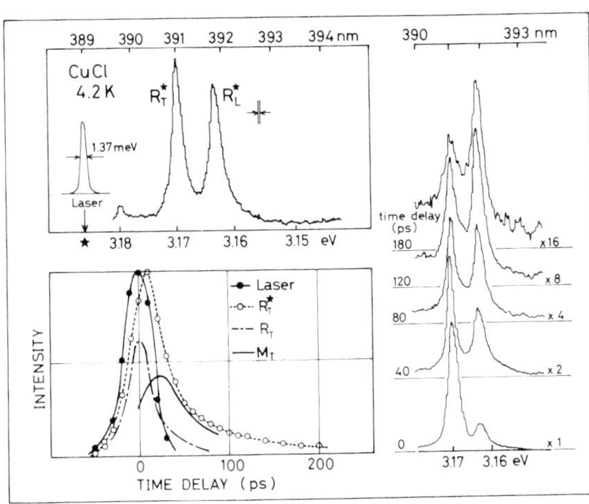

Fig.3 Time-integrated (upper-left) and time-resolved (right) spectra in the just resonance case and time dependence curve of the R_T^* line (lower-left)

[1] T. Kushida, Y. Tanaka, M. Ojima : Japan. J. Appl. Phys. 16, 2227(1977)
[2] N. Nagasawa, T. Mita, M. Ueta : J. Phys. Soc. Japan 41, 929 (1976)
[3] M. Ojima, T. Kushida, Y. Tanaka, S. Shionoya : Solid State Commun. 20, 847(1976); ibid. 24, 841(1977); J. Phys. Soc. Japan 44, 1294(1978); ibid. to be published

Dynamic Emission Spectroscopy Under Resonant Two Photon Absorption in CuCl and CuBr

Y. Aoyagi, Y. Segawa, and S. Namba

The Institute of Physical and Chemical Research
Wako-shi, Saitama, 351, Japan

O. Nakagawa

Faculty of Engineering Science, Osaka University
Toyonaka, Osaka, 560, Japan

K. Azuma

Faculty of Engineering, Osaka University, Suita, Osaka, 565, Japan

1. Introduction Recently, a question whether the optical process of the resonant two photon absorption of an excitonic molecule and its subsequent emission leaving a single exciton as a final state can be considered as two separate processes of absorption and emission or as an inseparable Raman process has been discussed theoretically [1], and experimentally [2]. The selection rule of the Raman process is the same as that of the luminescence under a resonant two photon absorption in CuCl and CuBr and we can not distinguish between these two processes by the selection rule. For the clear distinction between them, the time resolved spectroscopy under the resonant two photon absorption using a high intensity tunable picosecond pulse source is fruitful. We developed a high power tunable picosecond pulse dye laser source in UV and near UV regions, and we observed emission spectra and time resolved spectra under the resonant two photon exitation and discussed the Raman and luminescence processes in CuCl and CuBr single crystals.

2. Experimental The experimental set up is shown in Fig.1. The picosecond system contains a YAG Nd(3+) oscillator mode-locked simultaneously by a saturable dye and an acousto-optic mode-locker to stabilize mode-locking, two amplifires, a soft appature, and a dye cell to elliminate DC component of the output light. The dye laser was excited by the 3rd harmonics of the YAG laser train. Two etalons, E, were used to tune the oscillation wavelength and two lenses, L, were inserted at confocal points in the cavity to decrease the threshold power. The round trip time of light through the cavity was precisely adjusted to the period of the excitation pulse train. To change the spectral width of the excitation various transmission modes of the etalon were selected by rotating the etalon. Typical output performances of the tunable source were the peak power of 1 MW, the tunable range from 375 to 435 nm, the pulse width of 25 ps and the spectral width from 0.05 to 0.5 nm. The sample used was a flake type grown from a vaper phase in hydrogen atmos-

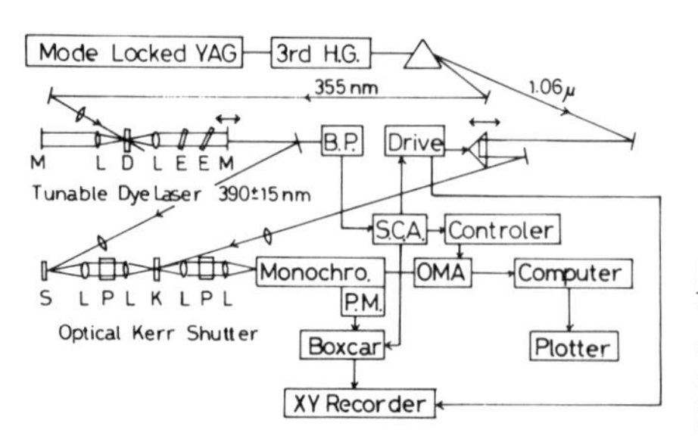

Fig.1 Experimental arrangement for the measurement of time resolved emission spectra. D; dye cell, P; polaryzer, K; CS_2 Kerr cell, B.P.; biplaner phototube.

phere after 100 times zone refining. The sample, S, at the temperature of 1.6°K was excited at the front surface and the emission from the sample was observed at the angle perpendicular to the surfase. The emission from the sample was observed through a Kerr shutter and analyzed by a 50 cm monochromator. Output spectrum was recorded by means of an optical multichannel analyzer (OMA). To improve the s/n ratio, the spectrum was accumulated by 300 to 1000 times. The dynamic behavior of the emission at various wavelengths was observed by a photomultiplier and a boxcar. The gate time of 1.06 μm laser light at the Kerr shutter was continuously scanned by moving a prism with a pulse motor. The output power of the dye laser was monitored by a single channel analyzer (SCA) and only when the laser power was in the window region of the SCA, the boxcar and the pulse motor of the prism were triggerd.

3. Results Typical emission spectra of CuCl and CuBr under the two photon excitation at various wavelengths are shown in Fig.2. The band width of the exciting laser was 0.8 meV. As shown in this figure we found two Raman lines in addition to two luminescence lines M_T and M_L, for CuCl and three Raman lines in addition to three luminescence compornents M_T, M_L and M_F for CuBr, where M_T, M_L and M_F lines show the luminescences due to the radiative recombination of the excitonic molecule with a transverse, a longitudinal and a triplet free exciton left in the final state, respectively. Peaks of the Raman lines in CuCl and CuBr shift by $2\Delta\nu$, when the wavelength of the exciting laser shifts by $\Delta\nu$. By approaching the wavelength of the exciting light to the resonant two photon absorption energy, intensities of these lines increase rapidly. These results apparently indicate that these sharp lines are resonant two photon Raman lines. When the wavelength of the exciting laser is at the resonant condition, the Raman and luminescence processes coexist and strong emission peaks were observed. Intensities of these lines become more than 20 times of those of the off resonant luminescence lines. These results are

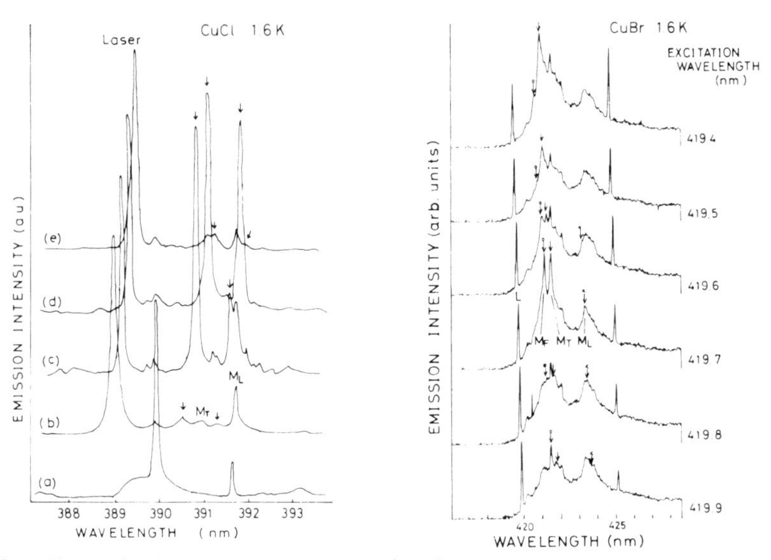

Fig.2 The emission spectra of CuCl (b-e) and CuBr under the two photon excitation at various excitation wavelengths. Arrows indicate two photon Raman lines. The spectrum (a) is the emission spectrum of CuCl under the Hg lamp excitation

similar to the behavior of the Raman and luminescence spectra under the excitation of a tunable dye laser of which the pulse width is several ns [2]. Since the decay time of the Raman process is very fast compared to that of the luminescence process (a few handred ps), one can distinguish the Raman process from the luminescence under the resonant two photon excitation by the time resolved spectroscopy.
Fig.3 and Fig.4 show dynamic behaviors of the M_T and M_L lines of CuCl under the resonant two photon excitation and dynamic emission spectra of CuBr, respectively. In the case of CuCl, the M_T and M_L lines have two different decay components. One is very fast and the other is slow. The decay of the fast component is the same as that of the exciting laser and this component is ascribed to be the Raman process. The slow component is the radiative decay process of the excitonic molecule. The assignment was confirmed by observing the decay of the Raman line separatly under the slightly off resonant excitation. In the case of CuBr three sharp lines M_T, M_L and M_F increase simultaneously with the exciting laser pulse and have the same decay time of that of the exciting laser light. Broad three emission bands remained during more than 300 ps.

4. Discussion In order to interpret the time behavior of the luminescence process in CuCl, we used a simple kinetic equation as follows;

$$dn/dt=(K/2)J(t)^2-n/\tau, \quad (1)$$

where n is the concentration of excitonic molecules, K is the two photon absorption coefficient, τ is the decay time constant of the excitonic molecule, J(t) is the exciting laser pulse and is given by $J(t)=J(0)\exp(-t^2/2\sigma^2)$, and σ=10 ps. In our analysis the recreation process of the excitonic molecule from excitons which appear from the radiative decay of the excitonic molecule is neglected. The calculated results are shown in dashed lines in Fig.3

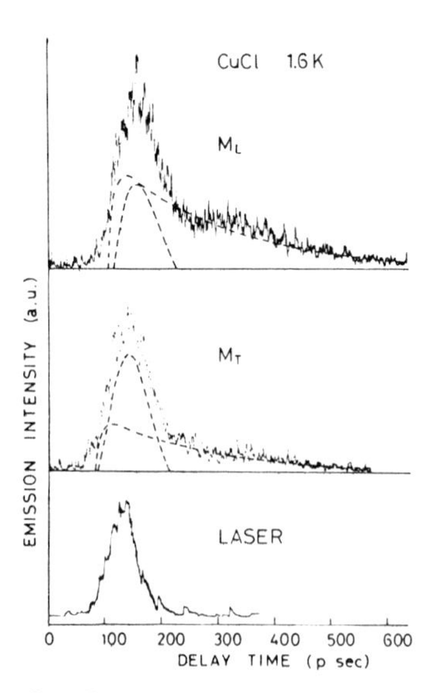

Fig.3 Decay profiles of the M_L and M_T lines under the resonant two photon excitation

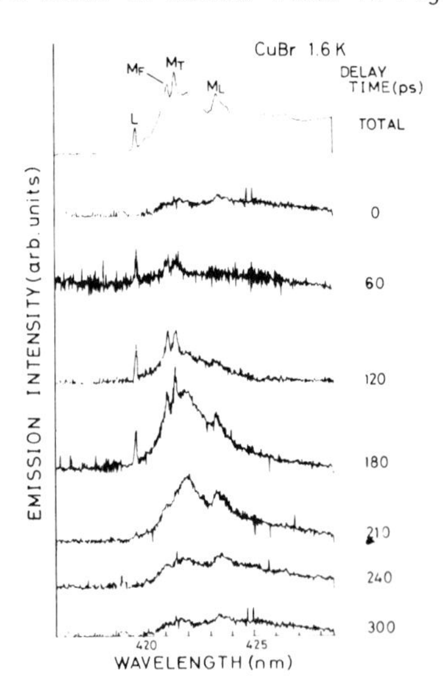

Fig.4 Dynamic emission spectra of CuBr under the resonant two photon excitation

Table 1 The ratio of the luminescence process to Raman process under the resonant two photon excitation of the excitonic molecule in CuCl

Band width	0.5 meV	0.8 meV	1.0 meV
M_T line	1.3	1.1	1.2
M_L line	1.0	3.6	5.0

with the value of τ=200 ps. Subtracting the luminescence parts from the whole spectra, the Raman parts are obtained for the transverse and longitudinal excitons, respectively. The F.W.H.M. of the Raman parts for M_T and M_L lines are 45 ps and they are same with that of the exciting laser light. The transition rates of the luminescence and Raman processes under the resonant two photon excitation are obtained from Fig.3 and are listed in Table 1 for different band widths of the exciting laser. Three values for the transverse exciton are constant for various excitation band widths in the experimental accuracy. On the other hand, the values for the longitudinal exciton apparently increase with the band width of the laser. The transition rates of the Raman and luminescence processes were evaluated using a damping theory [1]. Applying the theory of Ref.[1], moduration times of the excitons are roughly estimated. They are $1x10^9$/sec and $5x10^9$/sec for the transverse and longitudinal excitons, respectively under the excitation of $2x10^{11}$/sec (0.8 meV) band width laser light. In this estimation a value of $5x10^9$/sec was used for the radiative life time of the excitonic molecule. From the two photon absorption spectrum of the excitonic molecule [2], the moduration time of the excitonic molecule is estimated to be $8x10^{10}$/sec. These values suggest that the modulation times of the exciton and the excitonic molecule are in the same order.
In the time resolved spectra in CuBr, the broad bands with a long decay time are assigned to be emission spectra from the molecule created by the exciton which is generated by the exciton band-tail absorption [3]. In CuCl this broad band does not observed because the one photon absorption coefficient at the wavelength of the resonant two photon absorption is rather small in comparison with CuBr for the large binding energy of the excitonic molecule. The lack of the slow components in the sharp lines in CuBr suggests that the sharp lines under the resonant two photon excitation are almost due to the Raman scattering process and the luminescence process is small.

5. Conclusion

We separated the Raman and luminescence processes under the resonant two photon excitation of the excitonic molecule in CuCl and CuBr. In CuCl these two processes were found to coexist and simultaneously enhace their intensitics under the resonant condition. The transition rate of the Raman process for the longitudinal exciton rapidly decreased with increasing the band width of the exciting laser, but that for the transverse exciton was independent of the band width. In CuBr the Raman process is dominant under the resonant two photon excitation and the enhancement of the luminescence is small in our experimental condition.

References

1. E. Hanamura and H. Hang: Phys. Reports 33, 209 (1977)
2. N. Nagasawa, T. Mita and M. Ueta: J. Phys. Soc. Japan 41, 929 (1976)
3. Y. Nasu, S. Koizumi, N. Nagasawa and M. Ueta: J. Phys. Soc. Japan 41, 715 (1976)

VIII. High-Power Lasers and Plasmas

Saturation Effects in Fifth Harmonic Generation of Coherent XUV Radiation

C.Y. She

Department of Physics, Colorado State University, Ft. Collins, CO 80523, USA

J. Reintjes

Laser Physics Branch, Naval Research Laboratory, Washington, DC 20375, USA

1. Introduction

Optical frequency conversion in atomic and molecular vapors through third, fifth and seventh order nonlinearities has proven to be an important source of coherent radiation in the vacuum ultraviolet (VUV) and extreme ultraviolet (XUV). If only the frequency conversion process is considered, the theoretical conversion efficiency for many of these processes is very high, often in the range of pump depletion under conditions typically encountered in experiments. Measured conversion levels, on the other hand, often fall significantly below the theoretical ones, and the variation of conversion efficiency with vapor pressure or pump power may deviate from the simple forms predicted by theory. Such behavior indicates that additional nonlinear effects are playing a significant role in limiting the conversion efficiency that can be realized in the laboratory.

Several processes, such as dielectric breakdown, Stark shifts, depletion of ground state atoms or saturation of excited state populations due to multiphoton absorption or ionization, and changes in phase matching conditions due to nonlinear refractive indices have been suggested as being potentially important in limiting conversion efficiency [1-4]. Experimental results have been given in the literature indicating the importance of several of these processes in limiting third harmonic conversion. For example, measurements have been reported which show effects of the nonlinear refractive index on the phase matching conditions in both metal vapors and rare gases [2,3]. Evidence has also been reported of population saturation [5], and Stark shifts and Stark broadening [6], although some experiments may be open to more than one interpretation [4]. Proper identification of the effective limiting mechanisms is very important for making reliable estimates of the ultimate conversion efficiency that can be obtained in a given vapor system. Such knowledge can also be useful in identifying a proper choice of experimental parameters for obtaining the optimum conversion.

In this paper we discuss the nature of limiting processes which have been observed in experiments involving fifth harmonic generation of coherent radiation at 53.2 nm in helium [7]. Experimental measurements are presented in which the harmonic signals deviate from theoretical predictions with regard to their dependence on gas pressure, pump power and conversion efficiency. An analysis of these measurements indicates that two types of competing processes are present, involving both saturation of the nonlinear susceptibility for fifth harmonic generation, and changes in phase matching conditions due to a third-order nonlinear refractive index. Theoretical calculations of fifth harmonic conversion which include both of these effects are presented and the results are in excellent qualitative agreement with our measurements.

2. Fifth Harmonic Generation - Theory

Several treatments of harmonic conversion in vapors have been given in the literature [8]. In this section we summarize the results for fifth harmonic generation in a single component atomic vapor for a tight focused geometry. If we neglect pump depletion the fifth harmonic power, P_5, generated by a focused beam is given by

$$P_5 = \frac{2^4\pi^6}{5n_1n_5c^4} \left|\frac{N}{n_1-n_5} \chi^{(5)}\right|^2 \left(\frac{4P_1}{b\lambda_1}\right)^4 P_1 \left|b\Delta k F(b\Delta k)\right|^2 \tag{1}$$

where N is the vapor density, n_i the refractive index at the ith harmonic, $\chi^{(5)}$ is the susceptibility for fifth harmonic generation, P_1 is the pump power, λ_1 is the pump wavelength, $b = 2\pi\omega_0^2/\lambda_1$ is the confocal parameter of the focused fundamental wave, ω_0 is the radius of the pump wave, $\Delta k = k_5 - 5k_1 = 10\pi(n_5 - n_1)/\lambda_1$ is the wave vector dispersion between the fundamental and harmonic. F is an integral which describes the effects of phase mismatch and focusing and is given by

$$F(b\Delta k) = \int_{-\xi'}^{\xi} (1 + i\xi'')^{-4} \exp\left[-ib\Delta k(\xi''-\xi)/2\right] d\xi'' \tag{2}$$

where $\xi = 2(z-f)/b$, f is the focal length of the focusing lens and the interaction region extends from $-\xi'$ to ξ. When written in this form, the second term in (1) is independent of density for a single component medium since $(n_1-n_5)/N$ is the dispersion per atom, and all of the pressure dependence is contained in the last term. The strength of the fifth harmonic varies as the fifth power of the fundamental and increases with tighter focusing (smaller b). In an infinite medium fifth harmonic conversion is possible only for negative values of Δk, just as in the case of third harmonic generation. In a finite or semi-infinite medium (a geometry in which the interaction region is present only on one side of the beam waist) conversion is possible for both positive and negative values of Δk.

The theoretical variation of harmonic power with the parameter $b\Delta k$ is shown in Fig.1 for an infinite medium (curve b) and for a semi-infinite medium (curve a). For each geometry the harmonic conversion has a peak for a value of $b\Delta k$ ($b\Delta k_{opt}$) at which the dispersion in the medium is optimally adjusted to offset the phase slippage in the Gaussian focus [9]. At the peak of the curves, more conversion is obtained in the infinite medium, while at values of $b\Delta k$, that are much less or much greater than the optimum value, conversion is greater in the semi-infinite medium. For values of $\Delta k \ll \Delta k_{opt}$, the harmonic signal varies as $(b\Delta k)^2$ in the semi-infinite medium, and as $(b\Delta k)^8$ in the infinite medium. The stronger variation in the infinite medium is due to the near cancellation of the harmonic power generated in the front and back halves of the focus.

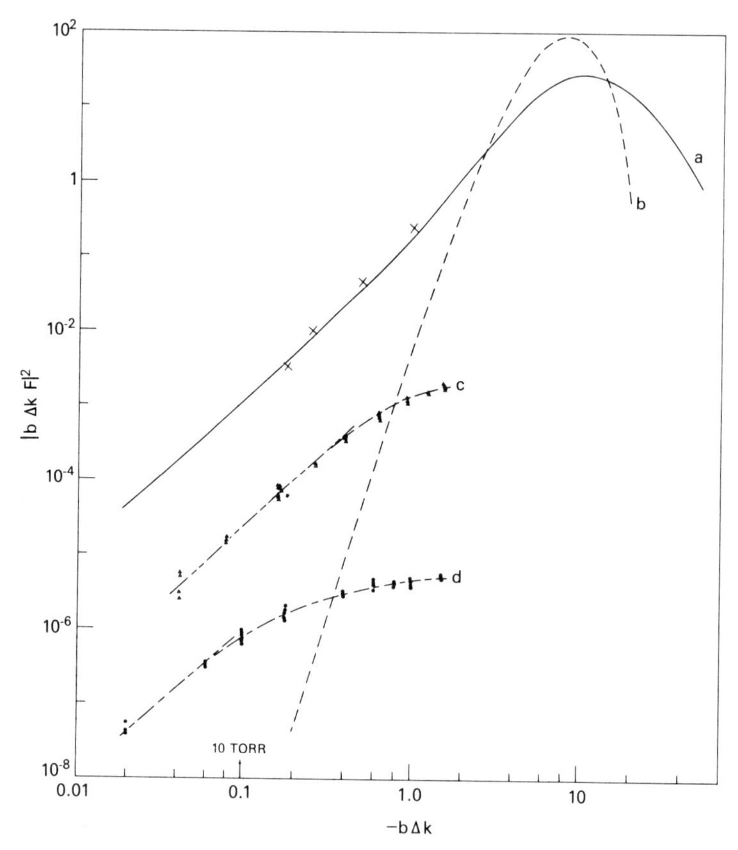

Fig.1 Comparison of measured fifth harmonic conversion in helium with theory. Curves (a) and (b) were calculated for a semi-infinite medium and an infinite medium respectively. Experimental data was obtained at three laser powers: 12 MW (crosses), 75 MW (triangles) and 300 MW (solid circles).

3. Fifth Harmonic Generation - Experiment

Experimental measurements of fifth harmonic conversion of pulses at 266 nm are compared with theory in Fig.1. The experiments were performed by focusing the pump pulses onto an aperture at the end of a cell containing helium. The gas was differentially pumped behind the aperture, and the generated radiation was analyzed with a VUV monochromator. Typical confocal parameters were of order of 0.5 mm for a 5 cm focal length lens, so the tight focusing approximation was well satisfied by the experiment. The fifth harmonic process is negatively dispersive in helium at our pump wavelength, so there is expected to be an optimum pressure for the harmonic generation, which for our experiments is calculated to be of the order of 1000 torr.

Measurements of the fifth harmonic are shown in Fig.1 as a function of gas pressure for three different pump powers. The horizontal scale is fixed by the calculated dispersion and the measured confocal parameter and is proportional to pressure. The vertical scale was fixed by the measured harmonic power and so there are no adjustable parameters in the comparison with theory. The estimated absolute accuracy of the measurements is about a factor 3, while the statistical accuracy is indicated by the spread in data points. The theoretical dependence of the harmonic signal on the fifth power of the pump strength has been divided out so that measurements taken at different pump levels can be compared with the same theoretical curve.

At low pump powers (12 MW) the experimental data are in excellent agreement with the calculated curve for the semi-infinite medium, both with regard to absolute agreement and to pressure scaling. The agreement with this curve indicates that the aperture at the end of the cell is acting as a jet, causing a shock front to develop in the gas which acts as a window near the beam waist. At higher values of pump power two types of deviation from theory are observed. At low pressures the absolute agreement with theory becomes worse as the power is increased, but the pressure dependence remains unchanged. Such behavior indicates a deviation from the ideal variation of harmonic signal with the fifth power of the pump strength. At higher pressures, the dependence of harmonic signal density deviates from theory as well, with the point of deviation moving to lower pressures as the pump power increases.

The behavior shown in Fig.1 indicates that one or more competing effects are present in our experiment. The processes mentioned earlier can affect harmonic conversion either by changing the value of the fifth order susceptibility or by changing the phase matching conditions through a nonlinear refractive index. A third order nonlinear refractive index (quadratic effect) can account for the observed deviation from the pressure dependence predicted by theory. To see this qualitatively we can calculate the total accumulated on axis nonlinear phase shift between the harmonic field and its driving polarization due to an intensity dependent refractive index

$$\Delta\phi^{NL} = N\left[k_{5\omega}n_2(5\omega,\omega) - 5k_{\omega}n_2(\omega,\omega)\right]\int_{-\infty}^{0} \langle E^2(z)\rangle dz \qquad (3)$$

where $n_2(q\omega,\omega)$ is the coefficient of nonlinear index per atom at frequency q_ω due to the intensity of the fundamental beam at ω, and $E(z)$ is the amplitude of the fundamental field. For a Gaussian beam of the form

$$\langle E^2\rangle = \langle E_0^{\,2}\rangle/(1 + (2z/b)^2) \qquad (4)$$

(3) has the form

$$\Delta\phi^{NL} = \frac{4\pi^2 k_{5\omega}(n_2(5\omega,\omega) - n_2(\omega,\omega))}{c\lambda_1} PN \qquad (5)$$

Thus the total on axis phase shift depends on the product of gas pressure and pump power and its effects would be visible at lower pressures for higher pump powers.

The nonlinear refractive index, however, cannot explain the deviation of the measurements from theory at low pressures when the proper density scaling is maintained. In order to explain this type of behavior, a saturation of the fifth order susceptibility is required. If such a saturation resulted from processes which involved interactions with single atoms, the reduction in susceptibility would be independent of pressure but would affect the dependence of the harmonic signal on pump strength. Saturation of the fifth order susceptibility would be expected to depend on the intensity of the pump beam in the focus while effects due to the quadratic Kerr effect would depend only on total pump power, as shown above, and not on how tightly the beam is focused.

4. Competing Processes

In order to investigate the effects of the competing processes, we have calculated the harmonic signal generated in the presence of both a nonlinear index and saturation of the fifth order susceptibility taking into account the variation of the pump beam in the transverse direction. The harmonic field generated by a focused beam can be expressed as

$$E_5(r,z) = \frac{-i\pi\omega_5^2}{2^2 k_5 c^2} \frac{N\,\chi^{(5)} E_0^5}{(1+i\xi)}\, b\, e^{\frac{-5k_1 r^2}{b(1+i\xi)}}\, e^{i5k_1 z} \int_{-\xi'}^{\xi} \frac{e^{-i\, b\Delta k\, (\xi''-\xi)/2}}{(1+i\xi'')^4}\, d\xi'' \qquad (6)$$

First we have investigated the effect of the quadratic Kerr effect alone. If we neglect the self-focusing effects of the nonlinear index and assume that its major effect is to change the phase matching between the harmonic and the fundamental, the nonlinear index can be accounted for in (6) by replacing the term $b\Delta k\xi/2$ in the integrand with a more general phase term

$$\phi(\xi) = \frac{b\Delta k_0 \xi}{2} + \frac{5\pi b}{\lambda_1} \underline{N}\, (n_2(5\omega) - n_2(\omega)) \int_{-\infty}^{\xi} \langle E_1^{\,2}(r,\xi')\rangle d\xi' \qquad (7)$$

where Δk_0 is the linear dispersion. The harmonic power can then be obtained by integrating the squared magnitude of (6) over the transverse coordinates.

Finally, integration over the temporal profile of the pulse can be performed to obtain the total fifth harmonic energy which can be directly compared with our experiments. The results of such a calculation showing harmonic conversion as a function of gas pressure are illustrated in Fig.2. Calculations were done for two different pump powers, chosen to correspond to the two highest values at which we obtained data and for three values of $\Delta n_2 = n_2(5\omega,\omega) - n_2(\omega,\omega)$. The values of Δn_2 used in the calculation are in approximate agreement with those calculated for helium and are of the same sign. Again the theoretical fifth power dependence on pump strength was divided out so that curves differing only in total pump power can be compared directly.

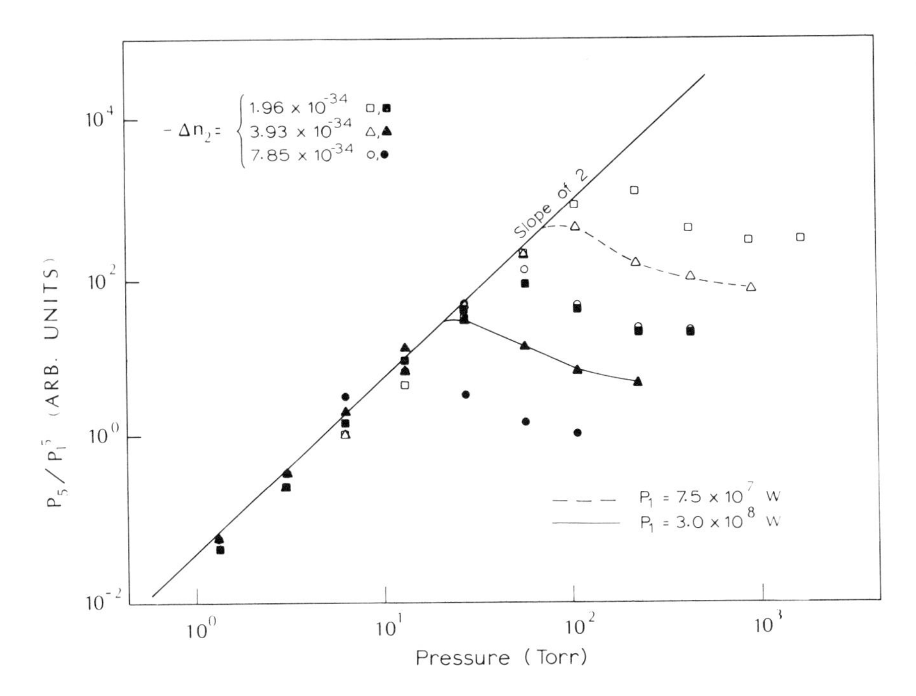

Fig.2 Calculated variation of fifth harmonic signal with helium pressure for two different laser powers and three values of nonlinear coefficient Δn_2.

The effect of the nonlinear index on the harmonic generation is illustrated by the calculations done for a value of $\Delta n_2 = -4 \times 10^{-34}$. At low pressures the nonlinear phase shift is negligible and the harmonic signal follows the same square-law dependence on gas pressure as in the absence of the nonlinear index. However, as the gas pressure is increased, the nonlinear phase shift becomes significant and the harmonic signal reaches a maximum and then decreases. The maximum occurs when the total effective phase mismatch, which now depends on pump power as well as density, is optimum to offset the phase slippage in the Gaussian focus. As the pump power is increased the effects of the nonlinear index becomes greater, and the deviation from the low-field behavior occurs at lower pressures, in agreement with the trends that are evident in our experiments. The pressure at which the optimum phase mismatch occurs is lower in the presence of the nonlinear index than in its absence because the sense of the nonlinear dispersion is such as to add to the linear dispersion. Again this behavior is in agreement with our experimental observations.

At low pressures, where the nonlinear phase shift is negligible, the calculated curves for different powers coincide. Thus this effect cannot account for our experimental observations in this region and saturation of the fifth order susceptibility must also be considered. To investigate this effect we can replace $\chi^{(5)}$ in (6) by a term of the form

$$\chi^{(5)} = \chi_0^{(5)} \; (1 + I(r,\xi)/I_s)^{-1} \tag{8}$$

where I_s is a saturation flux. Typical calculations of harmonic power vs. pressure which include the effects of both the saturation of $\chi^{(5)}$ and the nonlinear index are shown in Fig.3. At low pressures, the harmonic signal is seen to deviate from the fifth power dependence on pump strength but the square law dependence on density is maintained in agreement with our measurements.

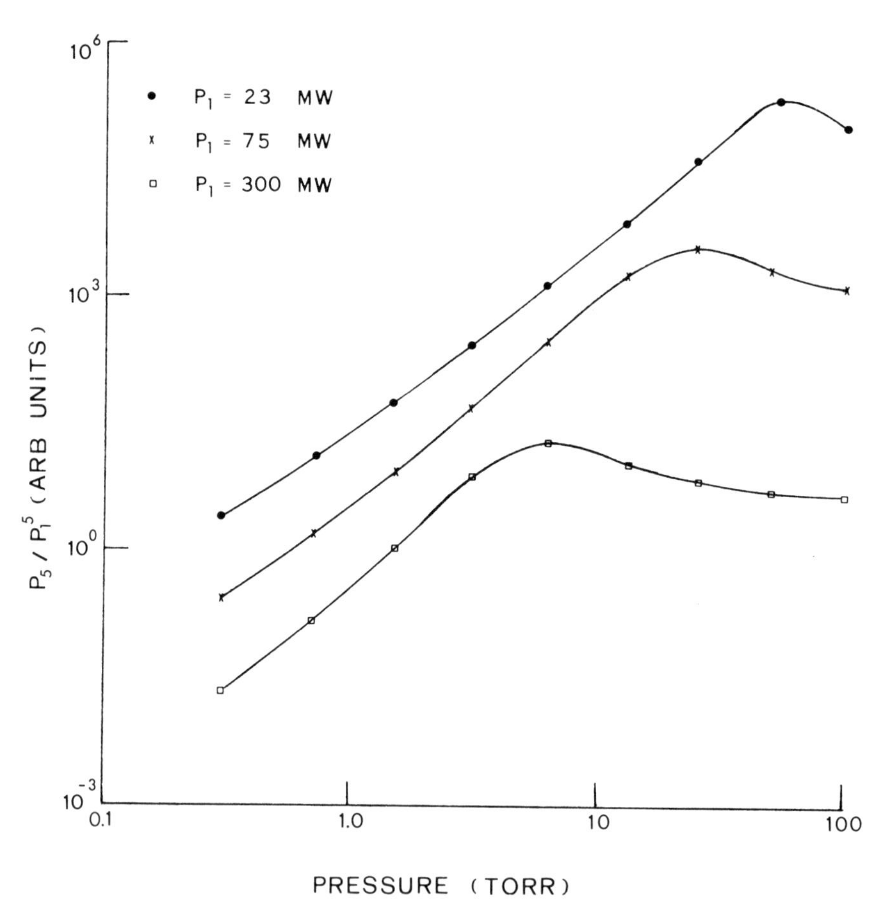

Fig.3 Calculated pressure variation of fifth harmonic, including effects of nonlinear index ($\Delta n_2 = -4 \times 10^{-34}$ esu) and saturation of the fifth order susceptibility of the form $\chi_0^{(5)}/(1 + I/I_s)$.

5. Conclusions

We have presented experimental observation of saturation of fifth harmonic conversion in helium. Our experiment implies the presence of two kinds of limiting mechanisms, one involving the saturation of the fifth harmonic susceptibility and the other affecting the phase matching conditions through a third order nonlinear refractive index. Calculations of fifth harmonic power which include both of these effects qualitatively account for all aspects of the saturation behavior which we have observed.

References

1. S. E. Harris, Phys. Rev. Lett. 31, 341 (1973).
2. H. Puell, K. Spanner, W. Falkenstein, W. Kaiser and C. R. Vidal, Phys. Rev. A 14, 2240 (1976).
3. L. J. Zych and J. F. Young, IEEE J. Quant. Elect. QE-14, 147 (1978).
4. A. T. George, P. Lambropoulos and J. H. Marburger, Phys. Rev. A 15, 300 (1977).
5. C. C. Wang and L. I. Davis, Jr., Phys. Rev. Lett. 35, 650 (1975).
6. J. F. Ward and A. V. Smith, Phys. Rev. Lett. 35, 653 (1975).
7. J. Reintjes, C. Y. She and R. C. Eckardt, IEEE J. Quant. Elect. QE-14 August, 1978.
8. J. F. Ward and G. H. C. New, Phys. Rev. 185, 57 (1969).
9. G. C. Bjorklund, IEEE J. Quant. Elect. QE-11, 287 (1975).

Picosecond Diagnosis of CO_2 Laser Produced Plasmas

M.C. Richardson, R. Fedosejevs, P. Jaanimagi, and G.D. Enright
National Research Council of Canada, Division of Physics
Ottawa, K1A 0R6, Canada

1. Introduction

The investigation of the interaction of intense 10 μm laser radiation with solid targets has in the past few years acquired increased significance due to the potential application of CO_2 lasers towards the achievement of laser fusion. In parallel to this effort, similar investigations are underway utilising 1 μm radiation from high power Nd:glass lasers. Although in many respects the stringent experimental demands on plasma and laser diagnosis are similar in both investigations, the difference in operating wavelength range has forced the development of alternative approaches toward the achievement of these aims. This has become particularly evident in those areas involving ultrafast diagnosis of the laser pulse characteristics and in the characterisation of the plasma by fast optical analysis. Whereas in the near IR and visible region the relative abundance of available sensitive ultrafast detectors and techniques has permitted accurate laser beam characterisation, comparable diagnosis in the CO_2 laser range has presented many challenging problems, which apparently will only be overcome by the adoption of novel techniques. In this paper, we wish to describe some of the approaches we have adopted to provide picosecond optical diagnostics in an investigation of the interaction of intense ($>10^{14}$W cm^{-2}) short ($\sim$1 ns) CO_2 laser pulses with solid targets. These investigations are centered around the 0.1 TW two-beam COCO-II CO_2 laser system presently operating at NRC-Canada.[1]

2. Ultrafast Diagnosis of Infrared Radiation

For these investigations with nominal 1 ns pulses, the demand for sensitive IR ($\sim$10 μm) detection with at least 10 ps resolution is imperative. It is necessary to accurately determine the intensity profile of the irradiating pulse. In addition many interesting features of the interaction itself may be obtained from analysis of the temporal and spectral nature of backscattered and side scattered infrared emission. The lack of suitable photocathodes has precluded streak camera development in this range, forcing reliance on other techniques. Although pyroelectric detectors with a temporal resolution of <100 ps exist,[2] their practical use is still limited by the bandwidth of available oscilloscope devices ($\sim$1 GHz).* Alternative approaches, employing nonlinear optical techniques, achieve greater resolution at the

*Recently some improvement in this area has been made with the development at L.A.S.L. of an oscilloscope with a bandwidth of $\sim$5 GHz.[3]

expense of flexibility and complexity. Optical Kerr effect photography where the source IR radiation is used to gate the output of a visible c.w. laser, in conjunction with an ultrafast streak camera can provide a theoretical temporal resolution of ~2 ps, but suffers from its inherent non-linear transmission function.[4]

In our investigations, linear picosecond infrared diagnostics are being implemented through the use of frequency upconversion to a wavelength for which ultrafast streak camera techniques can be employed. Infrared upconversion in proustite (Ag_3AsS_3)[5] is a well known technique, has high quantum efficiency, a large practical range of linearity,[6] and has already been used to analyse subnanosecond CO_2 laser pulses.[7]

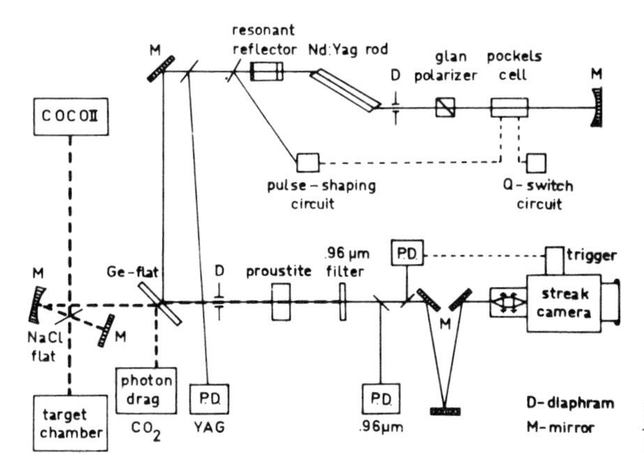

Fig.1 Experimental arrangement for IR upconversion

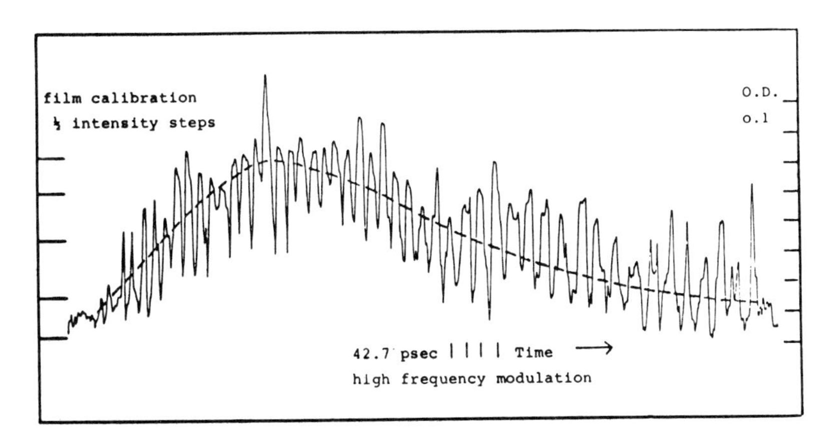

Fig.2 Densitometer trace of upconverted signal of amplified nanosecond pulse from the COCO-II laser system. Scales of intensity and optical density are indicated.

The experimental configuration is shown in Fig. 1. A small sample of the irradiating CO_2 laser (< 1 mJ) is optically mixed in a 5 mm long, 10 mm square aperture proustite crystal with the output of a synchronised Nd:YAG laser having a pulsewidth of $\sim$40 ns (FWHM) and peak power of 25 kW providing a power density of 150 kW cm^{-2}. The 0.96 μm upconverted radiation is isolated with an interference filter, monitored with a photodetector, and analysed with a picosecond (S1) streak camera.

A densitometer trace of a typical streak photograph of the upconverted signal is shown in Fig. 2. Although superimposed on the signal there is a periodic fine structure due to mode beating originating from the resonant reflector in the Nd:YAG laser, the profile of the CO_2 laser pulse can clearly be determined. It can be seen that the risetime of the leading edge of the pulse is $\sim$400 ps and the fall time $\sim$1 ns. However the depth of the 23.5 GHz substructure clearly indicates the system has a bandwidth approaching the theoretical limit (in the present case $\sim$200 GHz limited by phase matching conditions in the crystal). In the present system the range of useful operation is limited by the dynamic range of the streak camera. Thus with the use of spectrally pure radiation, eliminating the occurrence of substructure in the upconverted signal, this technique will have considerable application in the analysis of infrared phenomena associated with the interaction on a picosecond time scale.

3. Time Resolution of Other Infrared Phenomena

It has long been recognised by those investigating laser produced plasmas that valuable insight into the interaction processes can be gained through the analysis of the radiation scattered off the plasma, and of the various harmonics of the fundamental frequency emanating from the plasma. In particular, in recent experiments at very high intensities, (for values $I\lambda^2$ from 10^{15} to 10^{16} W $\mu m^2 cm^{-2}$) the existence of a complete series of integral harmonics has been observed; from the second through to the eleventh harmonic for CO_2 laser produced plasmas[8] and up to the fifth harmonic for plasma produced by Nd:glass lasers.[9] It is anticipated that the existence of these harmonics is a signature of the generation of strong resonant electromagnetic fields in the interaction zone, and therefore may well be an important diagnostic of resonant absorption. Thus a knowledge of their spatial location in the plasma and their temporal development should contribute considerably to their understanding. We have made an initial attempt at such a study by spatially and temporally resolving the tenth

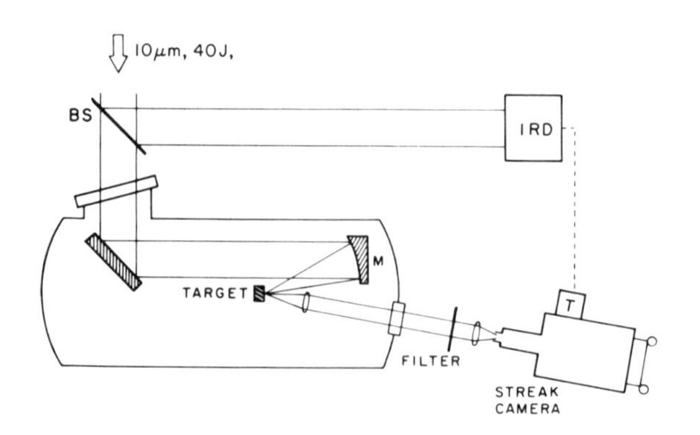

Fig.3 Time resolution of high harmonic emission

(1.06 μm) and eleventh (0.96 μm) harmonic radiation emitted by the plasma. [10] The general configuration of the experimental arrangement is shown in Fig. 3. The 10 μm nanosecond duration CO_2 laser pulse is focused onto plane Al targets with an f/2.5 20 cm focal length off-axis parabolic mirror in a 10^{-5} torr vacuum, the half energy focal spot diameter being ∿110 μm. The harmonic radiation emitted from the plasma close to the target normal, suitably isolated by a 100 Å bandwidth interference filter, is collected by an f/2 lens and imaged onto the slit of a picosecond streak camera. Typical densitometer traces of streak photographs of the tenth and eleventh harmonic emission taken on different shots are shown in Fig. 4. It can be seen that the duration of the emission is considerably shorter than the

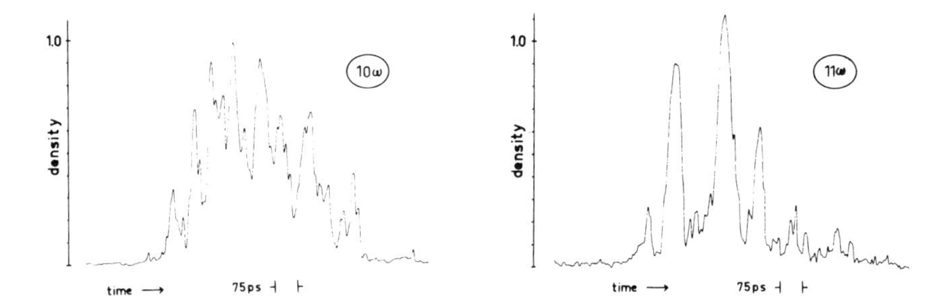

Fig.4 Densitometer traces of streak photographs of high harmonic emission. The CO_2 laser energy incident on target in each case was in 30 J. However, the laser pulse duration of the two shots were 1.6 ns (10 ω) and 1.0 ns (11 ω) (FWHM).

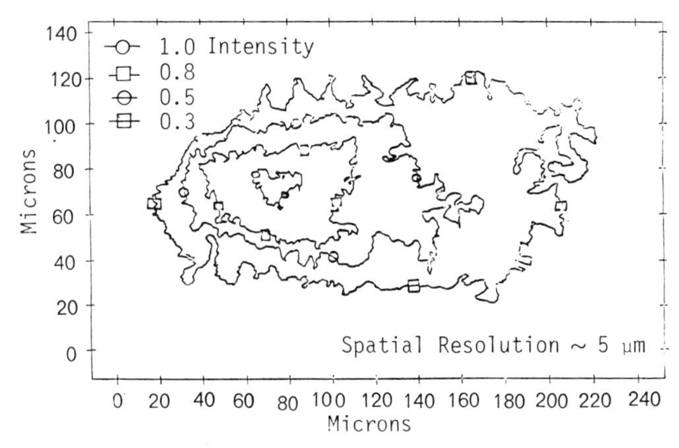

Fig.5 Iso-intensity contours of the spatial distribution of the 11th harmonic emission.

pulse duration (250 - 400 ps (FWHM)) and contains considerable temporal substructure. Although no direct measurement was made of the time of this emission relative to the incidence of the laser pulse on target, it is surmised from the time of the appearance of the lower level background Bremsstrahlung emission that the harmonic emission is generated early in the creation of the plasma. By utilizing high magnification optics, and the (S 1) streak camera as an image converter, spatially resolved photographs of the harmonic emission were obtained, (Fig. 5). It can be seen that the emission originates from a localized region which may well be associated with the high density region of the plasma.

4. Picosecond Optical Diagnosis

Optical diagnosis of CO_2 laser produced plasmas with coherent visible light is particularly desirable since through interferometry the electron density can be determined in the region of, and considerably greater than the critical density for 10 μm radiation (10^{19}cm^{-3}).[11] Knowledge of the latter, in so far as its gross morphology, spatial position and temporal development are concerned, is of prime importance to the illucidation of laser absorption and scattering processes in the plasma. In conjunction with other diagnostics, this permits the interpretation of the hydrodynamic and thermal transport behaviour of the plasma. In addition optical probing permits the measurement of self-generated magnetic fields in the plasma by means of Faraday rotation techniques[12] and the identification of various wave interactions using laser light scattering techniques. All of these studies require precisely[13] timed optical probe pulses. As reported in previous publications,[13] our approach to generating such diagnostic pulses was through the use of an actively mode-locked Nd:glass laser, which, through a common R.F. driver, is phase-locked to the master oscillator of the CO_2 laser system. This laser produces a single 1.06 μm pulse of 100 - 200 ps duration in synchronism to the high power CO_2 laser pulse to within measurement accuracy of ±200 ps. The single pulse is then upconverted to 530 nm and used to illuminate up to four sequentially delayed interferometric channels diagnosing the plasma at different times during the interaction. Analysis of these interferograms, utilizing Abel inversion techniques, then renders the gross morphology of the electron density distribution and its development during the interaction. In particular it has permitted the identification of a sharply steepened density profile through the critical density at CO_2 laser power densities on target of $\sim 10^{14}$W cm^{-2}[Fig. 6]. The existence of this steep density ramp, having a scale length ~10 μm at the

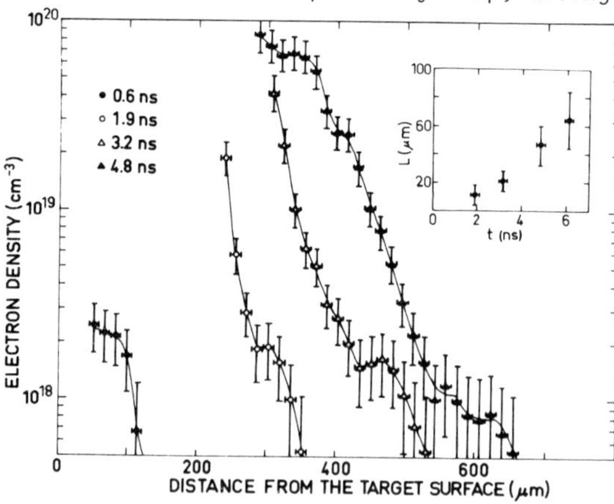

Fig.6 Variation of axial electron density profile as a function of time during the CO_2 laser pulse (t=0 ns being defined at 20% of the leading edge of the laser pulse), for a peak intensity of 2×10^{14}W cm^{-2}. The insert shows the variation of the scale length L in the critical density region as a function of time.

peak of the laser pulse is due to the effect of radiation pressure on the plasma at the critical density. As the intensity of the laser is varied, (Fig. 7), the height of the sharp density step is increased, its upper bound

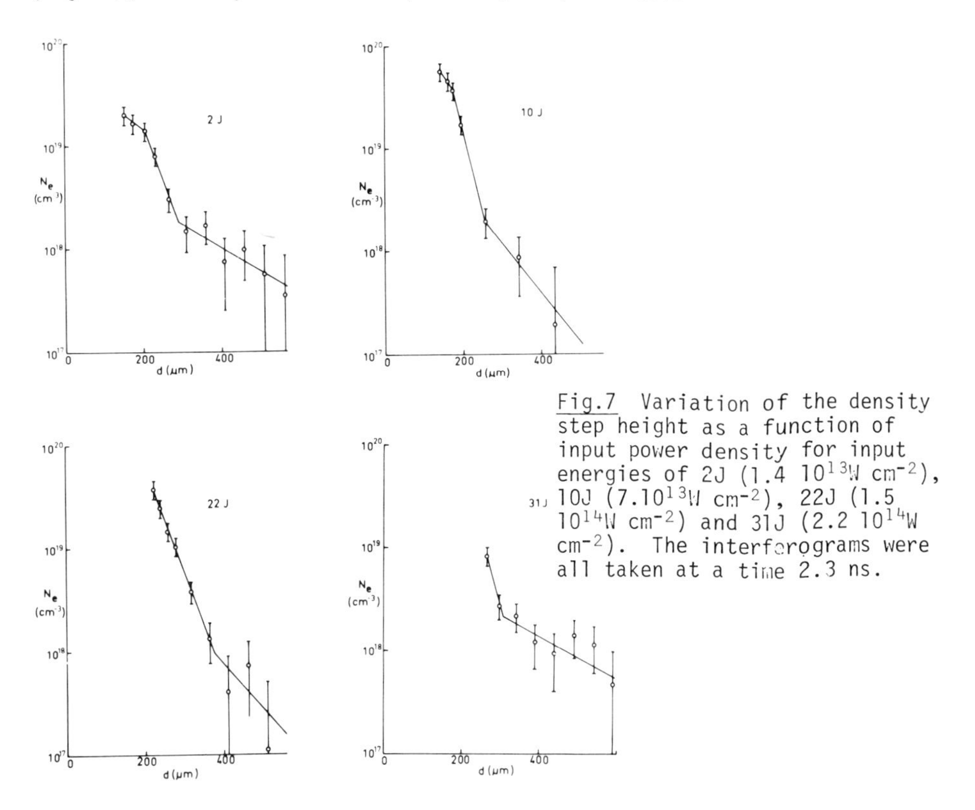

Fig.7 Variation of the density step height as a function of input power density for input energies of 2J (1.4 10^{13}W cm^{-2}), 10J (7.10^{13}W cm^{-2}), 22J (1.5 10^{14}W cm^{-2}) and 31J (2.2 10^{14}W cm^{-2}). The interferograms were all taken at a time 2.3 ns.

being determined by equality of the radiation pressure and the kinetic pressure at that density. These findings have important implications to the understanding of the interaction of high power CO_2 laser pulses with high density plasmas, and in particular to studies at present underway to determine the use of CO_2 lasers for inertially confined fusion.

5. Conclusions

It can be seen from this brief overview that the understanding of the interaction of high power CO_2 laser radiation with solid targets is a complex problem. It's solution requires stringent demands on high resolution diagnostic techniques. As is evident, it has, and unquestionably will continue to force the development of novel approaches and new technologies, which hopefully will find application in areas other than those associated with current laser-fusion studies.

References

1. M.C. Richardson, N.H. Burnett, H.A. Baldis, G.D. Enright, R. Fedosejevs, N.R. Isenor, and I.V. Tomov. Laser Interaction and Related Phenomena, Vol. 4A, p. 161, pub. Plennum Press (1977).

2. Molectron type P-5-00 pyroelectric detector.

3. E.J. McLellan and J.S. Lunsford. IEEE J. Quant. Electron. QE-13, 38D, (1977).

4. T.C. Owen, L.W. Coleman, T.J. Burgess, Appl. Phys. Lett. 22, 349 (1973).

5. J.F. Warner. Appl. Phys. Letts. 12, 222 (1968).

6. P. Jaanimagi, M.C. Richardson and N.R. Isenor. (To be published).

7. A.C. Walker and A.J. Alcock. Rev. Sci. Inst. 47, 915 (1976).

8. N.H. Burnett, H.A. Baldis, M.C. Richardson and S.D. Enright. Appl. Phys. Letts. 31, 172 (1977).

9. E.A. McLean, J.A. Stamper, B.H. Ripin, H.R. Griem, J.M. McMahon, and S.E. Bodner. Appl. Phys. Letts. 31, 825 (1977).

10. P. Jaanimagi, G.D. Enright and M.C. Richardson. (To be published).

11. R. Fedosejevs, I.V. Tomov, N.H. Burnett, G.D. Enright and M.C. Richardson. Phys. Rev. Letts. 39, 932 (1977).

12. J.A. Stamper, E.A. McLean, and B.H. Ripin. Phys. Rev. Lett. 40, 1177 (1978).

13. I.V. Tomov, R. Fedosejevs and M.C. Richardson. Appl. Phys. Letts. 30, 164 (1977).

The Regenerative Amplifier: A Source for Synchronized, Variable-Duration Pulses[1]

J.E. Murray and W.H. Lowdermilk

Lawrence Livermore Laboratory, University of California
Livermore, CA 94550, USA

Regenerative amplifiers (RA) are flexible and stable devices for amplifying and varying the pulsewidth of picosecond pulses. With a Nd:YAG - RA we have demonstrated amplifications up to 14 orders of magnitude with a stability of ± 2%. We have also demonstrated variable pulsewidths over the range 10^{-11} to 10^{-9} sec for a fixed injection pulsewidth. With these capabilities, the RA can be used in a dual-pulse system to provide stable, accurately synchronized pulses of widely different duration. Such a system greatly extends the range of pulse-probe experiments for studies of relaxation phenomena, photochemistry, and laser-generated plasmas.

An example of a RA dual-pulse system is illustrated in Fig. 1.

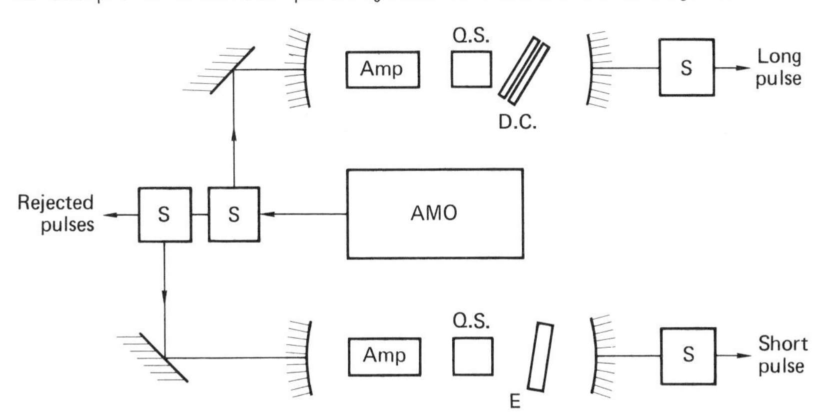

Fig. 1 Regenerative amplifier dual-pulse system: AMO — actively mode-locked oscillator, S — avalanche transistor switchout, QS — Q-switch, DC — dye cell, E — etalon.

Avalanche transistor switchouts [1] select single pulses from the actively mode-locked oscillator for injection into two RA's. The oscillator must be actively mode-locked to provide the necessary amplitude and pulsewidth stability for the injected pulses.

[1]Work performed by the Lawrence Livermore Laboratory under the auspices of the U. S. Department of Energy under contract W-7405-eng-48.

In one RA, the pulsewidth is stretched during amplification with an intracavity etalon; in the other, it is compressed with a saturable dye. Single pulses are selected from the RA pulse trains with two additional switchouts to obtain the synchronized output pulses.

Pulsewidth expansion is easily characterized analytically, and experiments have proven that pulsewidth and amplitude stability are independent of expansion. Pulsewidth compression, however, is much more complex and sensitive to initial conditions. Therefore, numerical analysis was used to optimize the compression process for minimum pulsewidth and maximum stability within the constraint that pulse fluence not exceed component damage levels.

Simultaneous pulse amplification and compression were modeled by numerical solution of three coupled equations. In the limit of small changes with each pass k, and assuming constant Gaussian pulse shape [2], these have the form:

$$\frac{dP}{dk} = P\left[G - \ell - \frac{d}{1+P} - 2\frac{G}{T^2}\right]$$

$$\frac{dT}{dk} = T\left[\frac{G}{T^2} - \frac{d}{2\,\ell n\,2}\,\frac{P}{(1+P)(2+P)}\right]$$

$$\frac{dG}{dk} = -\,s\,GTP,$$

where $P = I/I_S$, $T = \tau\Delta\omega_a/\sqrt{8\,\ell n\,2}$, $s = p\sqrt{2\pi}\,I_S\,R/(\Delta\omega_a J_S)$. Variables and parameters associated with the pulse are: FWHM pulsewidth τ and peak pulse intensity I measured at the dye cell; with the dye: unsaturated transmission exp(-d) and saturation intensity I_S; with the amplifier rod: gain exp(G), bandwidth $\Delta\omega_a$, saturation fluence J_S and the constant $p \leqslant 1$ whose value depends on level degeneracies; and with the resonator: loss $\exp(-\ell)$ and the ratio R of the beam area in the dye cell divided by the beam area in the gain medium.

In Fig. 2 are plotted the values of T and ℓnP at maximum pulse amplitude and the number of passes N required to reach the maximum. The two cases illustrate predicted behavior for two different dye concentrations. For large concentration (Fig. 2a), the variable values are found to be much more sensitive to the initial gain coefficient G_i than for small concentration (Fig. 2b). The transition between these two types of behavior occurs at a critical dye concentration d_c, whose value depends on ℓ, $\Delta\omega_a$, s, and the initial pulse parameters. The best predicted performance was found for minimum loss, dye concentration just below d_c and, with Nd:YAG and Eastman 9860 dye, the maximum value of R. An upper limit on R is dictated by damage at the Nd:YAG amplifier.

A Nd:YAG dual-pulse system has been developed for diagnosing laser-generated plasmas. It utilizes a pulse compression RA whose design is based on these results. An electro-optic cavity dump was used to maximize pulse extraction efficiency while holding ℓ to 0.10. A Brewster-angle dye cell, normal YAG rod and focusing cavity geometry gave R = 6.7. With the amplifier gain stabilized to ± 0.1% [3, 4], a pulsewidth compression from 120 ps to 15 ps was achieved with ± 5% amplitude stability. Predicted and measured output characteristics were in good agreement. Streak-camera photographs such as Fig. 3 verified synchronization of the long and short pulses within ± 5 ps.

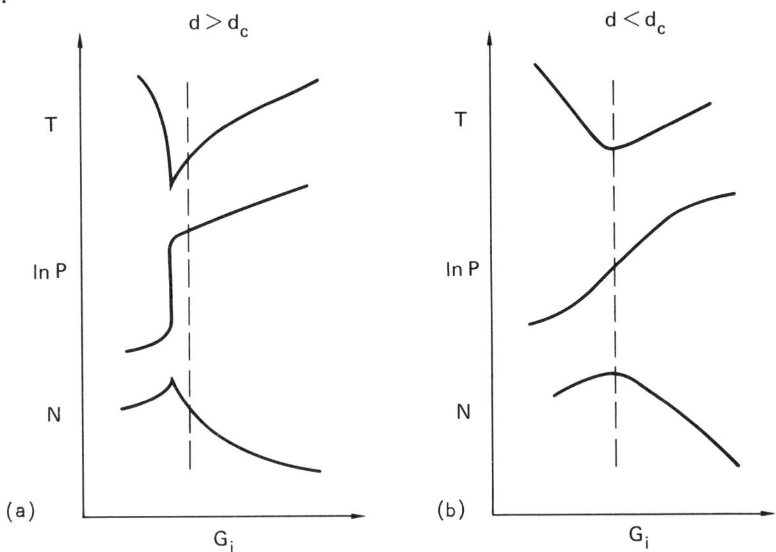

Fig. 2 RA variable values at the maximum pulse amplitude versus initial gain coefficient for dye concentration (a) above and (b) below the critical value d_C.

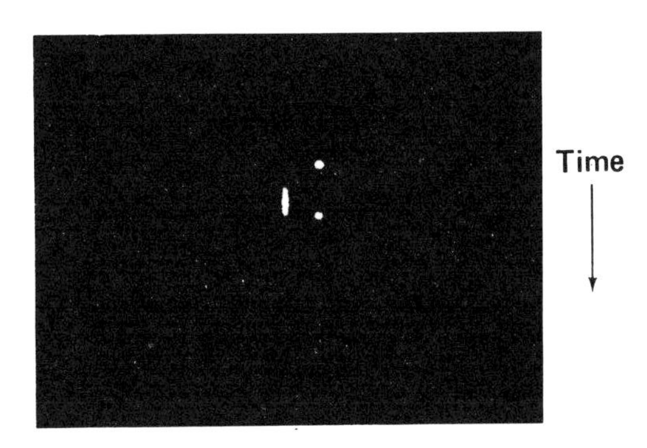

Fig. 3 Streak-camera photograph of 150-ps and 30-ps pulses synchronized within ± 5 ps.

References

[1] S. J. Davis, J. E. Murray, D. C. Downs, and W. H. Lowdermilk, Applied Optics (October 1, 1978).

[2] V. S. Letokhov, JEPT Lett. 7, 25 (1968).

[3] D. J. Kuizenga, Opt. Commun. 22, 156 (1977).

[4] D. C. Downs, J. E. Murray, and W. H. Lowdermilk, IEEE J. Quant. Electron. (August 1978).

Resonant Multiphoton Ionization of Cesium Atoms by Ultrashort Laser Pulses at 1.06 μm

L.A. Lompre and G. Mainfray

Service de Physique Atomique, Centre d'Etudes Nucléaires de Saclay
F-91190 Gif-sur-Yvette, France

1. Introduction

As is well known, in an intense laser field, the ionization of atoms can result from the simultaneous absorption of several photons. The multiphoton ionization rate is given by : $w = \alpha I^K$, where I is the laser intensity, K is the lowest integer greater than the ionization energy of the atom divided by the laser photon energy, and α is a function which emphasizes both atomic properties and laser properties namely coherence, frequency and polarization /1/. In particular, the multiphoton ionization rate of an atom, as a function of the laser frequency, exhibits a typical resonant character when the energy of an integral number of photons is close to the energy of an atomic level satisfying the selection rules. For example, the four-photon ionization of cesium atoms has been carefully studied by using a single-mode Nd-glass laser pulse of the order of 10^{-8} sec in length. This process gives evidence of a typical resonant character when the laser frequency is tuned through the resonant three-photon transition 6S ⟶ 6F /2/.

This resonant multiphoton ionization process has a characteristic time which is essentially the lifetime of the resonant 6F level under the influence of the laser field. This lifetime, which is governed mainly by the photoionization from the 6F level, is given by $T = 1/\sigma I$, where $\sigma = 1.4 \times 10^{-18}$ cm^2 is the photoionization cross section from the 6F level, and I is the laser intensity. For example, the characteristic time of the resonant multiphoton ionization of cesium atoms is expected to vary from 10^{-8} to 10^{-10} sec. when the laser intensity increases from 10^7 to 10^9 w/cm^2 which is the laser intensity range required to investigate the resonant multiphoton ionization of cesium atoms by a Nd-glass laser pulse.

The purpose of the present paper is to investigate whether or not resonance effects are observed in the four-photon ionization of cesium atoms, when the laser pulse duration is varied from 1-nsec to 10-psec, so that it becomes of the order of the characteristic ionization time of the atom. The simultaneous influence of resonance and coherence effects in multiphoton ionization of atoms is a difficult problem which is still far from being solved at the present time. Consequently, the investigation of a resonant multiphoton ionization process requires a bandwidth-limited laser pulse in order to avoid having to take into account the statistical properties of the laser radiation /3/. A bandwidth-limited pulse is defined as a pulse completely devoid of intensity or frequency modulation since its duration is less or equal to the coherence time.

2. Experimental Methods

The laser used in the present experiment is basically similar to the one described in detail elsewhere /4/. The laser rod is a Hoya LHG-5 phosphate glass. The Kodak 9740 saturable dye is in liquid contact with the output mirror. The laser bandwidth is narrowed by putting a very highly dispersive prism in the cavity. This prism reduces the spectrum to a width of 5 Å which corresponds to bandwidth-limited pulses of about 5-psec. By adding a Fabry-Perot etalon in the laser cavity, the pulse duration can be varied by changing the thickness of the etalon. However, when the thickness of the etalon is increased to produce pulse widths longer than 0.3 nsec, the free spectral range of the etalon becomes too small and several spectral lines are allowed to lase. Therefore, a second intracavity etalon is required to select a single line. By using a Pockels switch, a single pulse is selected from the early part of the pulse train generated by the mode-locked Nd-glass oscillator. This single pulse is amplified by two Nd-glass preamplifiers and then passes through a spatial filter before entering a three-stage Nd-glass amplifier. The laser pulse is focused into a vacuum chamber by a planocylindrical lens of 50 cm focal length. The atomic density is n_o (Cs) = 6×10^{10} cm^{-3} in the focal region. The focused laser intensity is known from surface determinations within the focal volume by using photometric measurements /2/. It should be pointed out that the focal surface varies as a function of the pulse duration. For instance, the focal spot area measured with a 1.5 nsec laser pulse is twice that for a 15-psec pulse. Verification of bandwidth-limited operation requires that careful and detailed diagnostics be applied to each laser pulse in both the time and frequency domains. The temporal characteristics of the single pulse are measured either by a 5-psec resolution streak camera or a photodiode connected to an oscilloscope, depending on the pulse time scale. Three different pulse lengths were successively used: 1.5- nsec, 50-psec and 15-psec. Detection is achieved either on a photographic film or on a TV screen /4/. The spectrum of the amplified single pulse is analysed by using a diffraction grating spectrograph with a dispersion of 4 Å/mm and a resolution of 0.1 cm^{-1} at 1.06μm. The usual photographic plate is replaced by the cathode of a TV pickup tube /4/. This method gives an direct measurement and control of the laser wavelength and bandwidth for each laser shot, by using a storage oscilloscope (Tektronix 7633).

3. Experimental Results

The experiment consists of measuring the number of ions formed as a function of the laser frequency in the neighbourhood of the resonance, for a fixed laser intensity. Fig.1 shows four resonance profiles obtained with four different laser intensities. The pulse duration is 15 psec, and the laser bandwidth is 1.4 cm^{-1}. These results exhibit a typical resonant character : enhancement in the number of ions, and a resonance shift. The resonance shift is mainly due to the shift of the ground state 6S and the resonant state 6F under the influence of the laser field. The variation of the resonance shift ΔE, expressed in terms of energy shift of the three-photon transition 6S$\rightarrow$6F is found to be linear with the laser intensity within the range 10^8-10^9 W.cm^{-2}. $\Delta E = \beta I$, with $\beta = 2 \pm 0.2$ cm^{-1}/GW.cm^{-2} in good agreement with the corresponding calculations /5-6/.

Resonance effects are still observed , even with a laser pulse as short as 15-psec. However, the short time behaviour in resonance effects can be observed by investigating the number of ions formed at the maximum of the resonance profiles : Nmax. res. The resonant multiphoton ionization probability of cesium atoms has been calculated as a function of the laser

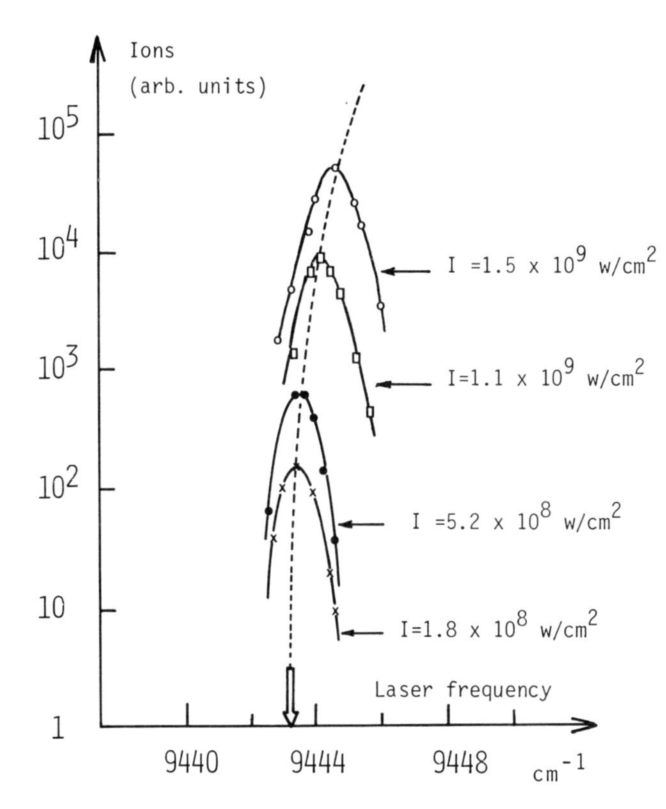

Fig.1 Resonance profiles in multiphoton ionization of cesium atoms. Pulse duration τ= 15-psec. The arrow indicates the frequency position of the three-photon transition 6S$\rightarrow$6F derived from spectroscopic tables. The dashed line shows the resonance shift for increasing values of the laser intensity.

intensity I /7/. Different interaction times have been chosen within the range of values used in the experiments. The laser pulse lengths change the slopes of the curves defined as $n = \frac{\delta \mathrm{Log}\ N_{max.res}}{\delta \mathrm{Log}\ I}$ These theoretical results show that the number of ions formed at the maximum of the resonance profile varies with the laser intensity I as I^2 when long pulse widths are used, and as I^4 when very short pulse widths are used (refer to Fig.2). The triangle point corresponds to the experimental value obtained with 15-psec laser pulses within the laser intensity range used in the present experiment. While the circle point is the experimental value obtained with a 35-nsec single mode pulse used in a previous experiment /2/. Thus, it is seen that temporal effects do play a role in the resonant multiphoton ionization of cesium atoms in good agreement with theoretical calculations /7-8/.

These temporal effects can be explained in terms of simple physical considerations. Under our experimental conditions, the one-photon 6F$\rightarrow$continuum transition rate is much larger than the resonant three-photons 6S$\rightarrow$6F transition rate, and is also much larger than the de-excitation rate of the 6F level towards the 6S ground state. In addition, the one-photon transition 6F$\rightarrow$ continuum is not saturated when ultra-short pulses are used, when saturation is defined as $w\tau \gg 1$, where w is the photoionization rate from the 6F level, and τ is the laser pulse length. $w\tau$ = 0.1 with τ=15-psec and the laser intensity I = 10^9 w/cm^2. Furthermore, the relation $w\tau = 1$ can also be expressed as τ=T, where T is the lifetime of the 6F level under the influence of the laser field. With our experimental parameters, the lifetime T of the 6F level is essentially governed by the photoionization rate from the 6F level.

Thus, when experiments are carried out with ultra-short laser pulses, the pulse width is much shorter than the lifetime of the resonant 6F level, and a slope n=4 is observed; while, when long pulses of 10^{-7} sec are used, the pulse duration is much larger than the lifetime of the resonant 6F level, and a slope n=2 is observed.

In conclusion, it is shown that temporal effects do play a role in the resonant multiphoton ionization of cesium atoms, in good agreement with theoretical calculations.

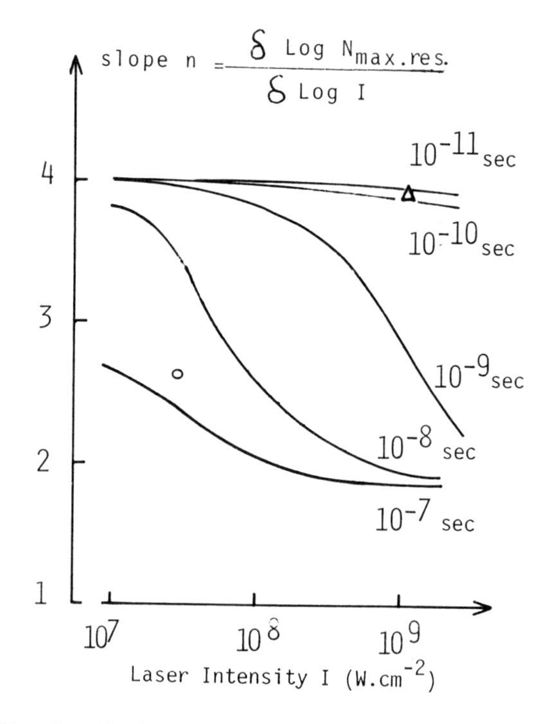

Fig. 2 Variation of the slope n as a function of the laser intensity, for pulse durations from 10^{-11} to 10^{-7} sec.

References

1. P. Lambropoulos. Topics on Multiphoton Processes in Atoms. Advances in Atomic and Molecular Physics 12, 87 (1976).
2. J. Morellec, D. Normand and G. Petite. Phys. Rev.A14, 300 (1976).
3. C. Lecompte, G. Mainfray, C. Manus and F. Sanchez. Phys.Rev.A11, 1009 (1975).
4. L.A. Lompré, G. Mainfray and J. Thébault. J. Appl. Phys.48, 1570 (1977)
5. Y. Gontier and M. Trahin. J. Phys. B 11, L131 (1978).
6. M. Crance. J. Phys. B (to be published).
7. Y. Gontier and M. Trahin. J. Phys. B Lett. (to be published).
8. C. Theodosiou, L. Armstrong, M. Crance and S. Feneuille (to be published).

Intense Picosecond Electrostatic Fields in Laser Generated Plasma

T.P. Donaldson, J.E. Balmer, P. Lädrach, P. Wägli, and H.P. Weber

Institute of Applied Physics, University of Berne
CH-3012 Berne, Switzerland

When plasma is formed by a p-polarised picosecond laser pulse focused obliquely onto a solid surface, resonance between the coherent electromagnetic waves and plasma waves can lead to the generation of an intense localised electrostatic field [1,2]. Such a field is very transient, persisting for a time period of the order of the laser pulse duration. Resonance with plasma waves has been demonstrated by the electron temperature difference seen between the case where a p-polarised rather than an s-polarised laser pulse is focused obliquely relative to the plasma electron density gradient [3]. The electrostatic field driven by this resonance can accelerate some electrons to energies well above the background electron temperature. These escape from the plasma but conserve charge neutrality, by adiabatically accelerating ions. The existance of an electrostatic field has been inferred from the behaviour of these non-thermal ions [4]. Experiments are in good agreement with the computer code ZWERG [5], that models resonance effects.

In the experiments, 1.06 μm, 35 picosecond pulses, from the laser system, OUTIS, were focused onto planar perspex $(C_5O_2H_8)_n$ slabs. Measured parameters were, incident laser intensity, I_o, X-ray electron temperature, T_e, and non-thermal ion emission energy, ε. Ion energies were measured using an electrostatic analyser with a Faraday cup, viewing the plasma along the target normal. The variable parameters were, laser intensity, I_o and laser incidence angle, θ.

For a p-polarised laser beam with θ non-zero and the target in focus, so that the E and k vectors of the beam are well defined, a group of fast, non-thermal, ions appears in addition to the thermal group. The non-thermal ions are those arising from acceleration of electrons by the electrostatic field. An ion trace obtained on the Faraday cup without the electrostatic analyser is shown in Fig.1(a). When measurements were made of the non-thermal energy distribution, the thermal ions were suppressed by a grid biased at 150 volts and placed over the entrance aperture of the analyser. The distribution function of the non-thermal ions, Fig.1(b), was obtained from combined analyser and time of flight measurements. These measurements identified the large peak in Fig. 1 as a proton peak and the non-thermal ion energy, ε, was taken as the value corresponding to the peak of the proton distribution function.

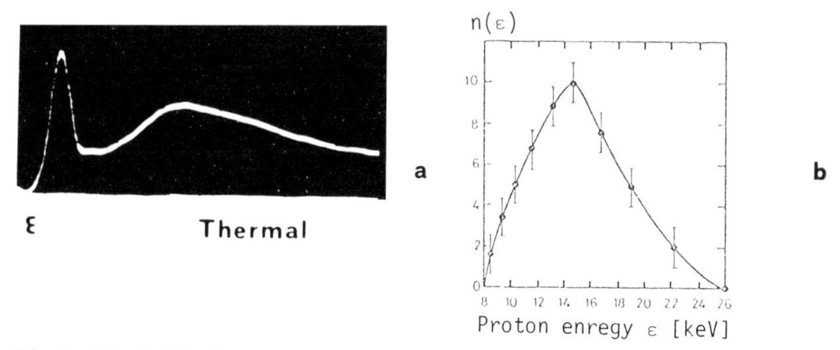

Fig.1 (a) Emitted ion current trace showing both the thermal and non-thermal ion peaks. (b) Non-thermal ion distribution.

For a laser beam of p-polarisation a resonance occurs at oblique incidence because, in this case, the laser beam has a finite electric field component in the direction of the plasma electron density gradient, see Fig.2

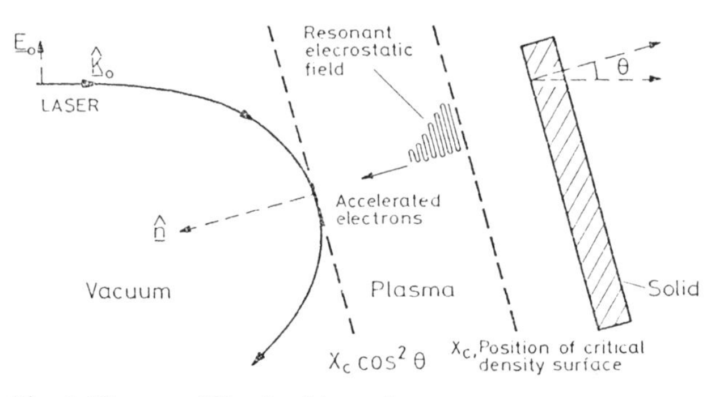

Fig.2 Diagram illustrating plasma resonance

An intense electrostatic field is then driven in the critical density region, where the laser frequency equals the plasma frequency. The minimum wavelength of the field will be equal to the Debye length at resonance, so that the phase velocity of the waves will be close to the electron thermal velocity. Thus the field will be Landau damped into electron thermal energy, resulting in increased plasma absorption of laser radiation as manifested by the enhancement of the cold electron temperature, T_e. This is shown in Fig.3(a), measured as a function of laser incidence angle, θ. The maximum at $15^\circ \pm 5^\circ$ is consistent with resonance absorption for a density scale length, $L = 3\ \mu m$, a value which was deduced independently from plasma reflectivity measurements [7].

The velocities of electrons in the extreme high energy tail of the Maxwellian are much higher than even the maximum phase velocity of the field components so that they travel through the electrostatic field region without being captured and fall through the maximum field potential, reaching energies of up to 14 keV. They then escape from the plasma accelerating protons to these energies, as seen in Figs.3(b) and (c). Note that ε peaks at the same angle as ΔT_e.

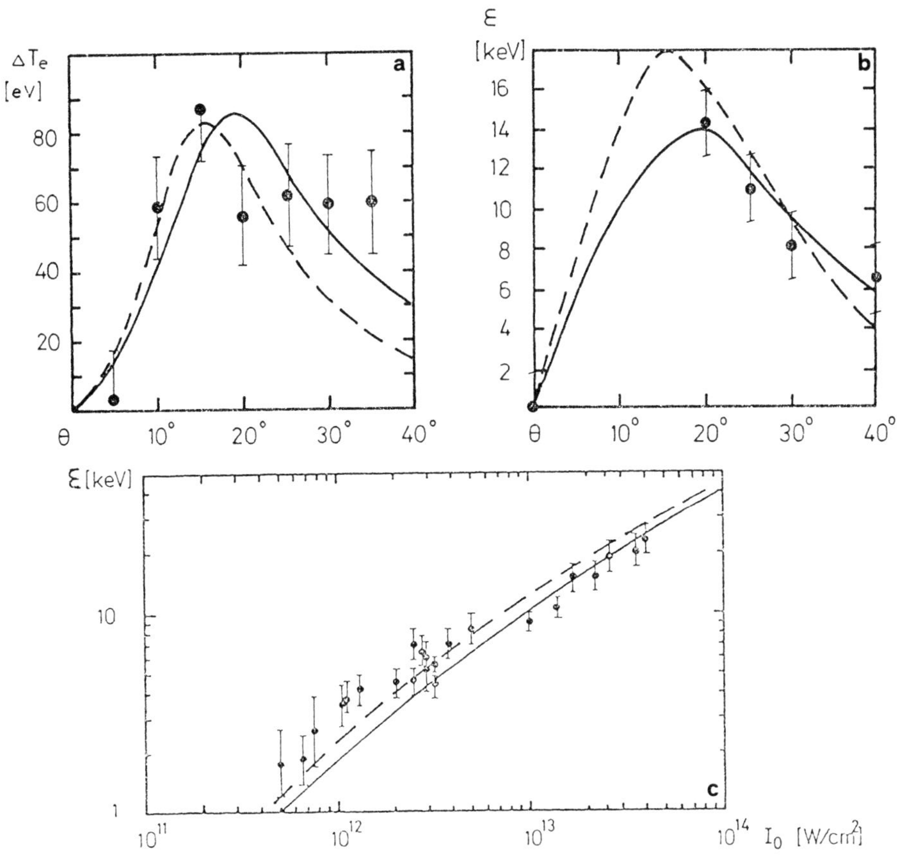

Fig.3 Measured behaviour of electron temperature enhancement and non-thermal ion energy. Solid lines represent the values predicted by the computer code with profile steepening included, while broken lines represent model predictions without profile steepening. (a) Electron temperature enhancement ΔT_e, against incidence angle. (b) Non-thermal ion energy against incidence angle. (c) Non-thermal ion energy scaling with laser intensity, for p-polarised laser incident at $\theta = 20^{\circ}$.

The computer code, ZWERG, was used to model the resonance effects. Absorption is by inverse bremsstrahlung and by plasma resonance, while the energy losses are by thermal and non-thermal ion convection. The resonance field grows to a value limited by wave convection and the electric field enhancement compared with the vacuum laser field is given by

$$(\phi(\tau) / (2\pi k_o L)^{1/2}) \ (L/\lambda_D)^{2/3} \simeq 10$$

where $\phi(\tau)$ is the Ginzberg resonance function. The electron density gradient is steepened by the ponderomotive force of the field, i.e electrostatic field pressure, which is included in the momentum balance equation. Code predictions of ΔT_e and ε with profile steepening included are close to the measured values. The good match between experiment and theory indicates that a resonant electrostatic field is present. For p-polarised laser radiation with $I_o = 2 \times 10^{13}$ watts cm^{-2} and $\theta = 20^o$, a field of 10^9 volts/cm and extent 10^{-5} cm, is generated. Protons are accelerated to energies of $\simeq$ 14 keV. Calculations show that the initial energy required by an electron to escape the field is $\simeq$0.5 keV and the energy gained by acceleration in the field is $\simeq$13.5 keV. By integration over a Maxwellian velocity distribution the energy carried away by the fast electrons was estimated to be $\simeq$ 15%, in agreement with the measured energy of the fast ion group. The field growth time is one oscillation period, while the decay is determined by the Landau damping time. These are in the subpicosecond regime. Thus the temporal behaviour of the field should follow the temporal envelope of the laser pulse.

The two main effects of profile steepening by ponderomotive force are (1) Reduction of the maximum non-thermal ion energy, due mainly to the shortening of the electron acceleration length. (2) Shifting of the absorption maximum to larger incidence angles. This is because the angle of incidence for optimum resonance is given by

$$\sin \theta_{opt} = (k_o L)^{-1/3}$$

Since L is decreased by profile steepening, θ_{opt}, is increased. Thus, for a given incidence angle, L is partially self-adjusting so that ΔT_e and ε are finite over a wider range of angles than expected without steepening. This means that the observed resonance effects are easy to control because the incidence angle is not very critical.

References

1. J.S.De Groot, J.E.Tull: Phys. Fluids 18, 672 (1975)
2. K.G.Estabrook, E.J.Valeo, W.L.Kruer: Phys. Fluids 18, 1151 (1975)
3. J.E.Balmer, T.P.Donaldson: Phys. Rev. Lett. 39, 1084 (1977)
4. P.Wägli, T.P.Donaldson: Phys. Rev. Lett. 40, 875 (1978)
5. P.Wägli, T.P. Donaldson, P.Lädrach: Appl. Phys. Lett.32, 638 (1978)
6. V.L.Ginzberg: Propagation of Electromagnetic Waves in Plasma (Pergamon, Oxford, 1970).
7. J.A.Zimmermann, J.E.Balmer, T.P.Donaldson, P.Wägli: Proceedings of the Eighth European Conference on Controlled Fusion and Plasma Physics, Prague, Czechoslovakia, 1977 Vol.1, p 55.

Picosecond Interferometry of Laser Fusion Targets[1]

D.T. Attwood, E.L. Pierce, D.W. Sweeney, J.M. Auerbach, and P.H.Y. Lee

University of California, Lawrence Livermore Laboratory
Livermore, CA 94550, USA

Introduction

Laser induced fusion is the forerunner of a class of inertial confinement schemes in which hydrogen isotopes are heated to thermonuclear conditions in a very short period. The process is characterized by such short time scales that fuel confinement is achieved through its' own finite mass and expansion velocity, approaching 1 μm/psec for ignition temperatures of order 10 keV (10^8 °K). With current laser powers limited to several terrawatts one readily estimates, on the basis of energy conservation, target mass, and expansion velocity, that target size and laser pulse duration are on the order of 100 μm and 100 psec, respectively. Within these constraints, targets have been heated and confined to the point where thermonuclear conditions have been achieved [1]. An important aspect affecting implosion dynamics is the process through which light is absorbed and apportioned between thermal and non-thermal particle distributions. In this paper we briefly review candidate absorption mechanisms, pointing out that the dominance of one or the other is strongly dependent on electron density gradient lengths in the surrounding plasma atmosphere. Since the gradient lengths are roughly proportional to sound speed times laser pulse heating time, we expect different mechanisms to dominate in short ($\sim$ 100 psec) and long ($\gtrsim$ 1 nsec) irradiation experiments. In this paper we review the use of a 1 μm, 15 psec ultraviolet interferometer to study electron densities in short pulse irradiation experiments. We expect significantly different results in future experiments with nsec plus laser heating pulses. In this paper we briefly review candidate absorption mechanisms and how ultraviolet interferometry resolved to 1 μm and 15 psec has been used to accurately measure electron densities in the plasma atmosphere where these processes take place.

Probing the Plasma Atmosphere

The plasma atmosphere is the relatively low density region surrounding the pellet where laser light is absorbed. It is formed by ablation of material from the target surface. In short pulse laser experiments, typically 100 psec duration the radial gradient of electron density is very sharp, e-folding on a scale of sub-micron to 10 microns (μm) or so. The density gradient in this region plays a very important role in the mechanism of light absorption, and through that plays an indirect role in the partition of absorbed energy between thermal and non-thermal particle distributions [2]. As outlined in Fig. 1, there are three basic mechanisms through which intense laser

[1]Work performed under the auspices of the U.S. Department of Energy by the Lawrence Livermore Laboratory under contract number W-7405-ENG-48.

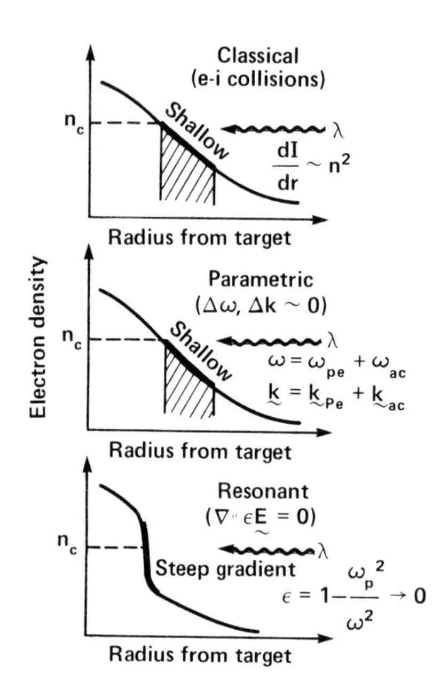

Fig. 1. Absorption of intense laser light is by three competing mechanisms.

light may interact with the plasma as it approaches the reflection or turning point in density where a given wavelength of light can no longer propagate. That density is referred to in plasma physics as the "critical density, n_c". The first mechanism is simple electron-ion collisions, sometimes called "inverse bremsstrahlung", where the electric field of the incident light rattles electrons, which then lose this energy in classical collisions with ions. The mechanism depends on both electron and ion density, thus increasing as n^2 as the wave approaches the critical density (or reflection point). This mechanism is obviously more important with slowly varying density gradients, as indicated in Fig. 1A, since the interaction volume for high n^2 is larger. The parametric processes of Fig. 1B also work most efficiently when the gradient is shallow. These are stimulated processes in which an intense (above threshold) laser field drives one or more acoustic waves out of the noise in a three wave parametric interaction. The dispersion properties of the acoustic waves are density dependent so that appreciable growth occurs only over fairly shallow density profiles. Both absorptive and reflective instabilities are possible depending on whether both or only one of the stimulated waves is acoustic. A typical absorptive process is the ion-acoustic or "parametric decay" instability in which laser light decays into a high frequency electron-plasma (acoustic) wave and a low frequency ion-acoustic wave conserving energy and momentum as indicated in Fig. 1B. An important reflective instability is stimulated Brillouin scattering in which the incident laser light decays into an ion wave and a slightly downshifted electromagnetic wave. As before, these processes are generally expected to play significant roles in shallow gradient plasmas formed by intense laser light in the nanosecond regime. For shorter pulse experiments the expanding plasma does not extend very far, tens of microns, before the pulse is over. An important short pulse, short gradient absorption mechanism is the rather poorly termed "resonance absorption" of Fig. 1C. In this case the gradient is sufficiently short that the field strength of the wave reaching critical density is quite large. For p-polarized light obliquely incident on the plasma surface, a radial component of electric field exists at the "plasma resonance" where the dielectric constant of the medium goes to zero, viz, $\varepsilon_p = 1 - \frac{n}{n_c} = 0$. With the boundary condition $\nabla \cdot \varepsilon_p E = 0$ at the critical surface there is indeed a resonance, and depending on gradients and loss mechanisms, the field amplitudes may be very large causing a large transfer of energy to electrons in this region. The mechanism is of course angle and polarization dependent, and will only play a significant role where refractive turning of the incident light permits a strong field component to tunnel to the critical surface.

In summary three different mechanisms are available for the absorption of light, contributions of each depending on the density gradient and temperature of the plasma, and on the intensity, polarization and angle of incidence of the laser light. With only inverse bremsstrahlung (collisional absorption leading to a thermal distribution of particle energies, the dominance of one mechanism over the others has a profound impact on implosion dynamics. We will see in the following that for short pulse experiments, finite expansion times, radiation pressure effects, and localized heating can combine to produce short scale lengths assuring the dominance of resonance absorption and the concomitant production of suprathermal electrons alluded to previously in the discussion of implosion dynamics.

To estimate expected scale lengths we consider typical plasma expansion velocities and pulse durations. For an SiO_2 plasma ablating from a pellet surface at 1 keV, the average ion mass is about 20 m_p so that on the basis of charge neutrality one expects an expansion velocity of order 0.1 μm/psec. For a pulse duration of 100 psec, typical of exploding pusher experiments, one estimates scale lengths to be of order 10 μm. Other factors conspire to reduce this value. For instance, radiation pressure due to the incident light itself may cause density profile steepening or critical surface deformation on both a small and large scale. Representative cases and nomenclature are described in Fig. 2. Both of these features have been observed in recent experiments [3]. The most successful technique used to date for studying the plasma atmosphere is that of optical probing in the classical sense of interferometry, polarimetry, etc., except that the densities and gradients demand

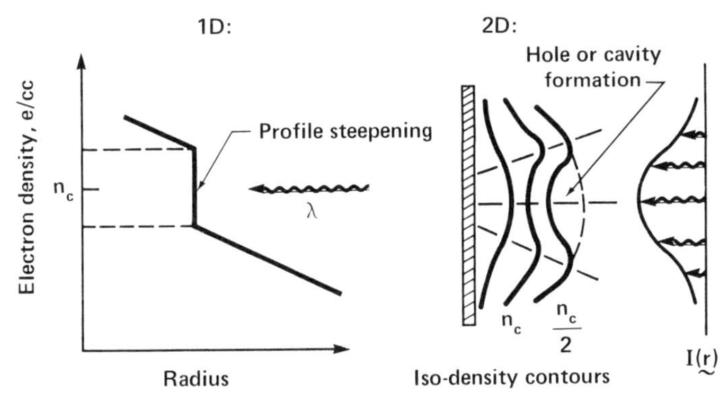

Fig. 2. Two manifestations of radiation pressure are described. Finite momentum transfer by a large photon flux push and distort the expanding plasma.

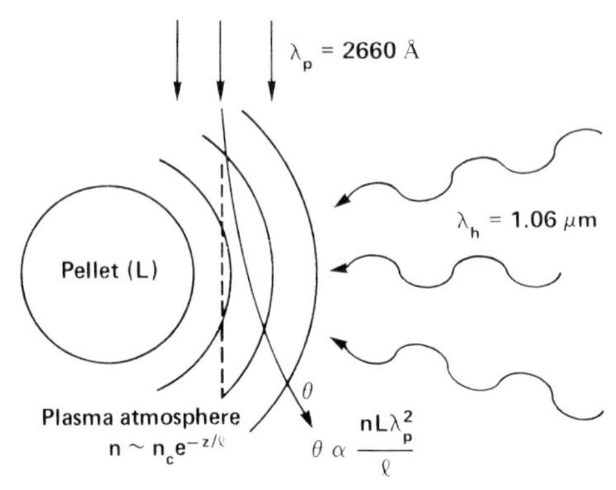

Fig. 3. The effect of refraction on optical probing is seen to depend on wavelength squared, and inversely on the axial scale length.

a wavelength in the ultraviolet. Figure 3 shows the general technique in which a probe pulse of wavelength λ_p passes transversely through the plasma, its phase and polarization being modified by the refractive index of this ionized medium. The pulse must be temporally short to avoid smearing as the plasma density contours move outward, typically with velocities of order 0.1 μm/psec. For an interferometer with fringe spacings of about 1 μm one then requires a probe duration of ∿ 10 psec. Refractive turning of the probe through an angle Θ, as in Fig. 3, must be minimized so that the ray is not lost to the collecting optics. For axial scale lengths of 1 μm and transverse dimensions of 20 μm, one requires a probe wavelength in the ultraviolet in order to probe critical densities.

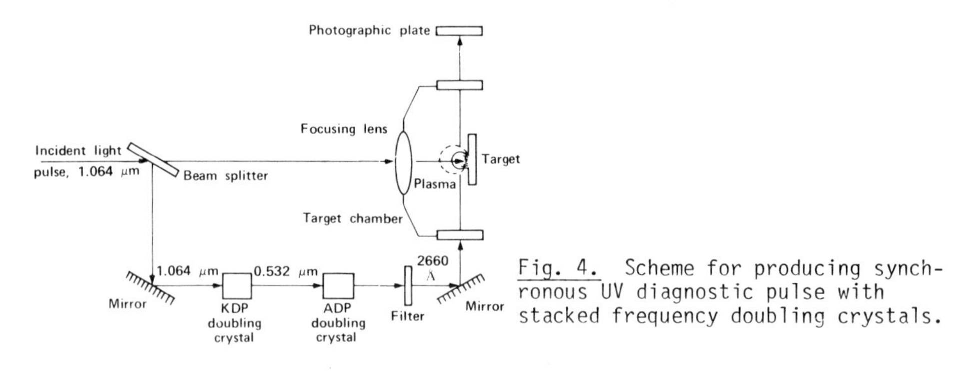

Fig. 4. Scheme for producing synchronous UV diagnostic pulse with stacked frequency doubling crystals.

The technique for obtaining a synchronous optical probe is shown in Fig. 4. Just after the oscillator switchout a portion of the main laser pulse is reflected into a separate beam filtering, shaping and amplifying chain. Nonlinear phase distortion is maintained below tenth wave. Amplification is followed by cascaded frequency doubling, for an overall wavelength change from 1.064 μm to 2660 Å. Conversion efficiency in each case is maintained at no more tnan 10% so as to maximize pulse shortening effects [4]. The 30 psec infrared pulse is thus converted to an approximately 15 psec ultraviolet pulse. The UV pulse is then directed towards the target chamber for diagnostic use. The interferometer used is shown in Fig. 5. Its holographic nature is particularly useful in obtaining near diffraction limited resolution (∿ 1

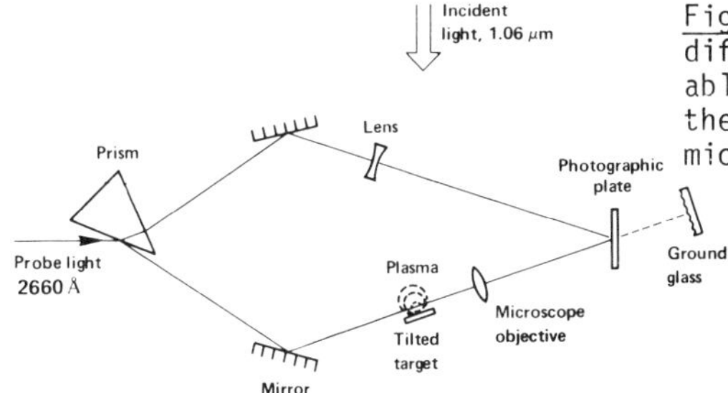

Fig. 5. Holographic system used for diffraction limited, accurately focusable interferometry. A key element is the 0.2 NA ultraviolet transmitting microscope objective.

μm), and providing accurate focusing in the central target plane. Focusing accuracy of several microns is desireable but difficult to obtain with a pulsed UV source and inherent mechanical drift. By imaging after the fact with CW He-Ne reconstruction procedures, these difficulties vanish. An example of focusing effects is shown in Fig. 6, where the effect of a 40 μm focusing "error" produces a different fringe pattern. This is not a difficulty here, but would be in a classical interferometer.

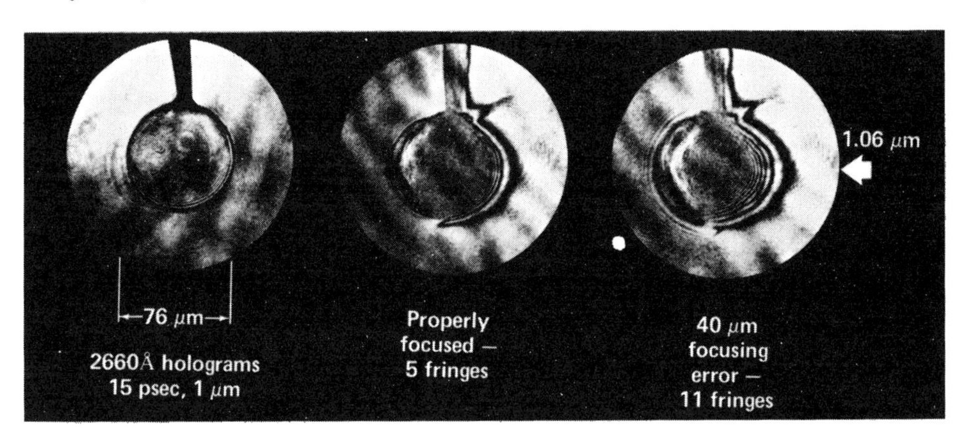

Fig. 6. Fringe pattern is a strong function of image plane position in a steep gradient plasma. This can lead to serious errors of misinterpretation if a variable focus holographic system is not used.

An example of a well focused interferogram, along with its analysis, is shown in Fig. 7. In this target experiment a 41 μm diameter glass microballon was irradiated with a 30 psec, 1.06 μm laser pulse at an intensity of 3×10^{14} W/cm^2. The resulting interferogram has been Abel inverted, with axial electron densities plotted to the right. These results demonstrate that radial density profiles are indeed steepened by radiation pressure effects. A simple but instructive model for this effect is to consider a static balance of partial pressures at the critical surface, where the high density side has only particle pressure, while the lower density laser side has both a thermal (particle) component and a portion due to momentum transferred by photon reflections, viz., $n_u\ \kappa T = n_L\ \kappa T + 2I_0/c$, where n_u and n_L are the upper and lower shelf density, κT is the thermal temperature, I_0 is

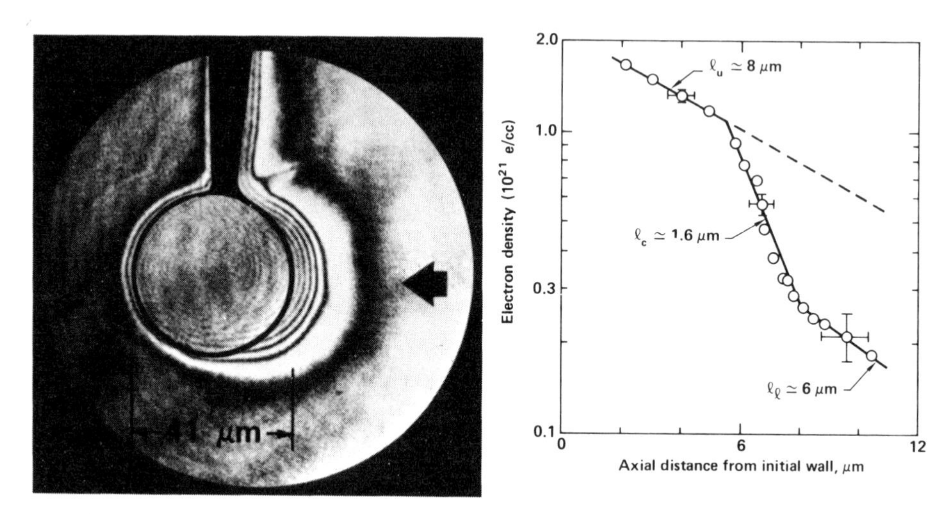

Fig. 7. Profile steepening due to radiation pressure is demonstrated in this target irradiation experiment. I ≈ 3 x 10^{14}W/cm^2.

the laser intensity and c is the speed of light. The model must then be modified for flow field dynamics, temperature differences across the interface, and finite absorption on a finite density profile. Typical numbers indicate an equivalence of thermal and radiation pressures at 10^{21}e/cc, 1 keV temperature, and a laser intensity of 3 x 10^{15}W/cm^2. The density step in Fig. 7 is thus somewhat larger than expected, probably due to temperature differences localized by inhibited energy transport in the critical region.

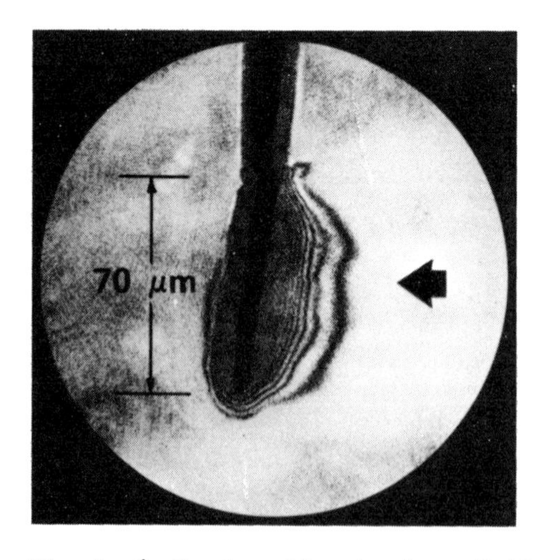

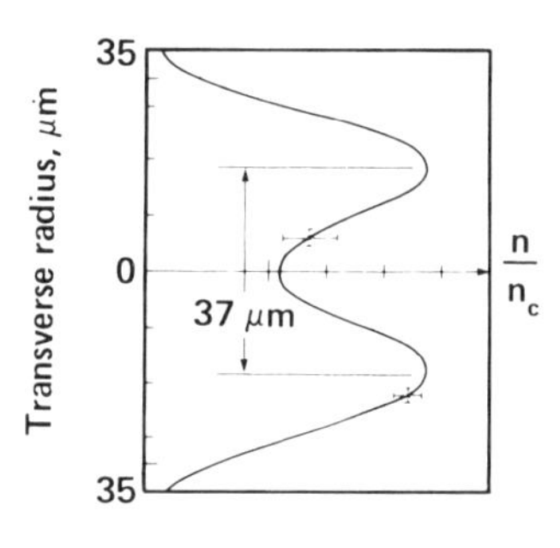

Fig. 8. Cavity formation due to radiation pressure is demonstrated in these flat disk experiments. I ≈ 3 x 10^{14}W/cm^2.

As a practical matter, these results show that for 30 psec experiments the scale lengths are everywhere less than 10 μm, and approximately 1 μm near the critical surface. Referring to fig. 1, this measured profile suggests that resonance absorption plays the major role in absorbing laser light. The shallow profile of 100 μm scale length or larger, required for efficient absorption by classical inverse bremsstrahlung or parametric decay, does not exist. Further confirmation of density contour deformation is observed in our experiments with flat disk targets, often referred to as "lollypops". An example is shown in Fig. 8, with the Abel inverted densities shown to the right for a plane transverse to the incident light. This is an example of the second case of radiation pressure described previously in Fig. 2B, that due to the transverse profile of the incident light. The density cavity or well formation shown in Fig. 8b has a transverse scale approximately equal to the incident light. Although the data here was numerically inverted, inversion is not required to appreciate the presence of a density well. The mere appearance of flat fringes is sufficient to draw this conclusion. Figure 9 sketches this reasoning by pointing out that ray paths are longer through the center of an axisymmetric plasma and therefore should provide larger phase shifts, which would give the appearance of outwardly bowed fringes. By a reversed reasoning, flat fringes imply a density well. In addition to large scale density cavities, we have also observed small scale rippling of the critical surface in our higher intensity experiments with parylene lollypop targets. An example is shown in Fig. 10. As argued in the lower left of

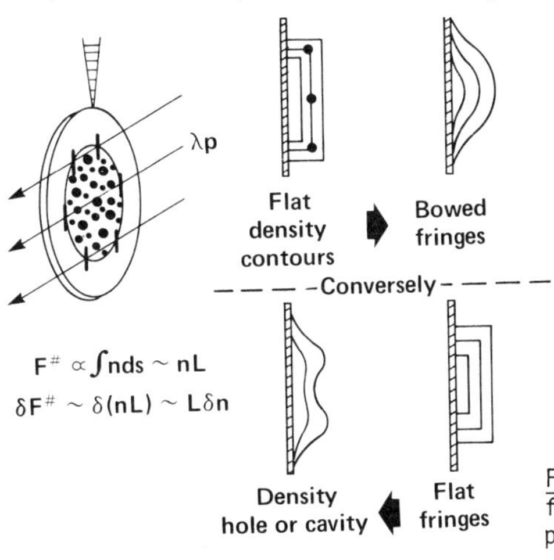

Fig. 9. Simple arguments show that flat fringes in an axisymmetric experiment indicate a density well.

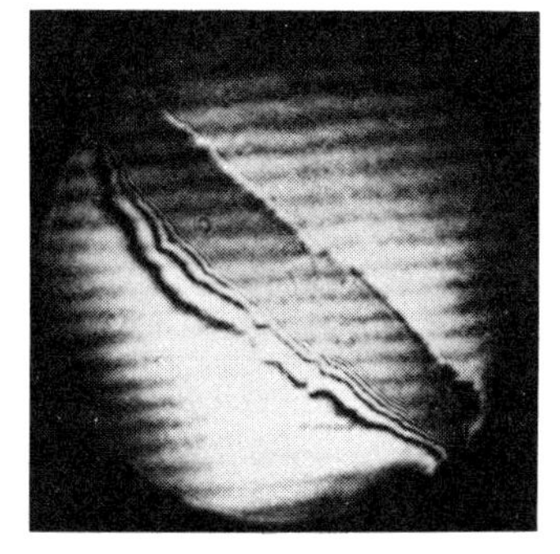

Fig. 10. Locally rippled fringes suggest density fluctuations of 20% amplitude and 10 μm scale size. $I \simeq 3 \times 10^{15}$ W/cm^2.

Fig. 9, localized fringe variations are primarily due to density changes since optical path lengths are essentially equal within small areas of the target. These density variations are most likely produced by hot spots in the incident light beam which push electrons out of their way through the increased radiation pressure. This process is unstable since the low density region is of higher refractive index, thus tending to focus the incident light further. This process of "filamentation" causes a localized mixing of angle and polarization of the incident wave, leading to a smoothing of measured absorption characteristics expected for the dominant resonance absorption process. In addition, these localized ripples very likely have strong surface currents due to the partial absorption of light (transverse momentum of the partially reflected waves is not conserved, a portion having been transferred to the electrons). These randomly oriented surface currents would then lead to randomly oriented magnetic fields, estimated by particle-in-cell computer simulations to be in the megagauss range [5]. These random microscopic fields are not to be confused with the somewhat larger scale toroidal fields due to full target size $\nabla n \times \nabla T$ gradients, whose magnitude is still a matter of discussion for laser fusion targets [6]. The great interest in strong microscopic fields is, of course, directed at understanding the reduction of energy transfer via electron conduction, which is observed indirectly in implosion experiments, a matter described in a companion paper [7]. Although simulations also suggest strong ion density fluctuations, which would further reduce electron conduction by strong scattering, measurements which demonstrate such effects have not yet been devised.

Summary

Holographic interferometry using a 2660 Å probe wavelength has been used to study the plasma surrounding laser irradiated targets with a resolution of 1 μm and 15 psec. Results indicated that for short pulse experiments gradient lengths are sufficiently short, and density depressions sufficiently large, that only resonant absorption plays a significant role, and that density fluctuations provide mixing of s- and p-polarization at the absorption surface. We anticipate different results in future experiments with nsec irradiation pulses. To obtain short probe pulses with long pulse irradiations, new techniques will be employed, as discussed at this conference by Kuizenga [8] and by Murray and Lowdermilk [9].

References

1. J. Nuckolls, L. Wood, A. Thiessen and G. Zimmerman, Nature 239, 119 (1972); P. M. Campbell, G. Charatis, and G. R. Montry, Phys Rev. Lett. 34, 74 (1975); V. W. Slivinsky, H. G. Ahlstrom, K. G. Tirsell, J. Larsen, S. Glaros, G. Zimmerman, and H. Shay, Phys. Rev. Lett. 35, 1083 (1975); N. M. Ceglio and L. W. Coleman, Phys. Rev. Lett. 39, 20 (1977).
2. F. W. Perkins and J. Flick, Phys. Fluids 14, 2012 (1971); C. S. Liu, M. N. Rosenbluth, and R. B. White, Phys. Fluids 17, 1211 (1974); J. P. Freidberg, R. W. Mitchell, R. L. Morse, and L. I. Rudsinski, Phys. Rev. Lett. 28, 795 (1972); K. G. Estabrook, E. J. Valeo and W. L. Kruer, Phys. Fluids 18, 1151 (1975); J. Dawson, P. Kaw and B. Green, Phys. Fluids 12, 875 (1969).
3. D. T. Attwood, D. W. Sweeney, J. M. Auerbach, and P. H. Y. Lee, Phys. Rev. Lett. 40, 184 (1978).

4. D. T. Attwood, E. L. Pierce and L. W. Coleman, Optics Commun. 15, 10 (1975); E. L. Pierce and D. T. Attwood, "Designing A Probe Beam And An Ultraviolet Holographic Microinterferometer For Plasma Probing", June 1978 (unpublished).
5. K. Estabrook, Phys. Fluids 19, 1733 (1976).
6. J. A. Stamper, E. A. McLean, and B. H. Ripin, Phys. Rev. Lett. 40, 1177 (1978) and reference 10 therein.
7. D. T. Attwood, "Diagnostics for the Laser Fusion Program-Plasma Physics on the Scale of Microns And Picoseconds", UCRL 81156 (May 1978), unpublished.
8. D. Kuizenga, paper FB7.
9. J. E. Murray and W. H. Lowdermilk, paper FB3.

Development of an Actively Modelocked and Q-Switched Oscillator for Laser Fusion Program at LLL[1]

D.J. Kuizenga

Lawrence Livermore Laboratory, University of California
Livermore, CA 94550, USA

The large Nd:glass laser-fusion systems such as Argus and Shiva require short-pulse oscillators that are very reliable and predictable. The requirements go well beyond what can reasonably be expected from a passively mode-locked laser. An actively mode-locked and Q-switched oscillator has now been developed that is extremely reliable and predictable, and satisfies all the requirements for the present Nd:glass laser systems. These systems require pulses that are adjustable from less than 100 ps to more than 1 ns, with less than 5% shot-to-shot variation in pulse energy and pulse width. Single-pulse energy from 100 μJ to 1 mJ is sufficient.

An actively mode-locked CW Nd:YAG laser can provide the range of short pulses required [1,2] and is also sufficiently stable. It has been shown that under steady-state conditions the pulse width in this laser, mode-locked with an amplitude modulator, is given by:

$$\tau_p = \frac{(2 \ln 2)^{1/2}}{\pi} \frac{(g_o)^{1/4}}{\theta_m^{1/2}} \frac{1}{(f_m \cdot \Delta f)^{1/2}} , \qquad (1)$$

where

θ_m = depth of modulation

g_o = round-trip amplitude gain

Δf = linewidth

f_m = modulation frequency

For typical laser parameters, pulses shorter than 100 ps can be obtained; and with the use of an etalon in the cavity, longer pulses of up to 1 ns or more can be obtained, depending on the cavity length.

---

[1]Work performed under the auspices of the U.S. Department of Energy, Contract W-7405-eng-48.

A typical CW mode-locked Nd:YAG laser gives only a few nanojoules in a single short pulse. This can be increased by more than four orders of magnitude by simultaneously Q-switching the laser. However, the buildup time in a typical Q-switched laser is not long enough for the short-pulse generation process to reach steady state. To obtain good mode locking in the Q-switched Nd:YAG laser, the buildup time of the Q-switching has to be increased. Several workers [4-7] have developed techniques to do this in an attempt to obtain steady-state mode locking before the Q-switched pulse train is obtained.

We have now further developed a method that allows the mode-locking process to go to its steady state condition before the laser is Q-switched.[8] This is done by pumping the laser quasi-cw for about 5 ms. During this time, the loss in the Q-switch is such that the laser will just slightly go above threshold. The active modulator is on during this time, and the laser oscillates quasi-cw for a period long enough to obtain stable transform-limited short pulses. At the end of this prelase period, the laser is Q-switched, and a train of stable, short pulses is obtained.

Three conditions have to be satisfied during this prelase period. First, the short pulse envelope has to approach a steady state value. It has been shown [3] that the pulses are within 5% of the steady state value after a number of round trips M given by:

$$M = \frac{0.38}{g^{1/2}\theta_m}\left(\frac{\Delta f}{f_m}\right), \tag{2}$$

where g is the roundtrip amplitude gain during the prelase period, and the other parameters are defined for (1). The dominating factor here is f/f_m, which is typically on the order of 10^3, and hence for a typical mode-locked Nd:YAG laser, it takes about 10 to 100 μs for the short pulses to approach their steady-state value. This is easily satisfied during the prelase period.

The prelase period must also be long enough for the pulses to become transform-limited. It may happen that even though the short pulse envelope has reached its steady state value, the pulse still has considerable substructure and may be chirped, etc. This type of substructure is smoothed out by repeated passes through the finite bandwidth of the active medium (or etalon). If one considers the narrowing of the spectrum on repeated round trips, it can be shown that the spectrum is within a factor of two from the steady state, transform limited spectrum after a number of round trips M given by:

$$M > \frac{0.5}{g^{1/2}\theta_m}\left(\frac{\Delta f}{f_m}\right). \tag{3}$$

This is essentially the same as the condition for the short pulse envelope to reach steady state. A much more severe condition to obtain perfect transform limited pulses is given by:

$$M > \frac{0.35}{g}\left(\frac{\Delta f}{f_{ax}}\right)^2, \tag{4}$$

where f_{ax} is the axial mode spacing of the cavity. This is the condition required to get single axial mode operation without modulation. With the modular on, this condition means that we get this single axial mode plus its sidebands produced by the modulator, and hence a perfect transform limited pulse. For a typical mode-locked Nd:YAG laser, it can take as much as several milliseconds to satisfy this condition. This condition is more severe than is actually required to get good pulses from a Nd:YAG laser.

Finally, when the laser goes above threshold at the start of the prelase period, there are the usual relaxation oscillations. These have to die out so that the laser can oscillate quasi-cw. It usually takes a few upper level lifetimes for this to happen and hence in a Nd:YAG laser, these relaxation oscillations die out in about 1 ms. In practice it is found that a prelase period of about 3 to 5 ms gives very good operation of this type of laser.

A typical arrangement for this type of oscillator now in operation on Argus and Shiva is shown in fig. 1. An acousto-optic modulator and Q-switch were used. The modulator, as well as the Q-switch, had direct water cooling of the transducer (35° Y cut $LiNbO_3$) to ensure stable thermal conditions in the modulator. The modulator and Q-switch, as well as the Nd:YAG rod, were all cut as Brewster's angle to completely eliminate any spurious etalon effects or multiple reflections between components in the laser cavity. This gave stable and predictable short pulses from this laser, and also reduced the noise level between pulses to a minimum.

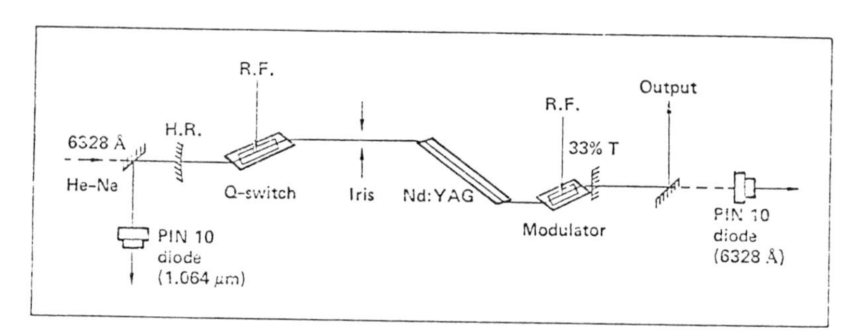

Fig.1 Typical arrangement of modelocked and Q-Switched Nd:YAG laser. Pulse repetition rate is 5 or 10 pps.

It was quite difficult to obtain stable, quasi-CW pumping of the lamps to satisfy the ± 5% stability of the pulse energy. The final and successful arrangement consisted of a bank of power transistors to regulate the voltage of a number of capacitors. In addition, a transistor bank in series with the lamps provided feedback regulation of the simmer and peak currents through the lamps adjustable from 1 to 5 ms and shot-to-shot stability of better than 0.3%. This was sufficient to give good operation of the laser.

We found that the pre-lase signal is a very good indicator of how well the laser is mode-locked [8]. It was found that with the modulator off, the initial relaxation oscillations are quite irregular; but with the modulator on, these relaxation oscillations become very smooth and regular. For a small change in cavity length ($\sim$25 μm) the relaxation oscillation continues through the entire pre-lase period. These driven oscillations occur for length changes longer and shorter than the optimum length. It is precisely the combination of these characteristics that allows one to easily adjust the laser for optimum mode locking, and one only needs a very slow detector and oscilloscope to do this. Many other laser problems can be diagnosed from this pre-lase period, such as higher-order transverse modes.

An optical correlator was used to measure the pulse widths from the Argus laser. Although in principle similar to that described before [3] this correlator was considerably improved to routinely do very accurate short-pulse measurements. Figure 2 shows the range of short pulses available from this oscillator. Note that with the aid of a 2.5 mm and 11 mm uncoated quartz etalon, the required pulsewidth range from 100 ps to 1 ns can be obtained. Figure 3 shows the energy in a single, selected pulse near the peak of the pulse train. For strong modulation, there is some reduction in output

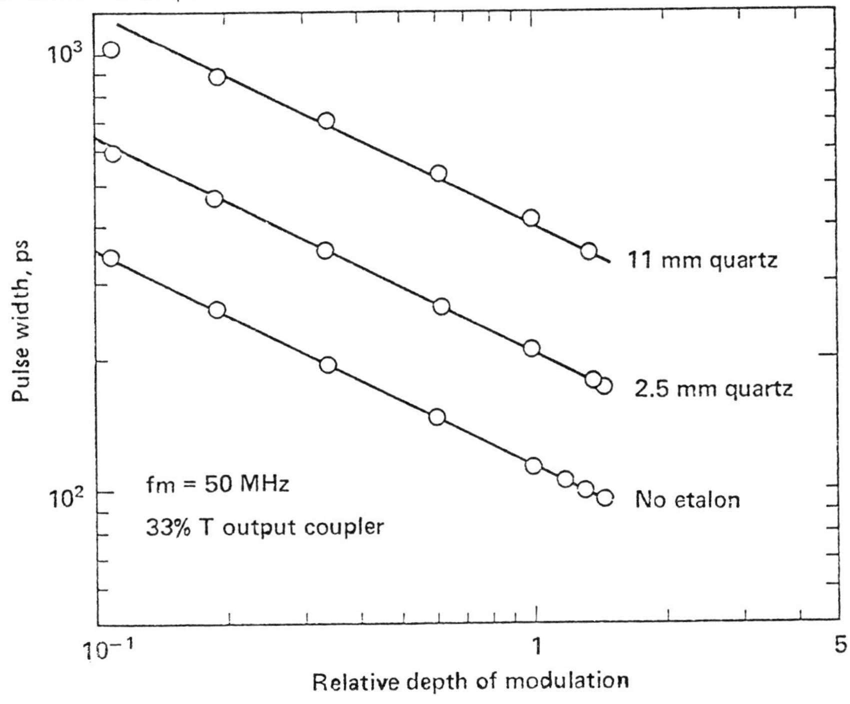

Fig.2 Range of pulses available from Argus Oscillator.

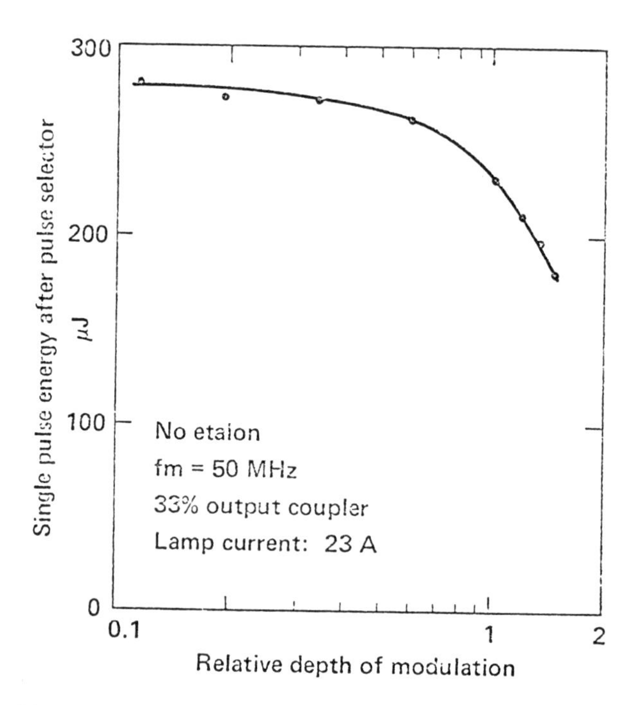

Fig.3 Single pulse energy from Argus Oscillator.

The stability of these oscillators is very good. Typically, over the duration of one hour, the variation of a single selected pulse is less than ± 2%. For the Shiva Oscillator, we have also introduced a new method to switch out a single pulse. We found that the time from when the laser is Q-switched, to a single pulse at the peak of the pulse train is stable to less than one roundtrip in the cavity. Hence we can select this single pulse after a fixed time delay from Q-switching the laser. An avalanche transistor stack is used to drive the KD*P Pockells Cell to select a single pulse. Typically, two or three stages are used in the pulse selector to obtain sufficient pre-pulse isolation.

REFERENCES

1. D. J. Kuizenga and A. E. Siegman, IEEE J. Quant. Electron. QE-6, 694 (1970)

2. D. J. Kuizenga and A. E. Siegman, IEEE J. Quant. Electron. QE-6, 709 (1970).

3. D. J. Kuizenga, D. W. Phillion, T. Lund, and A. E. Siegman, Opt. Commun. 9, 221 (1973).

4. G. V. Krivoshchekov, L. A. Kulevskii, N. G. Nikulin, V. M. Semibalamut, V. A. Smirnov, and V. V. Smirnov, Sov. Phys. JETP 37, 1007 (1973).

5. G. V. Krivoshchekov, N. G. Nikulin, and V. A. Smirnov, Sov. J. Quant. Electron. 5, 1096 (1976).

6. I. V. Tomov, R. Fedosejevs, M. C. Richardson, and W. J. Orr, Appl. Phys. Lett. 29, 193 (1976).

7. I. V. Tomov, R. Fedosejevs, and M. C. Richardson, Appl. Phys. Lett. 30, 164 (1977).

8. D. J. Kuizenga, Opt. Commun. 22, 156 (1977).

Generation and Amplification of Short 10 μm Pulses

P.B. Corkum and A.J. Alcock

Division of Physics, National Research Council of Canada
Ottawa, Canada K1A 0R6

Abstract

Ultrashort pulse of 10μm radiation, generated with a dye laser controlled semiconductor reflection switch, have been used to probe temporal pulse broadening during amplification in a multi-atmosphere TE CO_2 gain module. At approximately 4 atmosphere the pulse duration deviates considerably from the duration predicted for a single pressure broadened gain line and by 7 atmospheres pulses regeneratively amplified to 10^8 times the initial power have a measured duration of $\lesssim$100 ps.

The development of semiconductor reflection switching [1,2] has provided a means for generating very short 10μm pulses and it appears that the technique will be applicable over a wide range of infrared wavelengths. Calculations indicate [3] that in the case of a single germanium switch illuminated by 10 ps, .6μm pulses with an absorbed power density of 80 MW/cm^2, a semiconductor plasma is produced with its density remaining greater than the critical density for 10μm radiation for $\sim$30 ps. Thus reflected pulses, $\sim$30 ps in duration, should be produced with germanium while the use of other semiconductors (eg InSb) may permit the generation of even shorter pulses with single switches. Furthermore, with double switching arrangements it appears feasible to generate short infrared pulses whose duration is largely independent of the semiconductor properties. One such double switching arrangement has already been demonstrated to be capable of gating pulses of <10 ps in duration[4]

In parallel with the development of techniques for creating ultrashort infrared pulses it is of interest to use such pulses in picosecond probe experiments. Since development thus far has been restricted to switching 10μm pulses generated by CO_2 lasers an experiment has been performed to probe the pressure dependence of the gain bandwidth in a multi-atmosphere TE CO_2 laser.

Line broadening coefficients for the constituent gases of CO_2 lasers have previously been measured [5,6] and these yield an effective value of 3.5 GHz/atm for a 10:10:80::CO_2:N_2:He laser gas mix. Since the rotational lines are separated by $\sim$60 GHz significant line overlap is expected at pressures of $\sim$8 atm or greater [7] while at even lower pressures the sequence lines are expected to modify the spectral profile of the gain [8]. This is illustrated in Fig.1 where the gain profile of the 10.6μm p-20 transition and the nearest sequence transition (p-17) are plotted for both 1 atm and 4 atm operation assuming a Lorentzian line shape and a gain on the sequence line that is 40% of that on the regular line. There are, however, no measurements of the gain on the sequence lines for multi-atmosphere CO_2 lasers.

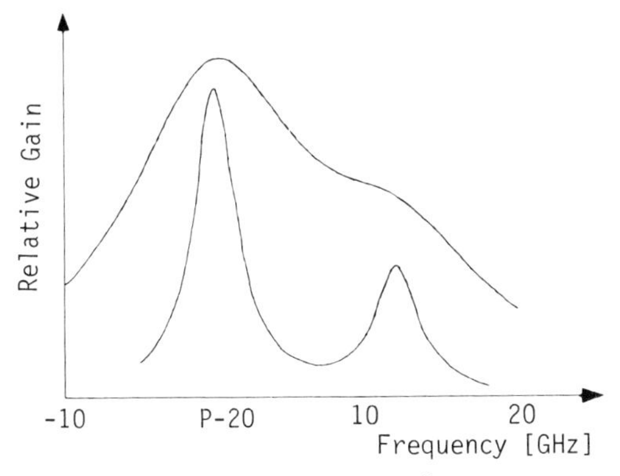

Fig.1 Gain profile for CO_2 laser near the p-20 10.6μm transition. The gain on the p-17 sequence line is assumed to be .4 gain on the regular (p-20) transition

If a sufficiently short low intensity probe pulse is amplified in a CO_2 laser gain medium, spectral narrowing due to the finite gain bandwidth will increase the duration of the amplified pulse. If the initial pulse duration is sufficiently small, then temporal broadening due to the gain medium will control the pulse duration (τ) of the amplified pulse. A simplified expression for this pulse duration

$$\tau = \frac{1}{\pi \Delta \nu} (8 \ln(2) g)^{\frac{1}{2}} \tag{1}$$

can be derived if it is assumed that the CO_2 gain profile can be approximated by a Gaussian with the same half width. In (1) $\Delta\nu$ is the full width at half maximum of the Gaussian gain profile and g is the gain coefficient at line center (Gain = e^g). Since the bandwidth $\Delta\nu$ is proportional to pressure, a scaling law

$$\tau \sim 1/p \tag{2}$$

follows. (Eq. (2) is valid for other line shapes as well.) Deviation from this scaling law is expected as pressure broadening causes significant line overlap and this should become important in the vicinity of the p-20, 10.6μm transition at pressures of ~3-4 atmospheres, Fig.1.

To probe the pressure dependent gain of a TE CO_2 laser short probe pulses are gated from the ~1 kw output of a grating tuned, low pressure CO_2 laser operated on the p-20, 10.6μm transition. The output of a passively mode-locked flashlamp pumped dye laser provided a train of ~10 ps pulses at 0.6μm to operate the gate. Both germanium and cadmium telluride have been used for switching. The train of 10μm pulses generated in this manner is injected [3] into a multi-atmosphere TE CO_2 oscillator [9] where the pulses are regeneratively amplified in the gain module. If the round trip cavity transit time of the 10μm pulse in the CO_2 laser (τ_C) is an integral multiple of the cavity transit time of the 0.6μm pulse in the dye cavity (τ_D), then the output of the TE CO_2 oscillator will appear as a train of mode-locked pulses Fig.2a,b. However, to avoid artificially increasing the duration of the regeneratively amplified pulse by incorrectly matching the cavity lengths, the cavities were adjusted so that

$$\tau_C = n\,\tau_D + \Delta T \qquad (3)$$

where ΔT, which can be varied, provides a streak camera calibration and n = 2 in this experiment. Due to the dynamics of the TE CO_2 gain buildup, only a few pulses are of any significance in the mode-locked output.

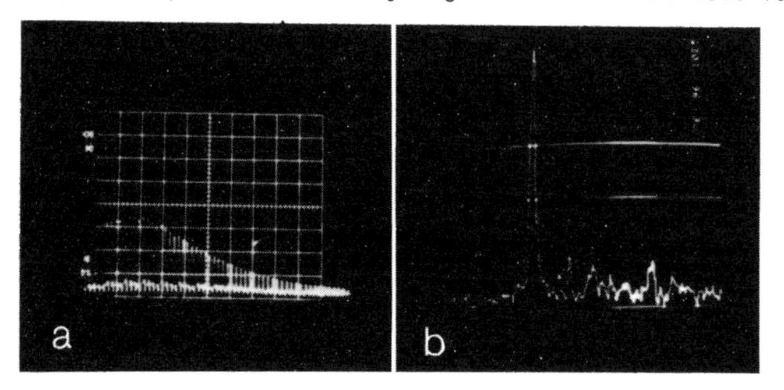

Fig.2 The regeneratively amplified output train of the TE CO_2 oscillator
A) Mode-locked train - 20 ns/div
B) Single regeneratively amplified pulse - duration $\sim$150 ps

This output was monitored on a photon drag detector and simultaneously the time dependence of one of the output pulses was recorded on an optical Kerr effect detection system [10] consisting of a 0.5 watt single longitudinal mode argon ion laser, a CS_2 optical Kerr effect shutter and a Hammatsu streak camera. Fig.3a and b illustrate the temporal characteristics of the regeneratively amplified 10μm pulses at 4 and 7 atm respectively (note: the pulse duration of the latter is limited by the streak camera resolution on the 3 mm/ns streak speed used) with cavity mismatch parameter ΔT = 100 ps and 110 ps respectively.

Combining the present data (labeled B), taken at 4, 6 and 7 atm with previous data (labeled A) taken at 2, 3 and 4 atmospheres (where the time response of the detection system was limited to $\sim$200 ps) the results displayed

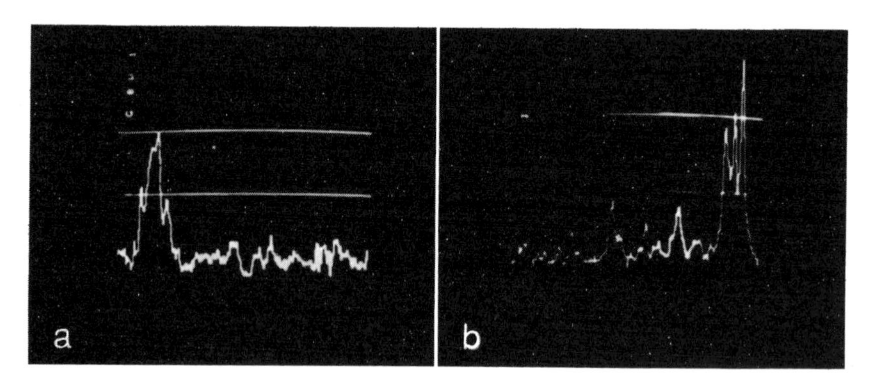

Fig.3 Streak camera measurement of the time duration of the regeneratively amplified pulses.
A) p = 4 atm, ΔT = 100 ps B) p = 7 atm, ΔT = 110 ps.

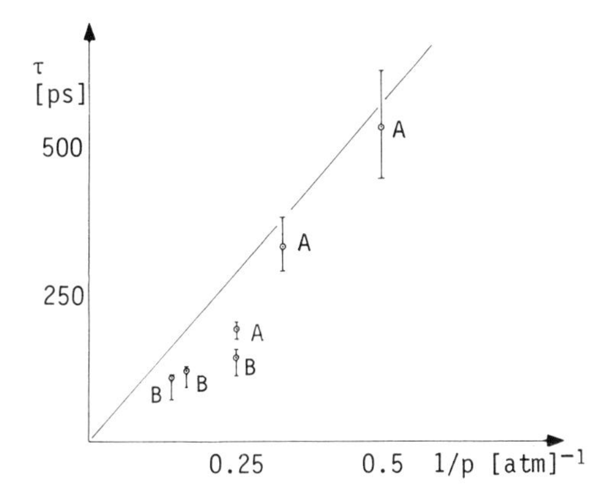

Fig.4 Pulse duration plotted as a function of 1/p. Points labeled A were measured previously with a system resolution of ~200 ps. The solid line shows the time duration as a function of 1/p as predicted by (1)

in Fig.4 are obtained. The solid curve in Fig.4 is calculated from (1) assuming $\Delta\nu/p$ = 3.5 GHz/atm, a net gain of e^{20} for the regeneratively amplified output pulse that was monitored and an additional gain of e^{10} to compensate for the linear losses experienced by the pulse between the time of injection and the time of the measurement (g = 30).

The large deviation of the experimental points in Fig.4 at pressures of $\gtrsim$4 atm from the curve computed by assuming a single Gaussian gain profile indicates that the sequence bands have considerable influence on the gain in multi-atmosphere TE CO_2 oscillator. It should now be possible to estimate the gain in the sequence lines by fitting the experimental data with the predictions of a more complete pulse broadening theory.

At pressures of 8 atm and above it is expected that the overlap of the regular lines will significantly influence the pulse duration and in future experiments the effect of regular line overlap on a regeneratively amplified pulse will be investigated.

References

1. A.J. Alcock, P.B. Corkum and D.J. James, Appl. Phys. Lett., 27, 680 (1975).

2. P.B. Corkum, A.J. Alcock, D.J. James, K.J. Andrews, K.E. Leopold, D.F. Rollin and J.C. Samson, Laser Interaction and Related Plasma Phenomena, edited by H.J. Schwartz and H. Hora (Plenum Press, New York, 1977) 143 ff.

3. P.B. Corkum, A.J. Alcock, D.F. Rollin and H.D. Morrison, Appl. Phys. Lett., 32, 27 (1978).

4. S.A. Jamison and V. Nurmikko. Quantum Electronics Conference, Atlanta, May 1978.

5. O.P. Judd, Hughes Res. Rep. 452, August 1971.

6. R.R. Patty, E.R. Manring and J.A. Gardner, Appl. Opt., 7, 2241 (1968).

7. A.J. Alcock, R. Fedosejevs and A. Walker, IEEE J. Quantum Electron., QE 11, 767 (1975).

8. J. Reid and K.J. Siemson, IEEE J. Quantum Electron., QE-14, 217 (1978).

9. A.J. Alcock, K. Leopold and M.C. Richardson, Appl. Phys. Lett., 23, 562 (1973).

10. T.C. Owen, L.W. Coleman and T.J. Burgess, Appl. Phys. Lett., 22, 272 (1973).

11. P.B. Corkum, A.J. Alcock, D.F. Rollin and H.D. Morrison, Optical Society Meeting, Toronto, Ont., Oct. (1977).

IX. Postdeadline Papers

Measurement of Singlet Lifetimes by Upper State Fluorescence Excitation

L. Hallidy, Horn-Bond Lin, and M. Topp

Department of Chemistry, University of Pennsylvania
Philadelphia, PA 19104, USA

1. Introduction

Increasingly sophisticated analysis of chemical and related phenomena in the subnanosecond time-range has imposed increasingly stringent requirements on the accuracy and reproducibility of experimental results obtained through fast laser experiments. Foremost among the problems being studied in this fashion are relaxation processes involving excited singlet states of small and medium-sized aromatic molecules, particularly in liquid environments. Much current attention is being turned towards weakly or effectively non-fluorescent states since their relaxation is dominated by strong, non-radiative interactions. Where the lowest states are fluorescent, there are several ways to time-resolve the transient populations. However, where the fluorescence is weak, two-photon gating techniques including transient absorption and fluorescence excitation spectroscopy have much to offer.

Picosecond flash photolysis experiments have provided substantial evidence for the rapid population of excited triplet states in several heterocyclic and substituted aromatic molecules [1-8]. However, almost without exception [4], published kinetic data have shown a growth of longer-lived spectra assigned to triplet states, but no decaying portion attributable to excited singlet state absorption. Further, there appears to be a rather poor correlation between results obtained for the same compounds, but in different laboratories. Problems such as these render it extremely difficult to interpret experimental results indicating the dependence of these rates on environmental factors.

The actual absorbance profile may be represented by:

$$D(t) = S_0(\alpha_T\Phi_T + (\alpha_S - \alpha_T\Phi_T)e^{-kt}) \tag{1}$$

where the α represent singlet and triplet absorption cross-sections, Φ_T, the triplet quantum-yield and S_0, the original excited singlet state concentration generated by a delta-function optical pulse. Thus, the conclusion on the basis of literature measurements would seem to be that $\alpha_T\Phi_T \gg \alpha_S$ over the sampling wavelength range.

This stresses the need to probe absorption changes with a wide-range spectral continuum, but there is another constraint. Although there should be no triplet absorption at t=0,a basic instrumental problem arises in determination of S_o and t=0. Thus, especially for short-lived states, one encounters a signal-to-noise problem having its origins in the intrinsic difficulty in specifying laser pulse-shape and duration. However, as we show, acridine has a singlet decay time considerably longer than the optical

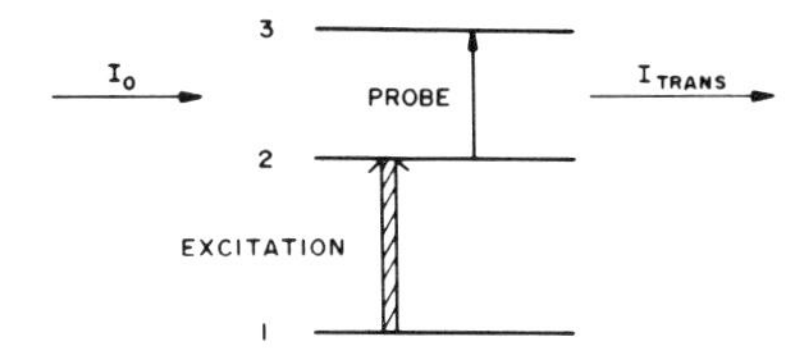

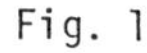
Fig. 1

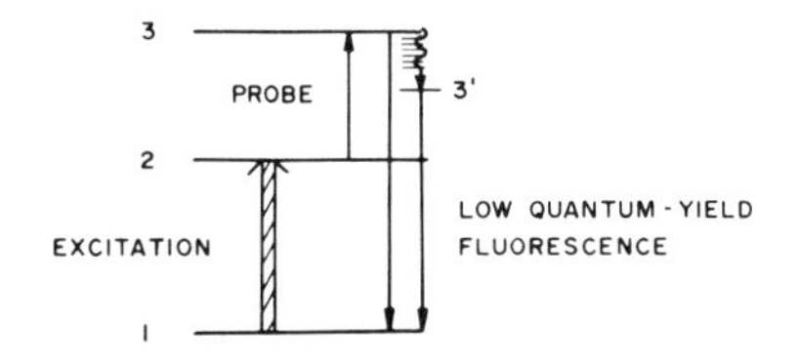

pulse and in this and similar cases, the experimental uncertainty encountered in transient absorption measurements must arise from an inability to distinguish between singlet and triplet absorptions.

This problem can be resolved by imposing additonal experimental constraints. In particular, we can require the terminal level of the secondary absorption process to be strongly radiatively coupled to the ground state. Thus, instead of measuring the actual absorption process, we can monitor fluorescence from the short-lived upper states to which the system is promoted, as in Fig. 1.

Thus, equation (1) becomes:

$$I_f(t) = \text{const} \times \Phi_{FS}(1-\exp(-S_o\alpha_s e^{-kt})) \quad (2)$$

where $I_f(t)$ is the intensity of upper state fluorescence for a particular sampling time t, and Φ_{FS} is the quantum-yield of emission from the upper state (usually 10^{-4} - 10^{-5}). It is assumed for the present that the upper triplet states give rise to negligible ultraviolet fluorescence. Thus, the time-differential of the signal changes sign following the pulse and the spontaneous decay can easily be distinguished from the excitation process. The presence of a short-lived singlet state will be detected as a pulse-limited "spike" whereas the triplet rise-time may not be separable from the time-integrated pulse-profile.

2. Time-resolution using consecutive two-photon excitation.

Over the past few years, we have introduced the use of consecutive two-photon laser methods to observe the spectra of highly excited states of aromatic molecules with lifetimes in the subpicosecond region [9-13]. The appearance of such emission depends on the coincidence of a secondary laser pulse with an excited state population and therefore it is possible to measure population fluctuations of the lowest excited states [14]. The method is independent of the fluorescence properties of the lowest state

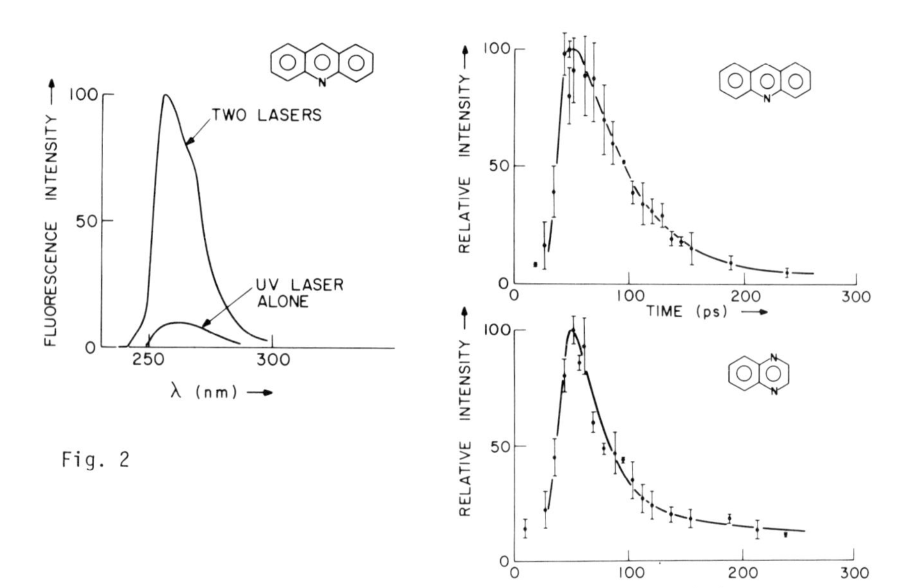

Fig. 2

Fig. 3

and, for example, in addition to the extremely weakly fluorescent aza compounds, we have measured the singlet absorption profile of BBOT scintillator using upper state fluorescence excitation. Parity restrictions apply only indirectly since, as we have recently shown for centrosymmetric molecules [13], fluorescent states can be populated following relaxation of the states directly excited by the laser.

This analytical technique uses a three-fold selection mechanism. Here, the molecules excited by the original laser pulse are further selected by a second pulse and still further by the spectrum of the weak upper state fluorescence. Therefore, while an impurity may dominate the S_1 emission spectrum by virtue of a favourable oscillator strength, these further selection criteria virtually eliminate the possibility of impurity emission, as we have shown in the spectra of numerous aromatic compounds.

3. Results

It has been known for some time that the fluorescence quantum-yield of acridine is strongly solvent-dependent, and this has been attributed to a level inversion between n-π* and π-π* lowest singlet states. The π-π* singlet state is weakly coupled to the triplet manifold and, if it is the lowest excited state, then the molecule is observed to fluoresce strongly, as is the case in aqueous solution. On the other hand, if the n-π* state is lowest (e.g. hexane solution) the combination of efficient coupling to the triplet manifold and the long radiative life-time to the ground greatly reducesthe fluorescence quantum-yield.

We have shown [12] that acridine has a strong upper state fluorescence band near 260 nm as, for example shown in Fig. 2. Therefore, we require only a small amount of singlet absorption at the probe wavelength in order to detect the singlet state population. Our experiment is designed to time-resolve the appearance and subsequent decay of this transient excitation spectrum. The apparatus has the same basic principles as for the nanosecond experiments [11] except for the use of picosecond laser harmonics and accurately synchronised optical paths.

We present, in Figs. 3 & 4, the first reported singlet absorption profiles of acridine and of molecules with related properties, quinoxaline and 9-fluorenone. The obvious background signal of 10-20%, also seen in Fig. 2, is due to multiphoton excitation of upper singlet state fluorescence by the two laser pulses separately. The effects of pulse synchronisation are quite clear (the absolute value of the time-scale is not important here), yielding high signal-to-noise profiles for all three molecules. (These represent true excited singlet state absorption profiles). The calculated decay times for acridine, quinoxaline and 9-fluorenone, all in hexane solution are 54 ± 4 ps, 35 ± 3 ps and 110 ± 20 ps respectively, the last two of which compare favourably with existing data [5,15]. The respective signal detection wavelengths were 265 nm, 247 nm and 265 nm.

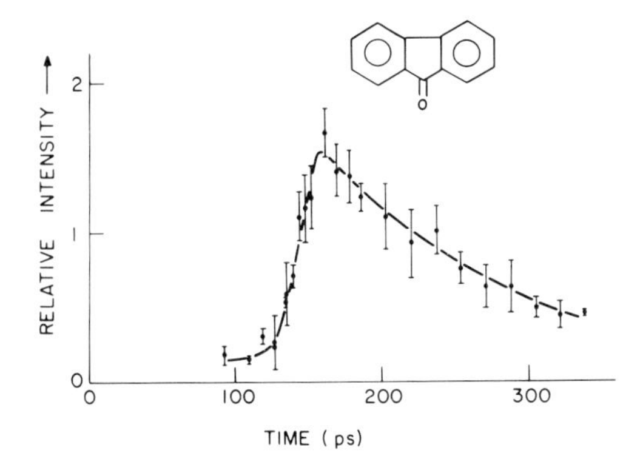

Fig. 4

We have also recorded the upper state emission spectra for acridine and quinoxaline with exactly synchronised pulses. The acridine spectrum agreed well with Fig. 2 (which was recorded by our nanosecond apparatus, in ethanol solution). The quinoxaline spectrum showed good agreement with the ground state absorption spectrum.

This very straightforward technique has virtually unrestricted application, since we can make use of at least three principal degrees of freedom: pulse-synchronisation and variation of both excitation and detection wavelengths. It forms an ideal complement of our frequency-conversion experiments on fluorescent states [16,17], which are currently being carried out using a high-repetition rate neodymium-YAG laser.

This work was supported by the National Science Foundation directly (CHE-76-10336) and through the Materials Research Programme (DMR-76-00678).

References

1. R. M. Hochstrasser, H. Lutz and G. W. Scott, Chem. Phys. Lett. 24 162 (1974).

2. R. W. Anderson, R. M. Hochstrasser, H. Lutz and G. W. Scott, Chem. Phys. Lett. 28, 153 (1974).

3. Y. Hirata and I. Tanaka, Chem. Phys. Lett. 41, 336 (1976).

4. R. W. Anderson, R. M. Hochstrasser and H. J. Pownall, Chem. Phys. Lett. 43, 224 (1976).

5. T. Kobayashi and S. Nagakura, Chem. Phys. Lett. 43, 429 (1976).

6. Y. Hirata and I. Tanaka, Chem. Phys. Lett. 43, 568 (1976).

7. V. Sundstrom, P. M. Rentzepis and E. C. Lim, J. Chem. Phys. 66, 4287 (1977).

8. G. W. Scott and L. D. Talley, Chem. Phys. Lett. 53, 431 (1977).

9. G. C. Orner and M. R. Topp, Chem. Phys. Lett. 36, 295 (1975).

10. H. B. Lin and M. R. Topp, Chem. Phys. Lett. 47, 442 (1977); 48, 251 (1977).

11. M. R. Topp and H. B. Lin, Chem. Phys. Lett. 50, 412 (1977).

12. K.J.Choi, L.A.Hallidy, H.B.Lin, M.R.Topp: In *Advances in Laser Chemistry*, ed. by A.H. Zewail, Springer Series in Chemical Physics, Vol. 3 (Springer, Berlin, Heidelberg, New York 1978).

13. H. B. Lin and M. R. Topp (in preparation).

14. P. M. Rentzepis, Chem. Phys. Lett. 2, 117 (1968).

15. R. M. Hochstrasser (private communication).

16. L. A. Hallidy and M. R. Topp, Chem. Phys. Lett. 46, 8 (1977); 48, 40 (1977).

17. L. A. Hallidy, H. B. Lin and M. R. Topp (in preparation).

Picosecond Studies of Excited State Proton Transfer Reactions: The Laser pH Jump[1]

A.J. Campillo, J.H. Clark[2], S.L. Shapiro, and K.R. Winn

University of California, Los Alamos Scientific Laboratory
Los Alamos, NM 87545, USA

1. Introduction

In terms of both the chemical insight gained and the predictive power provided, the concept of acids and bases must surely be one of the most general in all of chemistry. For Brönsted [1] acids and bases in protic solvents, acid-base chemistry becomes the chemistry of the proton. Since it is a naked charge bereft of any electron cloud, the proton has an ionic size of $\sim 10^{-6}$ nm, while all other cations are of the order of 0.1 nm or larger. Thus the proton is very reactive, since it can closely approach and highly polarize molecules and chemical bonds. While a detailed understanding of the mechanisms and kinetics of proton transfer reactions is a prerequisite for thoroughly explaining acid-base chemistry, the exceedingly rapid rates typical of these reactions has made them difficult to study using conventional techniques. Picosecond laser spectroscopy [2,3] provides a general method for the study of these rapid reactions.

As first observed by Weber [4] and later quantified by Föster [5] and Weller [6], the acid-base properites of many organic molecules change markedly upon electronic excitation. A number of examples of such behavior are given in Table I. The molecules studied in the present work, 2-naphthol and

Table.1 Ground (S_0) and first excited (S_1) singlet state pK_a values for various organic molecules (from Ref.7).

COMPOUND	$pK_a(S_0)$	$pK_a(S_1)$
2-NAPHTHOL	9.5	2.5
2-NAPHTHOL-6-SULFONATE	9.1	1.9
PHENANTHRENE	-3.5	19.5
PHENOL	10.0	3.6
AZOBENZENE	-2.9	13.7
CHALCONE	-5.0	7.1
FLUORENE	20.5	-8.5

[1]Work performed under the auspices of the U. S. Department of Energy.

[2]J. Robert Oppenheimer Research Fellow.

2-naphthol-6-sulfonate, become more than 7 orders of magnitude more acidic upon excitation. Other systems, such as phenanthrene and fluorene, display even larger pK_a changes upon excitation [7]. Furthermore, it should be noted that a molecule can be either more basic or more acidic in the excited state.

2. Kinetic Scheme

The reaction scheme pertinent to the experiments reported here is shown in Fig.1. Electronically excited states of the protonated species, ROH^*, and

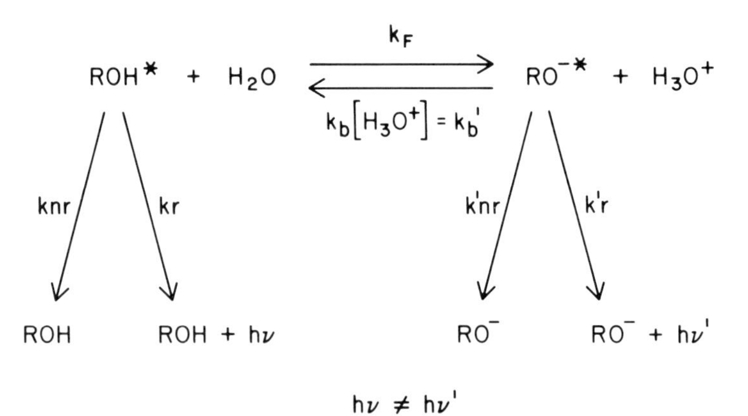

Fig.1. Reaction scheme for excited state proton transfer reactions.

the anion, RO^{-*}, may either fluoresce or be quenched non-radiatively back to their respective ground states. For 2-naphthol, 2-naphthol-6-sulfonate, and their respective anions, typical fluorescence lifetimes are ∿10 ns [8-10]. ROH^* may also transfer a proton to H_2O, while RO^{-*} may be protonated in the bimolecular reaction $RO^{-*} + H_3O^+$.

Changes in pK_a upon electronic excitation are simply a result of the difference between the absorption spectrum of the protonated species and that of its anion [11]. For fluorescent compounds such as 2-naphthol and 2-naphthol-6-sulfonate, the changes in the absorption spectra lead to concomitant changes in the fluorescence spectra. The fluorescence spectra of 2-naphthol and its anion, 2-naphtholate, are shown in Fig.2. While the 2-naphthol emission peaks at 360 nm, that of the 2-naphtholate is a maximum at 420 nm. Thus the two species may be distinguished by spectrally resolving the emission. The emission spectra of 2-naphthol-6-sulfonate and its anion are very similar to those of 2-naphthol and 2-naphtholate, respectively [12]. Since the ground state pK_a values for 2-naphthol and 2-naphthol-6-sulfonate are ∿9, only the protonated species is present in solutions whose pH < 7. Immediately following excitation, the only excited species is ROH^* (referring to Fig.1). Thus a measurement of the risetime of the emission from the excited anion (RO^{-*} in Fig.1) provides a direct measure of the proton transfer reaction rate, k_f. As the solution becomes more and more acidic, the protonation rate of RO^{-*} becomes faster and faster, and the risetime of the RO^{-*} emission becomes shorter and shorter. In the limit of very high acidity the temporal profile of the anion emission will mimic that of the protonated species.

Therefore by determining the risetime of the anion emission as a function of the pH of the solution, the complete protonation and deprotonation kinetics may be obtained.

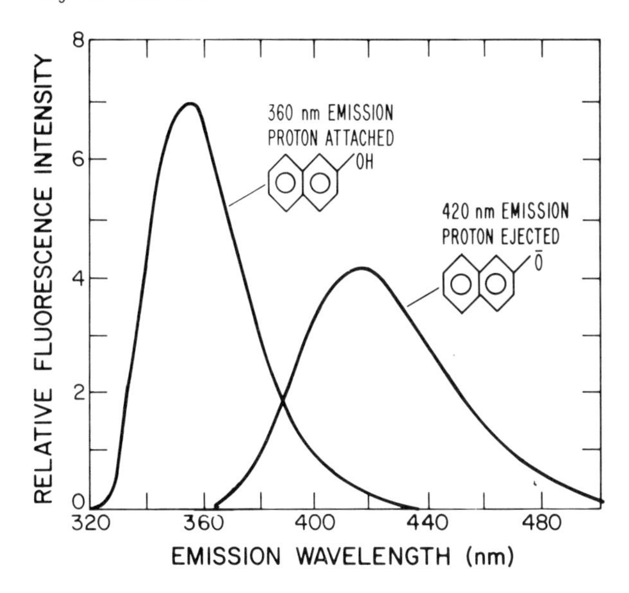

Fig.2 Emission spectra of 2-naphthol (proton attached) and its anion, 2-naphtholate (proton ejected).

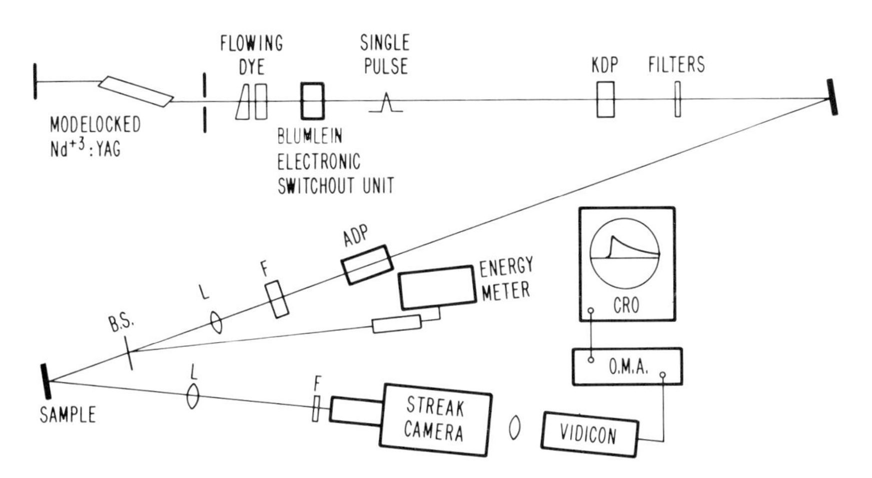

Fig.3 Schematic of experimental apparatus for measurement of excited state proton transfer reactions. (F = spectral filter, L = lens, B.S. = beam splitter).

3. Experimental

The experimental arrangement for measuring proton transfer rates is shown in Fig.3. A 1064 nm pulse is selected from the pulse train emitted by a mode-locked Nd:YAG laser by means of a longitudinal mode KDP Pockels cell, and is frequency shifted to 532 nm in KDP. The 532 nm radiation is redoubled in ADP to generate a pulse at 266 nm. The 1064 and 532 nm pulses are removed with filters, and the 266 nm pulse excites the sample. The fluorescence is collected onto the slit of an Electrophotonics ICC 512 (S-20 response) streak camera. A Corning 7-54 filter, which passes light below ∿380 nm, and a narrow band (12.5 nm FWHM) interference filter centered at 450 nm are placed in front of the slit of the streak camera to isolate the fluorescence from either the protonated species or the anion, respectively. Fluorescence streaks are imaged onto a silicon vidicon optical multichannel analyzer (PAR 1205D) and then displayed on an oscilloscope after each shot. The laser pulse duration at 266 nm is 20 ps. A beam splitting mirror directs a portion of the pulse to a Laser Precision energy meter. With knowledge of the energy and beam size the intensities could be determined. Samples of 2-naphthol and sodium 2-naphthol-6-sulfonate (Eastman) were used without further purification. Solutions were initially prepared in a N_2 purged glove bag using multiple-distilled, deionized water which was degassed by a series of freeze-pump-thaw cycles. However, since identical results were obtained with ordinary distilled water which was not degassed, subsequent solutions were prepared in this simpler manner. For all the experiments described here, the 2-naphthol concentration was 1.3×10^{-3} M, and that of the 2-naphthol-6-sulfonate was 1.0×10^{-3} M. The solution pH was varied by addition of reagent grade HCl or NaOH. The pH was measured with a Fisher Model 230 pH meter to an accuracy of ∿0.1 pH unit. Although samples were typically used immediately after preparation, the results obtained with freshly prepared and day-old samples were identical.

4. Results and Discussion

Figure 4 shows the time history of the emission for 2-naphthol-6-sulfonate from the protonated species at pH 1.6 (lower trace), and from the anion at pH 0.9 (upper trace) and pH 0.5 (center trace). Since it is directly pumped by the laser, the risetime of the emission of the protonated form is determined by the temporal profile of the laser pulse. At the streak rate used in Fig.4, this risetime is faster than the time resolution of the streak camera. The risetime of the anion emission, however, is much slower, and as expected from the reaction scheme of Fig.1, becomes more rapid with increasing solution acidity.

The points in Fig.5 are the experimentally determined risetimes (10-90%) of the anion emission for various solution pH values. The fact that the proton transfer rate does not vary between pH 3 and 7 confirms that the proton transfer is to H_2O, as shown in Fig.1, and not to OH^-, since the OH^- concentration changes by four orders of magnitude over this range of pH. The experimental points in Fig.5 may be fit with a kinetic model for proton transfer based on the reaction scheme of Fig.1. The equations for the ROH^* population, n_1, and the RO^{-*} population, n_2, may be written as:

$$\frac{dn_1}{dt} = -(k_r + k_{nr} + k_f)n_1 + k_b[H_3O^+]n_2 \quad \text{and} \tag{1}$$

$$\frac{dn_2}{dt} = -(k_r' + k_{nr}' + k_b[H_3O^+])n_2 + k_f n_1, \tag{2}$$

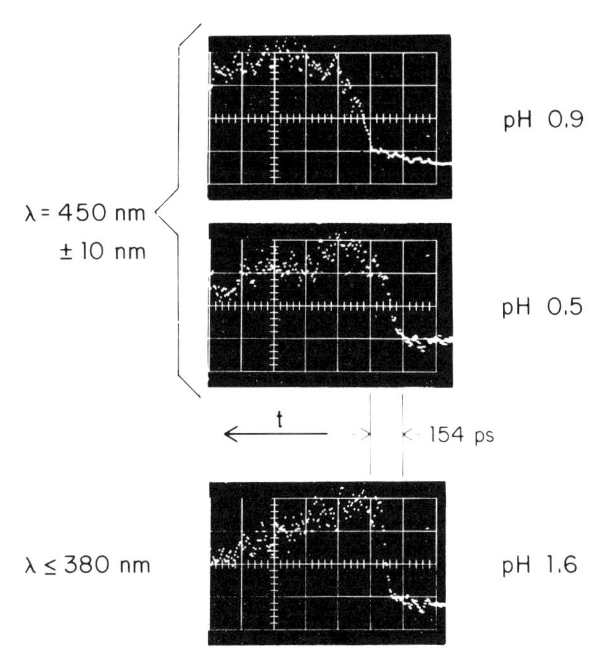

Fig.4 Time history of emission from 2-naphthol-6-sulfonate (lower trace) and its anion, 2-naphtholate-6-sulfonate (upper and middle traces) at various solution pH values.

where k_r and k_{nr} are the radiative and nonradiative loss rates for ROH^* and k'_r and k'_{nr} are the corresponding rates for RO^{-*}. In previous theoretical models [13,14], the backward rate is taken to be a diffusion controlled reaction between the excited anion, RO^{-*}, and H_3O^+ ions. Those analyses employ standard collision theory, modified by the electronic attraction between the ions. When $k_b[H_3O^+]n_2 >> k_f n_1$ (pH < 3), the risetime for the RO^{-*} emission becomes controlled by the back reaction, whereas in the intermediate pH ranges, (3 < pH < 7), the risetime is controlled by the forward reaction since the contribution from the backward reaction is small.

The solution of these equations for the initial condition $n_2 = 0$ at $t = 0$ is given by [7,8]:

$$n_1 = \frac{n_0}{\gamma_2-\gamma_1}\left[(\gamma_2-X)\,e^{-\gamma_1 t} + (X-\gamma_1)\,e^{-\gamma_2 t}\right] \qquad (3)$$

$$n_2 = \frac{k_f\,n_0}{\gamma_2-\gamma_1}\left[e^{-\gamma_1 t} - e^{-\gamma_2 t}\right], \qquad (4)$$

where $n_1 = n_0$ at $t = 0$,

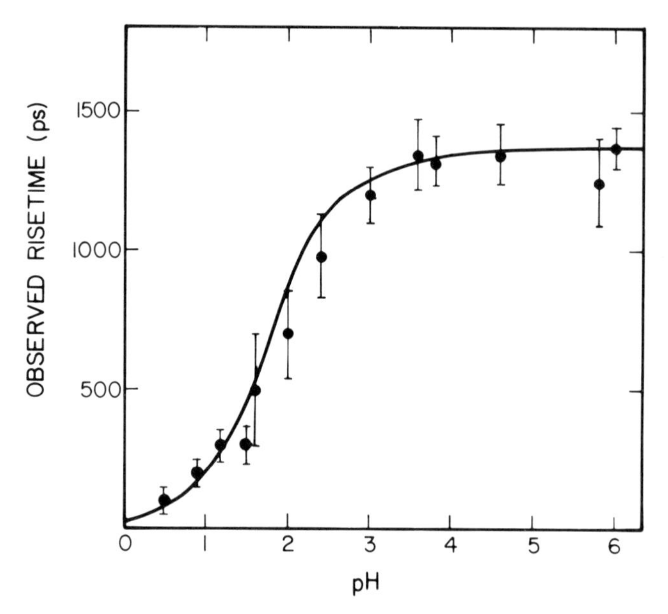

Fig.5 Risetime of emission (10-90%) for the 2-naphtholate-6-sulfonate anion vs. solution pH. Points represent the average of several experimental measurements. The solid line is a fit using a kinetic model based on the reaction scheme of Fig.1.

$$\gamma_{1,2} = 1/2 \left\{ (X+Y) \pm \left[(Y-X)^2 + 4\, k_f k_b\, [H_3O^+] \right]^{1/2} \right\},$$

$$X = k_r + k_{nr} + k_f, \text{ and } Y = k_r' + k_{nr}' + k_b\,[H_3O^+].$$

Using Eq.4 an excellent fit to the data for the risetime of the anion emission was obtained for 2-naphthol-6-sulfonate with the values $k_f = (1.02 \pm 0.2) \times 10^9\ s^{-1}$ and $k_b = (9 \pm 3) \times 10^{10}\ \ell\text{-mole}^{-1}\text{-s}^{-1}$. This fit is shown by the solid line in Fig.5. Use of the same fitting procedure resulted in rate constants $k_f = (7.0 \pm 2) \times 10^7\ s^{-1}$ and $k_b = (5 \pm 1) \times 10^{10}\ \ell\text{-mole-s}^{-1}$ for 2-naphthol in H_2O.

The proton transfer rate in 2-naphthol is sufficiently slow that it can be studied using nanosecond fluorimetry. The results reported here are in good agreement with previously reported values [8,9]. Comparison with values estimated from fluorescence titration work is also quite favorable [15,16,17].

In one direct measurement [9], it was necessary to invoke a very rapid ($k_f > 10^{10}\ s^{-1}$) proton transfer from "unrelaxed" excited 2-naphthol molecules in order to accurately model the experimental results. We have searched directly for evidence of such a transfer using very fast streak rates. Our failure to detect this rapid transfer confirms other nanosecond fluorimetry measurements [8,10], for which the results were adequately explained without invoking this mechanism.

The proton transfer rate in 2-naphthol-6-sulfonate is so rapid that it has defied previous attempts to measure it using nanosecond fluorimetry [8]. That such a measurement is easily carried out using picosecond techniques suggests that the method reported here will find general applicability in the direct measurement of fast proton transfer kinetics in a host of chemical and biological systems. The rapidity of the proton transfer in 2-naphthol-6-sulfonate has another important aspect. Since the proton transfer time is subnanosecond, while the fluorescence lifetime is $\sim$10 ns, acid-base equilibrium is achieved in the excited state. From Fig.1 it is seen that in addition to producing an excited anion, RO^{-*}, proton transfer also produces a hydronium ion, H_3O^+. With laser sources it is possible to produce substantial populations of excited state species, and hence substantial concentrations of H_3O^+. In the present experiments, knowing the beam shape, laser energy, and 2-naphthol-6-sulfonate absorption cross-section at 266 nm, we calculate that the H_3O^+ concentration was increased by 10^{-4} M. Thus a pH 7 solution was rapidly modulated, or "jumped" to a pH of 4 on a subnanosecond time scale.

This laser pH jump techniques offers a means for rapid initiation of the vast number of acid- or base-catalyzed, ground state reactions. The progress of reactions thus initiated can be followed by means of picosecond spectroscopy. The choice of a more appropriate jump molecule could provide both faster rising and longer lasting jumps [7] than those reported here. Such studies should allow the direct determination of the mechanism and kinetics of a wide array of condensed phase reactions. By providing a means with which such heretofore inaccessible aspects of these reactions can be obtained, the laser pH jump promises to bring a new, more detailed understanding to broad areas of chemistry and biochemistry.

Acknowledgments

Helpful discussions with W. C. Danen and J. L. Lyman are gratefully acknowledged.

References

[1] J.N.Brönsted, Rec. Trav. Chim. 42, 718 (1923).
[2] For a review of applications of picosecond techniques in chemistry, see K.B.Eisenthal: In *Ultrashort Light Pulses*, ed. by S.L.Shapiro, Topics in Applied Physics, Vol. 18 (Springer, Berlin, Heidelberg, New York 1977) pp. 275-315.
[3] For a review of picosecond measurement techniques, see E.P.Ippen, C.V.Shank: In *Ultrashort Light Pulses*, ed. by S.L.Shapiro, Topics in Applied Physics, Vol. 18 (Springer, Berlin, Heidelberg, New York 1977) pp. 83-122.
[4] K. Weber, Z. Phys. Chem. B15, 18 (1931).
[5] T. Förster, Naturwiss. 36, 186 (1949).
[6] A. Weller, Progr. Reaction Kinetics 1, 189 (1961).
[7] J. F. Ireland and P. A. H. Wyatt, Adv. Phys. Org. Chem. 12, 131 (1976).
[8] M. R. Loken, J. W. Hayes, J. R. Gohlke, and L. Brand, Biochem. 11, 4779 (1972).
[9] M. Ofran and J. Feitelson, Chem. Phys. Lett. 19, 427 (1973).
[10] A. B. Demjaschkewitch, N. K. Zaitsev, and M. G. Kuzmin, Chem. Phys. Lett. 55, 80 (1978).
[11] T. Förster, Z. Electrochem. 54, 531 (1950).

[12]R. M. C. Henson and P. A. H. Wyatt, J. Chem. Soc. Faraday Trans. II 71, 669 (1975).
[13]P. Debye, Trans. Electrochem. Soc. 82, 265 (1942).
[14]A. Weller, Z. Physik. Chem. N. F. 13, 335 (1957).
[15]L. Stryer, J. Am. Chem. Soc. 88, 5708 (1966).
[16]N. M. Trieff and B. R. Sundheim, J. Phys. Chem. 69, 2044 (1965).
[17]A. Weller, Z. Physik. Chem. N. F. 15, 438 (1958).

Measurement of Surface Recombination Velocity in Semiconductors by Diffraction from Picosecond Transient Free Carrier Gratings[1]

K. Jarašiunas[2,3], C. Hoffman[2], H. Gerritsen[2], and A. Nurmikko[4]

Abstract

Accurate measurements of surface recombination velocities at semiconductor interfaces (GaAs, InP) are accomplished using picosecond transient diffraction from an optically generated free carrier plasma grating. The method is insensitive to diffusion and bulk recombination processes.

Previous experimental studies of surface recombination velocity (S) in semiconductors have relied mainly on indirect methods [1,2]. We report the first application of transient diffraction from an optically generated free carrier plasma grating to the measurement of surface recombination at semiconductor interfaces. A significant advantage of this technique is its insensitivity, over wide ranges, to diffusion and bulk recombination processes. The method is based on picosecond optical resolution of a transient grating. The grating is generated at a semiconductor surface or interface by interband absorption of two coherent picosecond laser pulses coincident at a small angle. A delayed probe pulse, usually at a longer wavelength, monitors the diffraction efficiency as a function of time. The kinetics of the plasma grating following excitation are described by the diffusion equation

$$\frac{\partial N(x,z,t)}{\partial t} = D(N)\, \nabla^2\, N(x,z,t) - \frac{N(x,z,t)}{\tau_R} + g(x)\ell(z)$$

with the boundary condition $\left.\frac{\partial N(x,z,t)}{\partial t}\right|_{z=0} = (S/D)\, N(x,0,t)$ in

[1]Supported by National Science Foundation Grant DMR-76-81865 and International Research Exchange Program.

[2]Department of Physics, Brown University, Providence, RI.

[3]Division of Engineering, Brown University, Providence, RI.

[4]Permanent address: Department of Semiconductor Physics, Vilnius V. Kapsukas State University, Vilnius 232734, Lithuanian SSR, USSR.

the presence of surface recombination. $N(x,z,t)$ is the electron-hole density, $D(N)$ is the ambipolar diffusion coefficient, τ_R is the bulk recombination lifetime and $g(x)\ell(z)$ is the spatial generation function.

At the probe wavelength used in this work, the free carrier contribution to $\tilde{\varepsilon}(\omega) = \varepsilon_1(\omega) + i\,\varepsilon_2(\omega)$ leads to a thin periodic phase grating as the small conductive losses ($\varepsilon_2(\omega) \approx 0$) can be neglected. The first order diffraction intensity of the probe beam to which the sample is transparent, is [3]

$$I_1(t) = I_0\, J_1^2[(\pi n'/\lambda) \int_0^\infty dz\{N(\Lambda,z,t)-N(\Lambda/2,z,t)\}]$$

where I_0 is the probe pulse intensity, n' is the refractive index change per electron-hole pair, J_1 is the first order Bessel function, $\Lambda = \frac{\pi}{K\sin\theta}$ is the grating period, K is the wavevector of the incident excitation, λ is the probe wavelength, and 2θ is the angle between the two excitation beams symmetrically oriented with respect to the sample normal. Within the thin grating approximation [4], the integration of the plasma modulation into the bulk causes the diffraction efficiency to be relatively independent of diffusion in that direction, but dependent on the total number of carriers, determined by surface and bulk recombination processes. Therefore the variation of D with plasma degeneracy at high injected carrier densities plays only a minor role [5].

In our experiments the source of radiation was a modelocked Nd:glass laser producing 6 psec pulses at $\lambda = 1.06\mu$. A single pulse was frequency doubled in a KDP crystal and the residual 1.06μ radiation split off to serve as the probe pulse. The 0.53μ pulse was in turn divided into two beams of equal intensity which were recombined at the sample at a small ($\theta \approx 3°$) angle. One photon interband absorption of the green light generated a free carrier grating at the semiconductor surfaces under study with, typically, $\Delta N \lesssim 10^{20}$ cm^{-3} immediately after excitation. The picosecond $1.06\ \mu$ probe pulse, to which the samples were transparent, was delayed in controlled steps and the grating diffraction efficiency measured as a function of time delay.

We have initially performed experiments with GaAs and InP [6], both technologically important materials for which experimental characterization of surface recombination by other methods has often proven difficult. Some of our experimental results are shown in Fig.1 together with theoretical calculations. We find $S = 2 \pm 1 \times 10^4$ cm/sec for n-InP (Sn, 1×10^{17} cm^{-3}) and $S = 1.5 \pm 0.5 \times 10^5$ cm/sec for p-InP (Cd, 7×10^{17} cm^3). Both samples had (111) B surfaces chemomechanically polished with 1% Bromine in methanol. This rather unexpected difference agrees with the recent photoluminescence study by Casey and Buehler on InP [7]. For comparison we measured $S = 5 \pm 1 \times 10^5$ cm/sec for

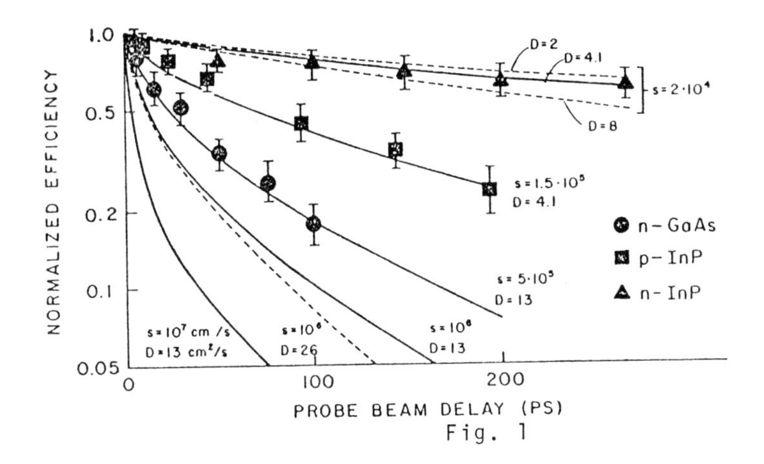

Fig. 1

a similarly prepared (100) n-GaAs (Te, 6×10^{16} cm^{-3}) surface in agreement with previous work [8]. As shown by the dashed curves in Fig.1, this method is indeed insensitive to relatively large changes in D. In addition, bulk carrier lifetimes of a few nanoseconds or longer will have negligible effect on the results.

The optically generated and probed picosecond grating provides a direct, sensitive and contactless method for studying recombination at semiconductor interfaces. The technique offers wide dynamic range for measuring recombination velocities and should be well suited to the investigation of interfaces in MIS devices and semiconductor heterostructures.

References

1. T. M. Buck and F. S. McKim, Phys. Rev. 106, 904 (1957).
2. D. B. Wittry and D. F. Kyser, J. Appl. Phys. 38, 375 (1967).
3. K. Jarašiunas and J. Vaitkus, Phys. Stat. Sol. (a) 44, 793 (1977).
4. J. P. Woerdman, Phillips Res. Rep. Suppl. 7, 1 (1971).
5. C. Hoffman, K. Jarašiunas, H. Gerritsen and A. Nurmikko (to be published).
6. We wish to thank E. Buehler af Bell Labs. for providing several InP samples.
7. H. C. Casey, Jr., and E. Buehler, Appl. Phys. Lett. 30, 247 (1977).
8. L. Jastrazebski, J. Lagowski, and H. C. Gatos, Appl. Phys. Lett. 27, 537 (1975).

On the Excited State Energy Surface of Rhodopsin and Bacteriorhodopsin

S. Rackovsky and A. Lewis

Department of Chemistry and School of Applied and Engineering Physics
Cornell University, Ithaca, NY 14853, USA

1. Introduction

Recently, LEWIS has proposed a molecular mechanism of excitation in visual transduction and bacteriorhodopsin [1]. There are three essential elements in this mechanism:

(1) Vertical excitation of the retinal chromophore produces electron density rearrangement.

(2) A bond rearrangement in the protein lowers the energy of this excited retinal-protein complex.

(3) Torsional motion and rearrangement of the chromophore couples the molecule back onto the ground state surface and stabilizes the new protein conformation.

Several experimental results on rhodopsin and bacteriorhodopsin have been accounted for in the original description [1] of the above excitation mechanism. In the present paper we discuss this mechanism of excitation in greater detail and show that recent observations on the temperature dependence of light induced picosecond absorption changes, photochemical quantum yields and fluorescence quantum efficiencies can also be accounted for by the above mechanism. These recent observations are the following: First, at low temperature the rate constant for the photochemical reaction exhibits non-Boltzman behavior of the type normally associated with tunneling [2]. Secondly, the number of molecules trapped in fluorescing states in bacteriorhodopsin increases strongly as the temperature is lowered and these states are apparently accessible only from rhodopsin and not bathorhodopsin [3]. Thirdly, quantum yield experiments indicate that rhodopsin and bathorhodopsin (the product of the photochemical reaction) have a common excited state [4].

To account for these experimental observations we will focus on the second element of the mechanism discussed above -- the protein bond arrangement. As has been noted [1] a very likely candidate for the rearranging species is a proton, for example in a hydrogen bond. It is well known [5] that protons in such bonds move in double-well potentials and exhibit tunneling behavior. Hydrogen bonds are present in large numbers in proteins and it seems reasonable that a transient electron redistribution is capable of inducing transitions between the sublevels of the doublets characteristic of double-well potentials. If the potential is asymmetric the sublevels of each doublet have different localization properties, which guarantees that such transitions are accompanied by translocation of the proton.

In this communication we demonstrate, using simple calculations on the above model, that proton translocation in the protein can indeed result from transient electron redistribution in the chromophore. Furthermore, we also demonstrate that this excitation mechanism can account for all the above experimental data.

2. The Theoretical Model

We consider a proton moving in a double-well potential. In the absence of any external interactions, the hamiltonian can be written as

$$H_o = K + V \tag{1}$$

where K is the kinetic energy operator and V is given by

$$V = \begin{array}{ll} 1/2\ \mu\omega^2\ (x + a)^2 & x<0 \\ 1/2\ \mu\omega^2\ (x - a')^2 + V_1 & x>0 \end{array} \tag{2}$$

Here μ is the proton mass, ω is the natural frequency of each potential well, and V_1 is the energy difference between the bottoms of the two wells. We have modelled the double-well potential by an asymmetric double harmonic oscillator potential; we do not anticipate any loss of physical generality from this mathematically convenient approximation.

It is known [5] that, if V_1 is not very large, the eigenstates of H_o occur as doublets, whose energy separation is considerably smaller than the separation between doublets. The eigenvalues and eigenfunctions of H_o will therefore be denoted by

$$H_o\phi_{n\sigma} = \varepsilon_{n\sigma}\ \phi_{n\sigma} \tag{3}$$

where n is the principal quantum number of those eigenstates of the separate wells from which the $\phi_{n\sigma}$ are constructed, σ = 1 or 2 and $\varepsilon_{n_2} > \varepsilon_{n_1}$. The $\phi_{n\sigma}$ are orthonormal.

To H_o we add a perturbation which represents the interaction of the proton with a transient charge density generated by photon induced electron redistribution. We model this charge density by a spherically symmetric charge density of total charge q, which is centered at r_o and decays with a characteristic lenth a_o. Thus,

$$\rho(|x-r_o|) = \frac{q}{8\Pi a_o^3}\ e^{-|x-r_o|/a_o} \tag{4}$$

Let q' be the (effective) charge associated with the proton. Then, it can easily be shown [7] that the potential due to $\rho(|x-r_o|)$ is

$$W(x) = \frac{qq'}{2|x-r_o|}\ [2(1-e^{-|x-r_o|/a_o}) - \frac{|x-r_o|}{a_o}\ e^{-|x-r_o|/a_o}] \tag{5}$$

We shall consider that this charge density is created at t=0 and acts for time $\hat{t}$ before decaying. As a simple approximation we can then write

$$H(t) = H_o + H'(x,t) \tag{6}$$

$$H'(x,t) = W(x)\ L(t) \tag{7}$$

where

$$L(t) = \begin{cases} 0 , & t<0 \\ 1 , & 0\leq t\leq \hat{t} \\ 0 , & T<t \end{cases} \tag{8}$$

We would like to calculate the probability that H' can induce a transition between the sublevels of the lowest doublet:

$$\lambda_{o,1\to 2} = |<\phi_{n_2}| \; e^{-i/\hbar \int_o^{\hat{t}} d\tau H(\tau)} \; | \; \phi_{n_1} >|^2 \tag{9}$$

The time-evolution operator in (9) can be expanded using a well-known identity [8]. To first order in the perturbation we have

$$e^{\frac{-i}{\hbar} \int_o^{\hat{t}} d\tau H\tau} = e^{\frac{-i}{\hbar} H_o \hat{t}} - \frac{i}{\hbar} \int_o^{\hat{t}} d\tau e^{\frac{-i}{\hbar} H_o(\hat{t}-\tau)} H'(x,\tau) e^{\frac{-i}{\hbar} H_o \tau} \tag{10}$$

When this expansion is inserted into the matrix element of (9), the first term does not contribute. Therefore, to first order

$$\lambda_{o,1\to 2} = \frac{2}{(\hbar\omega_o)^2} \; (|-\cos\omega_o \hat{t})| \; < \phi_{02}|W(x)|\phi_0 >|^2 \tag{11}$$

where $\omega_0 = \frac{1}{\hbar} (\varepsilon_{02} - \varepsilon_{01})$.

In (11) the term in paranthesis is an oscillatory function of ω_0 and $\hat{t}$. It is therefore necessary to ask whether, for physically realistic values of $\hat{t}$ and the potential parameters, $\lambda_{0,1\to 2}$ is of significant magnitude.

For a numerical estimate we chose $\hat{t} = 2(10^{-12})$ s and the following potential parameters:

$\mu = 1.67 \; (10^{-24})$ g

$\omega = 5 \; (10^{-13})/s$

$V_1 = 0.1$ Kcal/mole

V_0 (the potential barrier) = 3 Kcal/mole

These values are within the representative range for a hydrogen bond. We also chose $a_o = 3{\cdot}5A$, $q = 0{\cdot}5e$ and $q' = e$ and calculated $\lambda_{0,1\to 2}$ for three values of r_o. We find

$$\lambda_{0,1\to 2} = \begin{cases} 0{\cdot}114 , & r_o = 5{\cdot}0 \; \mathring{A} \\ 0{\cdot}107 , & r_o = 6{\cdot}0 \; \mathring{A} \\ 0{\cdot}098 , & r_o = 7{\cdot}0 \; \mathring{A} \end{cases}$$

Because of the simplicity of the model, these must be viewed as order-of-magnitude estimates. Nevertheless, they suggest that transient-charge-density-induced proton translocation is likely to be a process of some importance in rhodopsin and bacteriorhodopsin photochemistry. We shall now discuss the implications of this for the primary photochemical process.

3. The Excited State Energy Surface

In Figure 1 we show a proposed least-energy path along the excited state surface as a function of the transition coordinate.

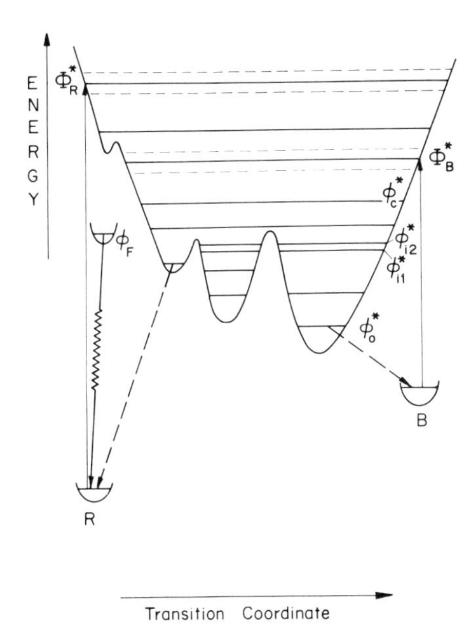

Fig.1 One possible least energy path for some combination of excited state protein (opsin) and retinal structural motion along the transition coordinate. Symbols are described in text.

This coordinate is some combination of the transition coordinates for protein/proton motion and retinal torsional coordinate. The presence of barriers due to various potentials including the hydrogen-bond potential provides the roughness in this surface suggested in [1] and causes at least in some cases the excited vibrational states of the electronically excited molecule to split into a doublet which we denote by the two states $\phi_{i_1}{}^*$ and $\phi_{i_2}{}^*$.

Vertical excitation from the ground electronic state of rhodopsin or bacteriorhodopsin without nuclear rearrangement leaves the molecule in a wave packet state ΦR^* which is a superposition of the vibrational levels ϕ_j of the electronic excited state (See Fig.1)

$$\Phi_R{}^* = \sum_j a_j \phi_j{}^* \tag{12}$$

Because of the nature of the excited state potential surface the higher energy vibrational levels will have a larger component in $\Phi R^*$and it is these states which are postulated to populate the state/states ϕF from which the ϕR^* populated (weak) fluorescence occurs. The vibronic state ϕc which is common to both R & B can be populated from both ΦR^* and ΦB^*.

The state ΦR^* evolves in time under the influence of a perturbation which can be thought of as involving two components: a specific vibrational relaxation mechanism and interaction with thermal modes of the system. At normal temperatures the only states which can interact through the thermal perturbation are split states such as $\phi_{i_1}{}^*$ and $\phi_{i_2}{}^*$. If we write

$$\Phi_R(t) = \sum_j a_j(t)\ \phi_j{}^* \tag{13}$$

then we expect

$$a_{i\sigma}(t) = a_{i\sigma}(t;T) \quad (\sigma = 1,\ 2) \tag{14}$$

In fact, this is possibly the origin of the strong T dependence of the rate of R → B photoreaction [2]. The system is prepared in an inverted state (in the sense of (13)). Thermal interactions will therefore have the effect of depleting the $\phi i_2{}^*$ component of ΦR (t) relative to the $\phi i_1{}^*$ component. Therefore, with decreasing temperature the lifetime of the $\phi i_2{}^*$ component of ΦR^* is increased and the rate of R → B decreases. A similar explanation may also apply to the increase in ϕF population as a function of temperature. In addition, the interaction between the two tunnel-split states may cause the characteristic [9] non-Arrhenius behavior at low temperatures, at which a T-independent rate is approached.

At long times, the only component which remains is that of $\phi_0{}^*$. It is from this state that relaxation to the B electronic ground state occurs in this model. Vertical excitation from the batho intermediate leaves the molecule in a different wave-packet state, ΦB^*. Because of the different geometry of ΦB^* we expect that, if we write

$$\Phi_B{}^* = \sum_\ell b_\ell\ \phi_\ell{}^* \tag{15}$$

we will have

$$b i_1 >> b i_2 \tag{16}$$

This condition, (16), could be responsible for the apparent lack of accessibility of the fluorescing state ϕF from B, but again the common vibronic state ϕc^*, indicated from the quantum yield measurements, results from relaxation of $\Phi_B{}^*$.

In summary we have discussed in some detail one possible version of the excited state vibrational energy surface for rhodopsin and bacteriorhodopsin. We have shown how the introduction of surface roughness, as first suggested in [1], can account for the known experimental features of these photochemical systems and can result in a multiminima excited state surface below the common excited state region ϕc. This multiminima surface, caused in part by movement of protons in opsin and bacterio-opsin hydrogen bonds, could cause a variety of phenomena including complicated picosecond behavior.

Acknowledgement - We would like to thank Professor Harold A. Scheraga for a most helpful discussion. This work was supported by a contract from the Naval Air Systems Command.

References

1. A. Lewis: Proc. Nat. Acad. Sci. USA, 75 549 (1978).

2. K. Peters, M.L. Applebury and P.M. Rentzepis: Proc. Nat. Acad. Sci. USA 74 3119 (1977).

3. a) A. Lewis, J.P. Spoonhower and G.J. Perreault: Nature 260 675 (1976).
   b) R.R. Alfano, W. Yu, R. Govindjee, B. Becher and T.G. Ebrey: Biophys. J. 16 541 (1976).

4. a) T. Rosenfeld, B. Honig, M. Ottolenghi, J. Hurley and T.G. Ebrey: Pure and Applied Chemistry 49 341 (1977).
   b) B. Becher and T.G. Ebrey: Biophys. J. 17 185 (1977).
   c) C.R. Goldschmidt, O. Kalisky, T. Rosenfeld and M. Ottolenghi: Biophys. J. 17 179 (1977).

5. a) S. Rackovsky: Chem. Phys. Letters 43 473 (1976).
   b) J. Brickmann and H. Zimmerman: Ber. Bunsenges. Physik. Chem. 70 157 (1966).

7. L. Pauling and E.B. Wilson, Jr.: Intro. to Quantum Mechanics, McGraw-Hill, (New York, 1935) p. 446.

8. A. Messiah: Quantum Mechanics, J. Wiley (NY, 1966) V.2, p. 722.

9. See, e.g., F.K. Fong: Theory of Molecular Relaxation, Wiley-Interscience (NY, 1975) Ch. 8.

Picosecond Self-Induced Transparency and Photon Echoes in Sodium Vapor

H.M. Gibbs and P. Hu

Bell Laboratories, Murray Hill, NJ 07974, USA

Abstract

Self-induced transparency breakup and narrowing of 200 to 500 ps pulses and photon echoes from two 20 ps pulses with picosecond pulse separation are reported.

Selection and amplification of one pulse from the modelocked output train of a synchronously pumped dye laser permit the extension of conventional coherent optical transient techniques to the picosecond regime. Studies of the dynamics of coherence are of interest in studies of collision processes in gases and relaxation processes in solids. The high transmission of self-induced transparency pulses permits analysis by real-time detectors or by correlation techniques such as second harmonic production with a delayed fraction of the output pulse or a shorter probe pulse. Angular separation of the photon echo from the two exciting pulses incident on the sample at slightly different angles permits the detection of the energy within the echo with slow detectors. It also permits the pulse separation to be made arbitrarily small since the detector is not subjected to the first two pulses. The conventional method of obtaining the two excitation pulses by splitting one pulse into a direct and a delayed pulse becomes especially simple in the picosecond regime. This paper illustrates these concepts in sodium vapor.

The apparatus is as follows. A homemade folded cavity Rh 6G dye laser is synchronously pumped by 200 ps, 68 MHz pulses from an actively modelocked Coherent Radiation 53 argon laser with extended cavity. Dye output pulse durations have been varied from 20 ps to 1 ns using intracavity etalons. An uncoated 0.8 mm etalon yields $\sim$ 20 ps pulses through the 26% transmitting output mirror for the photon echo studies. Addition of an uncoated 3-mm etalon and a 1-cm etalon either uncoated or with 20% coating lengthens the pulses to 200 to 500 ps for the self-induced transparency experiments. A single-pass dye amplifier pumped by a nitrogen laser selects [1] a single pulse by amplifying it by 10^3, resulting in near-transform-limited kW pulses with large electric field areas over millimeter diameters. Short pulses are studied via second harmonic correlation and spectral width determination. Real-time data are taken with a Rockwell International GaAs photodiode mounted directly on a Tektronix S-4 sampling head; the overall FWHM response time is about 100 ps.

The Na is contained in evacuated cells of 1 to 2.3 cm length and placed in a magnetic field of a few kG to lift the zero-field degeneracies. The pulses are incident perpendicular to the field and are linearly polarized along the field.

Self-induced transparency [2] (SIT) in Na is shown in Fig.1. A 5π 500 ps asymmetric input pulse (Fig.1a) is reshaped into two nearly symmetric output pulses by SIT breakup. [3] The first breakup pulse width is close to the detector response time. Narrowing of 500 ps 3π pulses to 200 ps and of 250 ps to 100 ps have also been observed. Uniform plane wave conditions are approximated by imaging the output of the cell onto an aperture to select the central portion of the beam. The frequency widths of these pulses is approaching the inhomogeneous width of about 3 GHz in high field. Pulses undergo similar reshaping in the sharp-line SIT limit provided the effective absorption is kept constant by increasing the density as the pulse duration is shortened. [4] Consequently, one should be able to observe SIT in Na with very short input pulses. The second harmonic detection scheme should be easily applicable since the output pulses are nearly as intense as the input pulses. Observations of 3π peaking and compression and large π to 2π delays should be easily studied by the second harmonic technique. Deconvolution of 4π breakup will be more difficult but may be possible for well-separated breakup pulses.

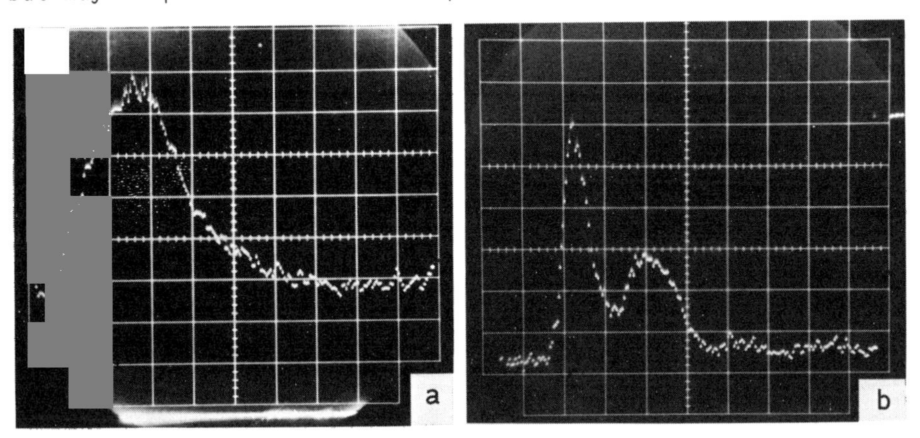

Fig.1 Self-induced transparency in Na. 200 ps per major horizontal division. (a) Input. (b) Output.

Photon echoes [5] have also been observed using 20 ps pulses separated by 450 ps. With the two excitation pulses separated by 0.5°, the echo is observed at 0.5° from the second pulse and is typically 10^3 times smaller. The angular separation between the echo and excitation pulses made it easy to reduce the leakage from the first two pulses to no more than a few percent of the echo signal. In systems with short dephasing times one should then be able to reduce the pulse separation to arbitrarily small time intervals and still use relatively slow, sensitive detectors to monitor the echoes. In contrast, the frequency switching technique [6] appears limited to approximately 50 ps by the crystal transit time. Picosecond photon echoes in pentacene, detected by optical mixing, are being reported by Hesselink and Wiersma at this conference.

References

1. T. Urisu and K. Kajiyama, Opt. Commun. 20, 34 (1977).
2. S. L. McCall and E. L. Hahn, Phys. Rev. 183, 457 (1969).
3. Compare with breakup of 2 ns pulses: B. Bölger, L. Baede, and H. M. Gibbs, Opt. Commun. 18, 67 and 19, 346 (1976).
4. H. M. Gibbs and R. E. Slusher, Phys. Rev. A 6, 2326 (1972).
5. I. D. Abella, N. A. Kurnit, and S. R. Hartmann, Phys. Rev. Lett. 141, 391 (1966).
6. R. G. DeVoe and R. G. Brewer, Phys. Rev. Lett. 40, 862 (1978).

Picosecond Spectroscopic Study on Excimer Formation in Solution and in Crystalline Phase

T. Kobayashi

Bell Laboratories, Murray Hill, NJ 07974, USA
(on leave from The Institute of Physical and Chemical Research
Wako, Saitama, 351, Japan)

1. Introduction

Complexes formed between excited and ground state aromatic hydrocarbons in solution [1] were named excimers [2]. Most aromatic compounds form excimers in solution which are usually characterized by their fluorescence [3]. No excimer emission has long been reported for anthracene under ordinary experimental conditions, although the self-quenching of its fluorescence is very efficient. Recently McVey et al. [4] interpreted a broad fluorescence from 0.2 - 0.1 M anthracene in chloroform solution to be due to anthracene excimer without confirming the coincidence of the monomer decay time and the excimer formation time.

Anomalously broad fluorescence was observed by Schneider and Lippert for 9,9'-bianthryl in many kinds of polar solvents [5]. The maximum positions fluorescence intensity depend on the solvent polarity, and therefore the emitting state of the fluorescence was assigned to the charge separated excited state (intramolecular exciplex) of two anthracene moieties of a 9,9'-bianthryl molecule. Excimer fluorescence from crystals of aromatic compounds were observed for dimeric crystals such as pyrene [6] and perylene [7]. The formation time constant of an excimer in the pyrene crystal at room temperature was measured to be shorter than 1 ns [8].

In the present study, the formation process of anthracene excimer was for the first time observed. Intramolecular exciplex formation is also observed for 9,9'-bianthryl. These processes have not been observed because of the low time resolution of the ordinary nanosecond spectroscopy. The excimer formation process in crystalline phase was also studied in relation with the excimer exciton transfer.

2. Experimental

Pulsed light source used for the fluorescence lifetime measurements is second harmonic (wavelength 347 nm, pulse width 20 ps, and peak power 300 - 500 MW) generated by the phase matched ADP crystal (2 cm thick) from an amplified single pulse (wavelength 694 nm, pulse width 20 ps, and peak power 1-5 GW) from a mode-locked ruby laser with a Pockels cell. Fluorescence intensity in the picosecond region was detected with a combined system of a monochromator, an image converter streak camera (ICSC, John Hadland IMACON 600), an image intensifier (II, EMI model 9914), an intensified silicon intensified target tube (ISIT, PAR model 1205I), an optical multichannel analyzer (OMA, PAR model 1205A), a desk-top computer (YHP model 9825A). The time resolution of the detecting system is about 20 ps.

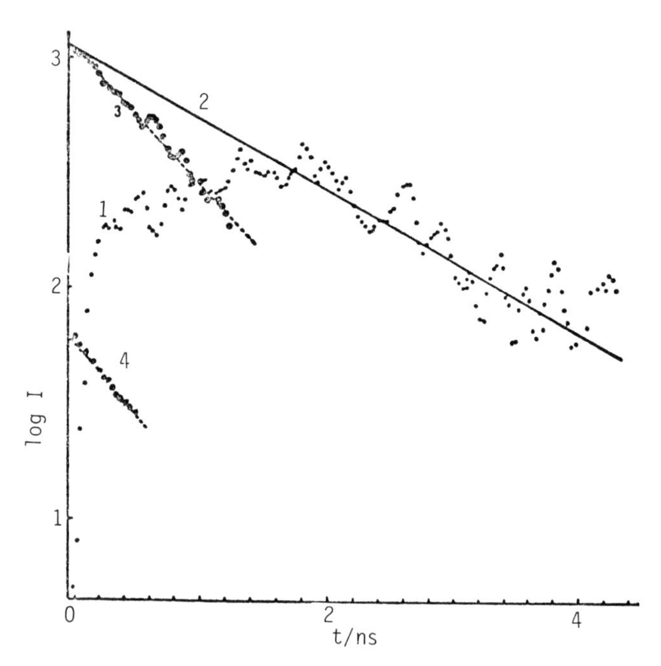

Fig.1 The logarithmic plot of fluorescence intensity of anthracene (0.1 M) in chloroform observed at 546 ± 10 nm (curve 1) and that at 400 ± 5 nm (curve 4). Line 2 and curve 3 are explained in the text.

3. Results and Discussion

3.1 Formation Process of Anthracene Excimer in Solution

Highly concentrated anthracene (0.1 M) in chloroform at room temperature exhibits very weak broad featureless fluorescence in the longer wavelength region than that of the ordinary characteristic fluorescence [4]. The peak of the broad fluorescence is located at 530 ± 20 nm while the wavelength of the 0-0 band of ordinary fluorescence is located at 385 nm. The energy difference between them is 7100 ± 700 cm^{-1}. The value is among the various values of the reported energy separation between monomer and excimer fluorescences.

In order to confirm the existence of anthracene excimer in fluid solution, the time dependence of the intensity was measured by the streak camera. The results at 0.1 ns/mm are shown in Fig.1. The time dependence of broad fluorescence intensity was observed at 546 ± 10 nm for the anthracene (0.1 M) in chloroform at 303 K and the result is given by dotted curve 1. Line 2 in Fig.1 has inclination corresponding to the decay time constant of 1.4 ns measured at the streak speed of 0.6 ns/mm and it is fitted to the time dependence of fluorescence intensity at 546 nm at the streak speed of 0.1 ns/mm. Curve 3 is obtained by substracting line 2 by curve 1. From curve 3 the rise time of the broad fluorescence was determined to be 710 ± 30 ps. The time dependence of the intensity of ordinary monomer fluorescence was observed at 400 ± 5 nm, the fluorescence was found to decay exponentially with a time constant of 690 ± 30 ps, as is shown by curve 4 in Fig.1. The decay time is in good agreement with the rise time of the broad fluorescence within the limitations of experimental error. This means that the excited species which emits broad fluorescence originates from the excited monomer state. From the viscosity (η=0.514 cP) of chloroform at 303K and the concentration of anthracene in solution ([A]), the time constant of the excimer formation (τ_f) is estimated to be 740 ps, by the equation $\tau_f=(k_{DF}[A])^{-1}=3000\eta/RT[A]$, where R and T have ordinary meanings. It is in agreement with both the decay and rise times within experimental error. From the energy separation between the intensity maximum of broad fluorescence and 0-0 band of monomer fluorescence and from the agreements of three time constants, i.e. decay time of monomer, rise time of broad fluorescence the calculated excimer formation time, the broad fluorescence is conclusively attributed to excimer. This gives the first direct evidence of excimer formation of anthracene in solution.

3.2 Intramolecular Exciplex Formation Process in 9,9'-bianthryl (C_0)

The time dependence of the intensity of fluorescence emitted from exciplex of C_0 in the glycerol-methanol solution was observed at 476 ± 5 nm. (Fig.2) The rise time of the exciplex fluorescence was obtained to be 105 ± 5 ps. The time dependence of monomer fluorescence was observed at 397 ± 10 nm. The decay time was found to be 100 ± 5 ps and it is in good agreement with the rise time of the exciplex fluorescence. This gives a strong evidence that the excited species which emits broad fluorescence is generated from locally excited state of anthracene moiety and most probably it is intramolecular exciplex.

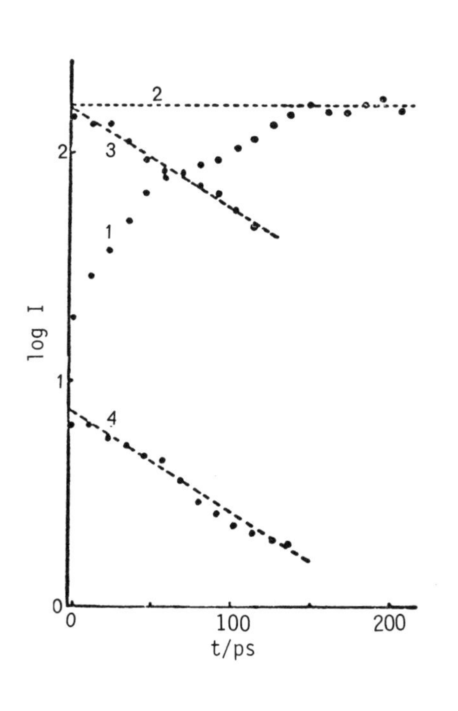

Fig.2 The logarithmic plot of fluorescence intensity versus time for C_0 (1.7×10^{-5}M) in glycerol-methanol (9:1) at 292K observed at 476 ± 5 nm (curve 1) and 397 ± 10 nm (curve 4). Line 2 represents the fluorescence decay at 476 ± 5 nm. Curve 3 is the difference between line 2 and curve 1.

3.3 The Formation Time of Excimer in Crystalline Phase

With the pyrene single crystal, a short-lived fluorescence (8 ns) which was attributed to monomer defect emission [6] was observed around 375 nm in addition to the long-lived excimer fluorescence (195 ns) around 465 nm. The rise curve of the fluorescence intensity of pyrene was observed at 395 nm and 478.5 nm at room temperature (298K) and 77K, the data at 77K being presented by circles (o), in Fig.3. The data were obtained by the accumulation of 8 signals.

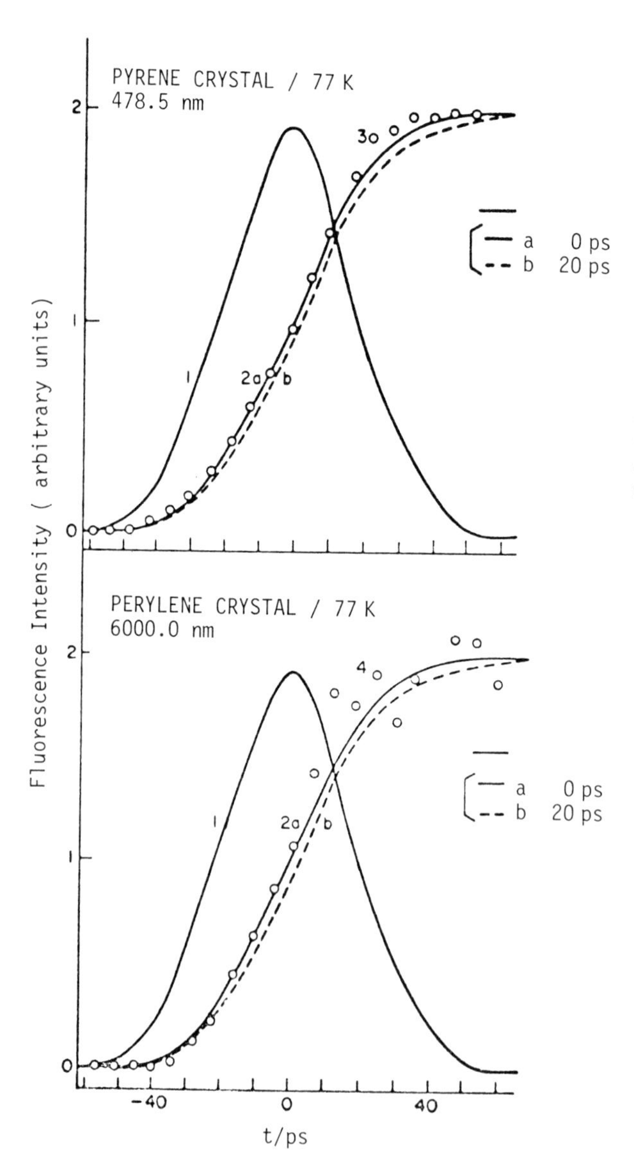

Fig.3 Curve 1 represents the observed laser pulse shape with the detection system shown in Fig.1. Curves 2a and 2b represent the convolution curve of exciting pulse shape, the time resolution function of the detection system, and the formation curve with time constant of 0 ps and 20 ps, respectively. Circles 3 and 4 indicate the time dependence of excimer fluorescence intensity from pyrene crystal observed at 478.5 nm and that from perylene crystals observed at 600 nm, respectively at 77K.

The calculated convolution curves of the excimer formation function (1-exp (-t/τ) with τ=0, and 20 ps), exciting pulse shape function, and the time resolution of the detection system are shown in Fig.2.

Comparison of the calculated convolution curves with the experimental results in Fig.2a, shows that the time constant of excimer formation in crystalline pyrene is shorter than 20 ps. This result indicates that the excimer in pyrene crystal is formed within 20 ps after excitation. The rise curve of pyrene excimer fluorescence at 293K was quite similar to that at 77K, and therefore the rise time of excimer in pyrene crystal was shorter than 20 ps both at 77K and at 293K.

The observed rise time was again shorter than 20 ps at 77K. The rise curve of excimer fluorescence of perylene crystal at 77K and at 293K were measured, and the results at 77K are given in Fig.3. The time constant of excimer formation in perylene crystal was found to be considerably shorter than 20 ps both at 77K and at 293K. The observed rise time of monomer fluorescence at 77K was also shorter than 20 ps.

The experimental results above described support the mechanism proposed by Inoue et al. [9], i.e., monomer fluorescence observed around 395 nm is emitted from a monomer defect and excimer fluorescence is emitted from the excimer exciton band, both being generated from a common excited state. Let us designate the rate constants for energy transfer of the processes from a monomer exciton band (A_m) to a monomer defect (D_m) and from A_m to an excimer exciton band (B_e) by k_M and k_E, respectively. From the results obtained in the present experiment the sum of k_M and k_E is considerably larger than $5 \times 10^{10} s^{-1}$ at both 293K and 77K. In the case of perylene crystal in the temperature range 96-135K, k_E is much larger than k_M. Since k_M is temperature independent and k_E has an activation energy of 280 cm^{-1} for perylene crystal [9], the condition $k_E \gg k_M$ is satisfied in the temperature range, 96-293K. Since the fluorescence intensity ratio of excimer to monomer defect is much larger in pyrene than in perylene, the condition $k_E \gg k_M$ is also satisfied for pyrene crystal in the temperature range 96-293K. Therefore k_E is concluded to be considerably larger than $5 \times 10^{10} s^{-1}$ in pyrene and perylene crystals at both 293K and 77K.

Since excimer exciton is localized, motion of the excimer through the lattice must involve sequential closing and opening of the intradimer distance as it propagates. Therefore the excimer formation in a crystalline lattice is closely related to the transport of excimer exciton.

In principle, the excimer state could be transferred from one pyrene pair to a neighboring one via the following three mechanisms [10].

Mechanism (a): A pair adjacent to the excimer is compressed by a lattice vibration to the position P_g^{ne} where the distance between two pyrene molecules is equal to the excimer equilibrium distance (d=3.34 Å), followed by excimer-energy transfer to the pair. The donating pair then is expanded from position P_g^{ne} to position P_g^{e}. (Fig.4a)

Mechanism (b): There is still excimer interaction in the ground-state equilibrium position P_g^{e} of the molecules. The separation of molecular planes is 3.53 A [11]; position P_e^{ne} is reached by thermal activation of the excimer followed by excimer-energy transfer to a neighboring pair in its equilibrium position P_g^{e}. (Fig.4b)

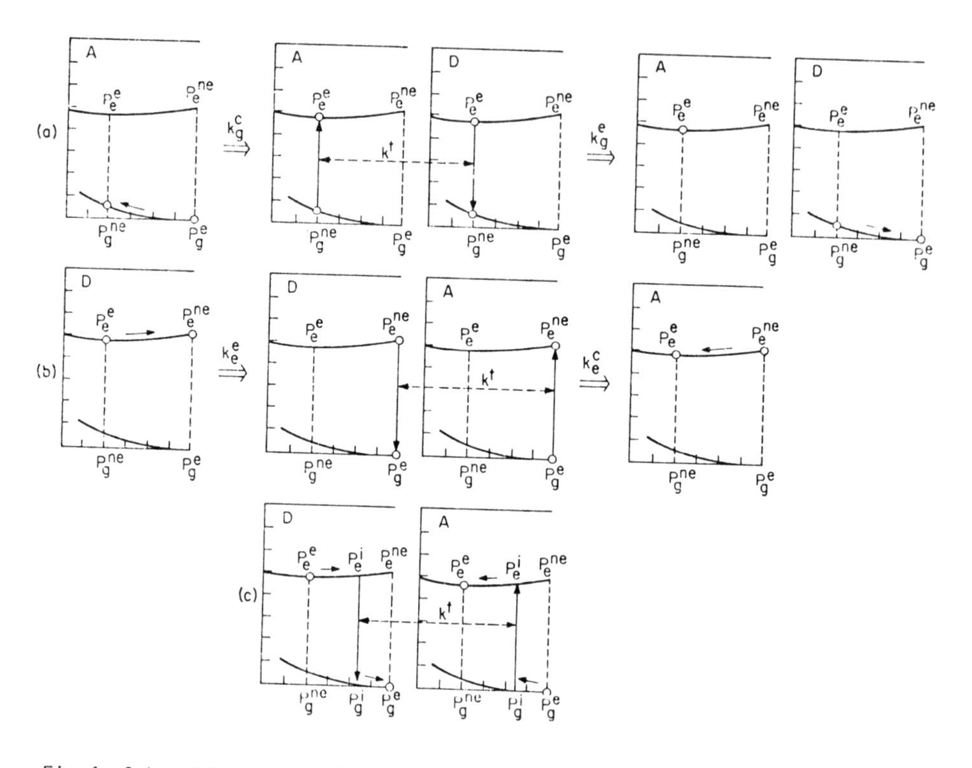

Fig.4 Schematic energy diagrams representing the three proposed mechanisms (a), (b), and (c) for the excimer exciton migration. D and A represent the energy-donating and -accepting pair molecules, respectively. P_g^e and P_e^e represent the points in the equilibrium configuration on the potential curve for the ground state and excimer state, respectively of the pairing molecules of pyrene in crystalline phase. P_g^{ne} on the ground state potential curve and P_e^{ne} on the excimer state potential curve represent the points which have the same configuration with P_e^e and P_g^e, respectively. k_g^c, k_g^e, k_e^c, k_e^e, and k^t represent the rate constants of the corresponding processes.

Mechanism (c): This is an intermediate case between mechanisms (a) and (b). The excimer is expanded from position P_e^e to position P_e^i and the adjacent pair in the ground state is compressed from position P_g^e to position P_g^i simultaneously, followed by excimer-energy transfer to the adjacent pair. Then the energy donating and accepting pairs change their conformation from position P_g^i go position P_g^e and from position P_e^i to position P_e^e, respectively. (Fig.4c)

Using the activation energy (2870 cm^{-1}[6]) necessary for the compression of a pair of pyrene molecules in the ground state from the position P_g^e to the position P_g^{ne} and the usual frequency factor of the compressive motion ($10^{13-14}s^{-1}$), the rate of the compression between 77K and 293K is estimated to be smaller than $10^{7-8}s^{-1}$. Therefore, mechanism (a) can not apply to excimer formation in crystalline pyrene and perylene. If the exciton hopping takes place by mechanism (b), the jump time of excimer exciton can be expressed approximately in terms of the sum of the expansion time $((k_e^e)^{-1})$ of an energy-donating excimer from position P_e^e to position P_e^{ne}, the excimer energy transfer time $((k^t)^{-1})$, and the compression time $((k_e^c)^{-1})$ from position P_e^{ne} to position P_e^e of an energy-accepting pair. Since the compression time $((k_e^c)^{-1})$ is included in the excimer formation time which is considerably shorter than 20 ps in the present study, $(k_e^c)^{-1}$ must also be considerably shorter than 20 ps. Therefore it appears that there is no potential barrier or a very low potential barrier in the potential curve from position P_e^n to position P_e^e. From the discussion given above, the sum of $(k_e^e)^{-1}$ and $(k^t)^{-1}$ is approximately equal to the exciton-transport time (20 ps) [12]. If the excimer exciton transport takes place by mechanism (c), position P_g^i is located higher than position P_g^e by appreciably less than 1400 cm^{-1}. If the position P_g^i is higher in energy than position P_g^e by 1400 cm^{-1} than $(k_g^c(P_g^e \rightarrow P_g^i))^{-1}$ would be longer than 20 ps.

The real exciton transport takes place by both mechanisms (b) and (c) with various location of positions P_g^i and P_e^i. The excimer energy transfer time $((k^t)^{-1})$ has been found to be shorter than 20 ps in both mechanisms.

We have observed for the first time that the formation times of excimer in pyrene and perylene crystals are considerably shorter than 20 ps both at room temperature and at liquie nitrogen temperature. In mechanism (b) there is still excimer interaction in the ground-state equilibrium position. In this sense, excimer in crystal may not be a suitable term. The study of sub-picosecond spectroscopy on the excimer formation and transport processes may be the subject of future research.

References

1 T. Förster and K. Kasper, Z. Physik. Chem. NF., 1, 275 (1954).
2 B. Stevens and E. Hutton, Nature (London) 186, 1045 (1960).
3 T. Förster, Angew. Chem. (Int. Ed. Engl.) 8, 333 (1969).
4 J. K. McBey, D. M. Shold, and N. C. Yang, J. Chem. Phys. 65, 3375 (1976).
5 F. Schneider and E. Lippert, Ber. Bunsenges. 72, 1156 (1968).
6 J. B. Birks, A. A. Kazzaz, and T. A. King, Proc. Roy. Soc., 291A, 556(1966).
7 J. Tanaka, Bull. Chem. Soc. Japan, 36, 1237 (1963).
8 J. B. Birks, Photophysics of Aromatic Molecules (Wiley, New York, 1970).
9 A. Inoue, K. Yoshihara, and S. Nagakura, Bull. Chem. Soc. Japan, 45, 720 (1972).
10 W. Klopffer, H. Bauser, F. Dolezalek, and G. Naundorf, private communication
11 J. M. Robertson and J. G. White, J. Chem. Soc., 358 (1947).
12 W. Klöpffer and H. Bauser, Chem. Phys. Letters, 6, 279 (1970).

Characterization and Application of a Combined Picosecond Streak Camera, Silicon-Intensified-Target Vidicon, and Image Analyzer System

W. Yu
Hamamatsu Corporation, Middlesex, NJ 08846, USA

A streak camera system consisting of a picosecond streak camera, a silicon-intensified-target vidicon, and a microprocessor-based image analyzer has facilitated the operation and readout of the streak camera (Hamamatsu Models C979, C1000-12, and C1098, respectively).

The microprocessor-based analyzer measures the streak image on the SIT Vidicon for each video line by integrating the intensity across the slit image of the streak camera. The output intensity as function of time formatted in 256 intensity levels in 256 time channels, is available instantly on a TV monitor, and can be transferred to a chart recorder, teletype writer, or a computer. The analyzer is designed to have capabilities for dark current subtraction, sensitivity shading normalization, and to look into information in any portion of the slit image.

To calibrate the linearity and dynamic range of the combined streak camera, SIT, and analyzer system, in its static (non-streak focussing mode), a voltage stabilized tungsten lamp light was projected through a calibrated neutral density step filter, and collected into the combined system. It was found the γ factor ($(\text{output}) \sim (\text{input})^{\gamma}$) is within 1% of unity for a dynamic range of better than 300. In Fig. 1, a log-log plot of the output from the combined system versus the input light inten-

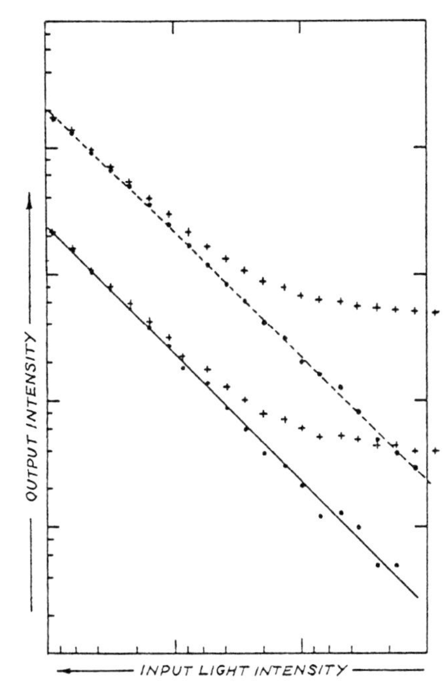

Fig.1 The output from the streak camera under focus (non-streaking) mode and SIT camera measured by the Temporalanalyzer as a function of the input light intensity in a log-log plot. The input light intensity is calibrated using calibrated Kodak step filter. +, raw data; •, after subtraction of background and dark current noise. Lower curve in solid line: one video frame; upper curve in dark line, 256 frame integration. The γ factor of both curves is 1% within unity.

sity is shown for one video frame measurement. After subracting the background dark current value, the linear dynamic range is determined to be better than 300. By integrating 256 frames, the linear dynamic range is further extended to greater than 700.

To calibrate the dynamic characters of the combined streak camera system, a single 6ps, 530nm laser pulse was passed through an etalon of transmission coefficient for each mirror T, and round trip time spacing Δt. The calibrating pulses produced is a train of pulses separated in time by Δt, and with its intensity profile decaying in a single exponential and each peak intensity reduced by $(1 - T)^2$. This pulse train was used to calibrate the linearity, dynamic range, time resolution, and streak speed linearity in a single shot. The result shows the streak camera system has better than 10ps time resolution, a linear dynamic range better than 100

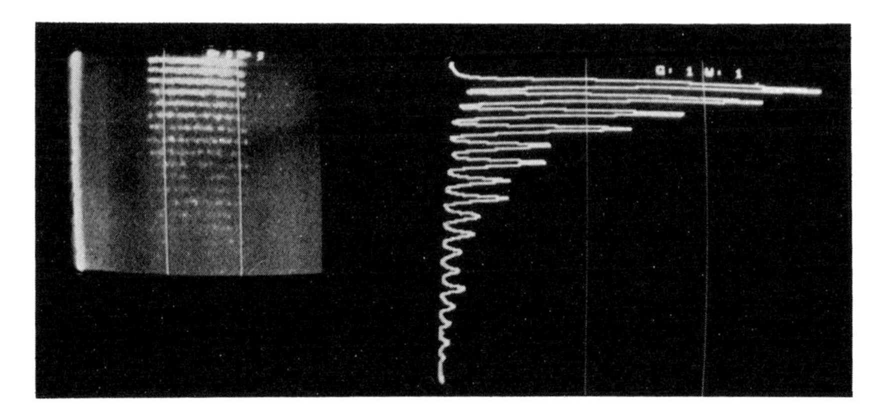

Fig.2 A photograph of the output from the streak camera. At left a video image of the streak. The time axis is vertically downward.Time between pulses: 30 ps; intensity ratio between pulses: 75%.On the right is the output from the Temporalanalyzer.

for all streak speeds, and a streak speed linearity well within 10% of the mean speed. In Fig. 2, an example of streak image captured by a one video frame grabber (Hitachi) is shown on the left. The streak image analyzed is shown on the right on the TV monitor. The graph is a single exponential decay with each peak intensity reduced by a factor of 0.75 for 17 peaks to cover a linear range of $\sim$100. The time spacing between the peaks is 30 ps.

To test the time jitter of the combined system, an electrical voltage pulse was splitted to trigger the streak camera and to excite a GaAlAs diode laser. The laser diode output of a 40ps, 820nm pulse was recorded by the system. By integrating streak image of 20 pulses, the time broadening of the output of the system was interpreted as the time jitter from the total streak camera system and laser diode. The total time jitter was found typically on the order of 50ps.

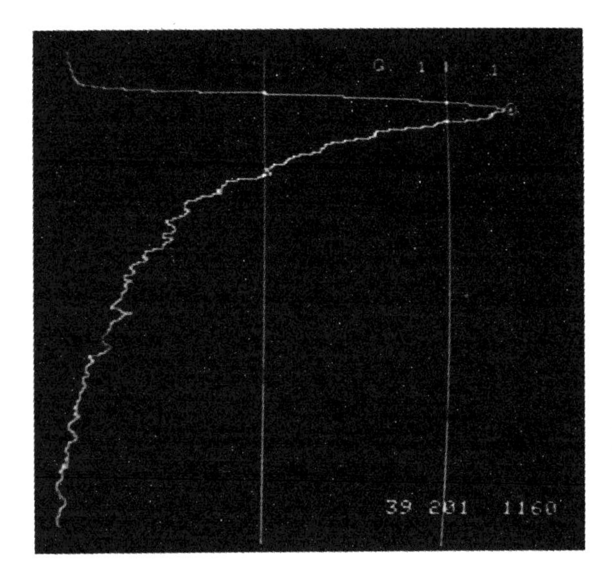

Fig.3 Typical fluorescence decay of CN7 in glycerol: η=1.000 cp, T=840 ps, λ>580nm.

The combination of a reliable ready to use streak camera and automatic data acquisition system has greatly facilitated the application of the streak camera to time resolved studies. We have used the system to measure the fluorescence decay of a malonitril substituted merocyanin dye ($(CH_3)_2N-[CH=CH]_3-CH=C(CN)_2$, CN7) in glycerol as a function of solvent viscosity. The only parameter varied for obtaining various solvent viscosity is the temperature of the solution (1). The fluorescence excited by a single pulse of 6ps, 530nm at 10^{13} photon/cm^2 was collected into the streak camera after passing through three Toshiba 0-58 filters. The fluorescence decay was observed to be single exponential. The lifetime was observed to vary as $\eta^{2/3}$ for $5 < \eta < 870$cp. For $\eta >$ 870 cp, the lifetime begins to level off. The $\eta^{2/3}$ dependence of fluorescence lifetime has been previously observed in malachite green in glycerol and glycerol and water mixture (2). The result is also consistent with the $\eta^{2/3}$ dep-

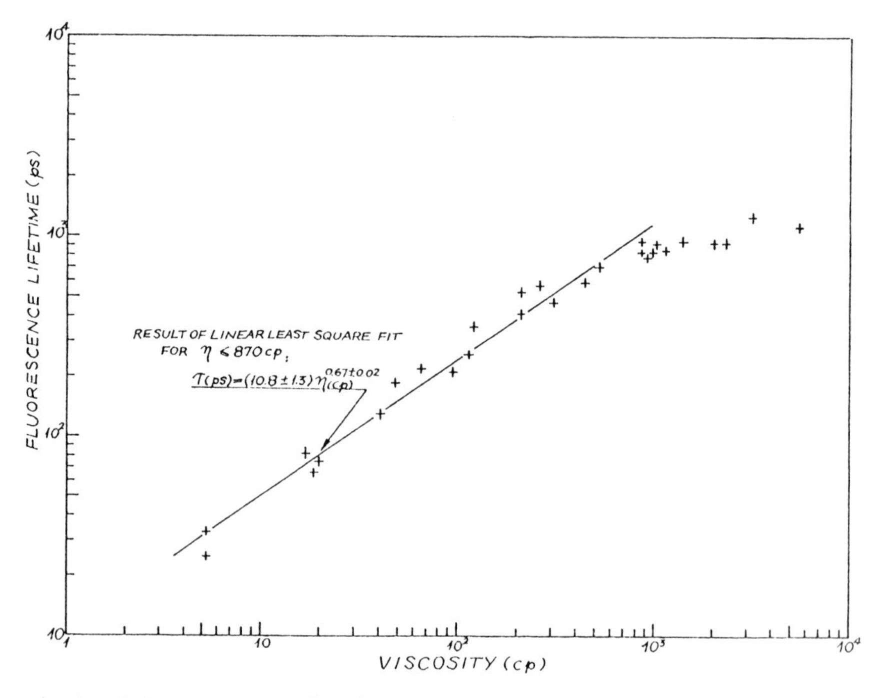

Fig.4 CN7 fluorescence lifetime vs. viscosity of the glycerol solution.

endence measured with the quantum yield measurement of triphenyl-methane dyes in various solutions (3). Merocyanin dye is known to have internal motions by cis-tran isomerization (4). Whether the isomerization and its behavior under solvent viscosity are the mechanisms of fluorescence quenching is an interesting one. Further experiments to elucidate such questions is in progress (5). I would like to thank Dr. S. Schneider for providing the merocyanin dye for the experiment.

References

1. Landolt-Bornstein, Zahlenwerte und Functionen, II Band, 5.Teil, p.213, Springer-Verlag, Berlin 1969.
2. Yu, W.Pellegrino, F.,Grant, M.,Alfano, R.R., J.Chem.Phys.67,1766 (1977)
3. Forster, Th., Hoffmann, Z.Phys.Chem. N.F.75, 63 (1971)
4. Jorges, E., Schneider, S., Dorr, F., Daltrozzo, E., Ber. Bunsenges Physik, Chem. 80, 639 (1976)
   Wirth, P., Schneider, S., Dorr, F., Ber. Bunsenges, Phys. Chem.81, 1127(1977)
5. Yu, W., Schneider, S. to be published.

List of Participants

Ackley, Donald C.
Brown University
Engineering Box D
Providence, RI 02912, USA

Anderson, Robert W.
Xerox Corporation
Rochester, NY 14580, USA

Aoyagi, Yoshinobu
Inst. Phys. Chem. Res.
Wako-smi, Saitama 356, Japan

Attwood, David
Lawrence Livermore Laboratory
L-482
Livermore, CA 94550, USA

Ausschnitt, Kit
Bell Laboratories
Crawford Corners Road
Holmdel, NJ 07733, USA

Auston, David H.
Bell Laboratories
126, New Providence Road
Mountainside, NJ 07092, USA

Baker, Fred
Galileo Electro-Optics
Galileo Park
Sturbridge, MA 01518, USA

Bates, Harry E.
Towson State University
Dept. of Physics
Baltimore, MD 21204, USA

Becker, Michael F.
University of Texas
Electrical Engineering Dept.
Austin, TX 78712, USA

Beddard, Godfrey
Davy Faraday Laboratory
21, Albemarle Street
London W1X 4BS, England

Berg, A.D.
Coherent
3210, Porter Drive
Palo Alto, CA 94304, USA

Berry, M.J.
Allied Chemical Corporation
P.O. Box 1021R
Morristown, NJ 07960, USA

Bloom, David M.
Bell Laboratories
4D-531
Holmdel, NJ 07733, USA

Boghosian, Charles
Army Research Office
Research Triangle, NC 27709, USA

Bohandy, Joe
Johns Hopkins
Johns Hopkins Road
Laurel, MD 20810, USA

Bolger, B.
Philips
Bld Wy 354
Eindhoven, The Netherlands

Bosacchi, Bruno
Optical Sciences Center
University of Arizona
Tucson, AZ 85721, USA

Bourkoff, Etan
University of California
Electronics Research Lab.
Berkeley, Ca 94720, USA

Bradley, D.J.
Imperial College
Blackett Laboratory
London, England

Brewer, Richard G.
IBM Research Laboratories
5600, Cottle Road
San Jose, CA 95193, USA

Brueck, Steven
Lincoln Laboratory
Lexington, MA 02173, USA

Campillo, Anthony J.
MS567
Los Alamos Scientific Lab.
Los Alamos, NM 87545, USA

Chan, C.K.
Spectra Physics
1250 W. Middlefield
Mt. View, CA 94042, USA

Clark, John H.
Los Alamos Scientific Lab.
P.O. Box 1663 MS 565
Los Alamos, CA 97545, USA

Clark, William G.
Spectra Physics
2 Harding Terrace
Morristown, NJ 07960, USA

Cone, R.L.
Physics Department
Montana State University
Bozeman, MT 59717, USA

Cooper, David E.
MIT
Bldg. 12-2017
Cambridge, MA 02139, USA

Corkum, Paul B.
National Research Co.
M23A Montreal Road
Ottawa, Ontario, Canada

Cox, A.J.
University of Redlands
1365 Pacific
Redlands, CA 92373, USA

Cramer, William H.
National Science Foundation
Washington, DC 20550, USA

DeFonzo, A.L.
Naval Research Laboratory
Washington, DC 20375, USA

Delmare, Claude
98, Rue Rakoczi
F-91330 Yerres, France

DeMaria, A.J.
United Technologies
Silver Lane
E. Hartford, CT 06118, USA

DeMartini, F.
Istituto di Fisica
G. Marconi Universita
Piazza delle Scienze, 5
I-00185 Roma, Italy

Devoe, Ralph G.
IBM Research Laboratories
K01-281
5600, Cottle Road
San Jose, CA 95193, USA

Diels, J.-C.
Center for Laser Studies
University of Southern California
University Park
Los Angeles, CA 90007, USA

Ducuing, Jacques
Ecole Polytechnique,
F-91120 Palaiseau, France

Duguay, Mike A.
Bell Laboratories
Holmdel, NJ 07733, USA

Eisenthal, Kenneth B.
Columbia University
New York, NY 10027, USA

Elliott, Richard A.
Oregon Graduate Center
19600, NW Walker Rd.
Beaverton, OR 97005, USA

Falcone, Roger
Stanford University
Ginzton Laboratory
Stanford, CA 94305, USA

Faust, Walter L.
Naval Research Laboratory
Washington, DC 20375, USA

Fayer, Michael D.
Stanford University
Chemistry Department
Stanford, CA 94305, USA

Fisher, Edward S.
Argonne National Laboratory
Argonne, IL 60439, USA

Fisher, Robert A.
Los Alamos Scientific Lab.
MS 535
Los Alamos, NM 87545, USA

Fleming, Graham
Royal Institution
21, Albemarle Street
London, England

Fleurot, N.
78, Avenue Laferrier
F-94000 Creteil, France

Fork, Richard L.
Bell Laboratories
Holmdel, NJ 07733, USA

Friedman, Joel
Bell Laboratories
Murray Hill, NJ 07974, USA

Frigo, Nicholas J.
Cornell University
Physics Department
Ithaca, NY 14853, USA

Gallant, Michel
Erindale College
Toronto, Ontario, Canada

Genack, Azriel
Exxon Research
1600, Linden Avenue
Linden, NJ 07036, USA

Gerritsen, H.J.
Brown University
Providence, RI 02912, USA

Gibbs, H.M.
Bell Laboratories
9, Stage Drive
Warren, NJ 07060, USA

Glasser, Lance
MIT
RLE 36-345
77, Mass Avenue
Cambridge, MA 02139, USA

Glenn, William H.
United Technologies
Research Center MS92
E. Hartford, CT 06108, USA

Gold, David
Center for Laser Studies
11715 Bellagio H3
Los Angeles, CA 90049, USA

Goldberg, Lawrence S.
Naval Research Laboratory
Code 5570
Washington, DC 20375, USA

Green, William R.
Stanford University
Ginzton Laboratory
Stanford, CA 94305, USA

Greene, Benjamin
University of Pennsylvania
Department of Chemistry
Philadelphia, PA 19104, USA

Gustafson, Ture Ken
University of California
Berkeley, CA 94720, USA

Harris, Charles B.
University of California
Berkeley, CA 94720, USA

Harris, Stephen E.
Stanford
880 Richardson Ct.
Palo Alto, CA 94303, USA

Harrison, Kiko
Los Alamos Scientific Lab.
Los Alamos, NM 87545, USA

Hartmann, Sven
Columbia University
New York, NY 10027, USA

Haus, Hermann A.
MIT
Massachusetts Avenue
Cambridge, MA 02139, USA

Heer, Clifford V.
Ohio State University
174 West 18th Avenue
Columbus, OH 43210, USA

Heisel, Francine
C.N.R.S.
Centre Nucléaire
F-67000 Strasbourg, France

Heller, Donald F.
Allied Chemical
MRS Box 1021R
Morristown, NJ 07960, USA

Hemenway, Carl P.
Cornell University
Chemistry Department
Ithaca, NY 14853, USA

Heritage, Jon
Bell Laboratories
Crawfords Corner Road
Holmdel, NJ 07733, USA

Hermann, J.P.
Ecole Polytechnique
Route de Saclay
F-91128 Palaiseau, France

Hesselink, William H.
University of Groningen
Nyenborgh 16
Gronigen, The Netherlands

Hirotoshi, Soma
University of Tokyo
Minato-ku Roppongi
Tokyo, Japan

Hirsch, Mitchell
University of Pennsylvania
Chemistry Department D5
Philadelphia, PA 19103, USA

Ho, Ping-Tong
MIT
77, Mass Avenue
Cambridge, MA 02139, USA

Hochstrasser, Robin
University of Pennsylvania
Philadelphia, PA 19104, USA

Hoffman, C.
Brown University
Providence, RI 02912, USA

Hopkins, James H.
Spectra Physics
1250 W. Middlefield
Mt. View, CA 94042, USA

Hyer, Ronald C.
MS 567
Los Alamos, Scientific Lab.
Los Alamos, NM 87545, USA

Ippen, Erich P.
Bell Laboratories
Holmdel, NJ 07733, USA

Jain, Ravi
Hughes Research Laboratories
Malibu, CA 90272, USA

Jarasiunas, K.
Brown University
Providence, RI 02912, USA

Johnson, Richard
Spectra Physics
1250 W. Middlefield
Mt. View, CA 94040, USA

Johnson, T.S.
Hadland-Photonics
Bovingdon
Herts, England

Kaiser, Wolfgang
Technische Universität
Arcisstr. 21
D-8000 München, Fed. Rep. of Germany

Kenney-Wallace, G.A.
University of Toronto
Toronto, Ontario M531A1, Canada

Kishida, Shunji
Nippon Electric Co.
4-1-1 Miyazaki Takatsu
Kawasaki, Japan

Kobayashi, Takayoshi
Bell Laboratories
Murray Hill, NJ 07974, USA

Kogelnik, H.
Bell Laboratories
Holmdel, NJ 07733, USA

Kryukov, P.G.
Institute of Spectroscopy
Moscow, USSR

Kuizenga, Dirk J.
Lawrence Livermore Laboratory
P.O. Box 808
Livermore, CA 94550, USA

Langelaar, Jan
University of Amsterdam
Nwe Prinsengracht 126
Amsterdam, The Netherlands

Langhoff, Charles
Ill. Institute of Technology
Department of Chemistry
Chicago, IL 60304, USA

Laubereau, Alfred
Technical University
Arcisstraße 21
D-8000 München, Fed. Rep. of Germany

Laval, Suzanne
C.N.R.S.
University Paris 11, Bât. 220
F-91405-Orsay, France

Lee, Chi H.
University of Maryland
College Park, MD 20742, USA

Levy, Roland
C.N.R.S.
5, Rue Université
F-67000 Strasbourg, France

Lewis, Aaron
Cornell University
Clark Hall
Ithaca, NY 14850, USA

Lieber, Albert
Los Alamos, Scientific Lab.
Los Alamos, NM 87544, USA

Lin, Cheng-Huei
EG and G
130, Robin Hill Road
Goleta, CA 93017, USA

Lindle, James Ryan
North Texas State University
Box 5368 N.T.S.U.
Denton, TX 76203, USA

Litster, David
MIT
13-2030
Cambridge, MA 02139, USA

Liu, Yung S.
GE Research Center
P.O. Box 8
Schenectady, NY 12301, USA

Lontz, Robert
Army Research Office
Research Triangle, NC 27709, USA

Lowdermilk, W. Howard
Lawrence Livermore Laboratory
Livermore, CA 94550, USA

Lukasik, Jacques
Ecole Polytechnique
Lab. D'Optique Quantique
F-91128 Palaiseau, France

Lutz, Stephen S.
EG and G
P.O. Box 682
Goleta, CA 93017, USA

Mainfray, G.
CEA Cen Saclay
F-91190 Gif sur Yvette, France

Magde, Douglas
University of California
at San Diego
La Jolla, CA 92092, USA

Mahr, Herbert
Cornell University
Ithaca, NY 14850, USA

Malley, Michael M.
San Diego State University
College Avenue
San Diego, CA 92124, USA

Mayer, Ulrich
Universität Stuttgart
Pfailenwaldring 54
D-7000 Stuttgart
Fed. Rep. of Germany

McAfee, Sigrid R.
Bell Telephone Lab.
600, Mountain Avenue
Murray Hill, NJ 07974, USA

McMullen, J.D.
Rockwell International
3370, Miraloma Avenue
Anaheim, CA 92714, USA

Merkelo, H.
University of Illinois
155EEB
Urbana, IL 61801, USA

Mialocq, Jean-Claude
CEN-Saclay
BP - No 2
F-91190 Gif sur Yvette, France

Migus, Arnold
Bell Laboratories
Holmdel, NJ 07733, USA

Moss, Steven C.
North Texas State University
Department of Physics NTSU
Denton, TX 76203, USA

Mourou, Gerard
University of Rochester
East River Road
Rochester, NY 14623, USA

Munro, Ian H.
Dept. Struct. Biology
Stanford Medical Center
Stanford, CA 94305, USA

Murray, James E.
Lawrence Livermore Laboratory
P.O. Box 808, L-479
Livermore, CA 94550, USA

Mysyrowicz, A.
C.N.R.S.
GPS-ENS Tour 23 Jussieu
F-75005 Paris, France

Nicolai, Van O.
Office of Naval Research
800 Quincy
Arlington, VA 22217, USA

Nicholson, Richard
National Science Foundation
1800 G, NW
Washington, DC 20550, USA

Nurmikko, Arto
Brown University
Division of Engineering
Providence, RI 02192, USA

O'Connor, Paul
Brown University
Physics Dept. Box 1843
Providence, RI 02912, USA

Ortega, J.D.
Naval Research Laboratory
Washington, DC 20375, USA

Owren, Joseph
Marco Scientific
1031, H. East Duane Avenue
Sunnyvale, CA 94086, USA

Penzkofer, Alfons
University Regensburg
D-8400 Regensburg
Fed. Rep. of Germany

Pilloff, H.
Office of Naval Research
Arlington, VA 22217, USA

Proffitt, Will
Coherent
3210, Porter Drive
Palo Alto, CA 94304, USA

Rabson, T.A.
Rice University
P.O. Box 1892
Houston, TX 77001, USA

Reintjes, John F.
US Naval Research Laboratory
Code 5540
Washington, DC 20375, USA

Ricard, Daniel
Lab. D'Optique Quantique
Ecole Polytechnique
F-91128 Palaiseau, France

Richardson, Martin
National Research Council of Canada
Bldg. M23A
Division of Physics
Ottawa, Ontario, Canada

Roach, Joseph Fr.
US NARADCOM
Kansas Street
Natick, MA 01760, USA

Robinson, C. Paul
Los Alamos Scientific Lab.
Box 1663
Los Alamos, NM 87544, USA

Royt, Terry
Naval Research Laboratory
Code 5570
Washington, DC 20375, USA

Saari, Peeter
Academy Sciences Fl
Riia 142
Tartu ESSR 202400, USSR

Sabersky, Andrew
Stanford University SLAC
P.O. Box 4349
Stanford, CA 94305, USA

Salamo, Gregory J.
University Arkansas
Physics Department
Fayetteville, AR 72701, USA

Salcedo, Jose R.
Stanford University
Ginzton Laboratories
Stanford, CA 94305, USA

Salour, Michael
100, Memorial Drive 1118C
Cambridge, MA 02139, USA

Schneider, Siegfried
Technical University München
Lichtenbergstraße 4
D-8046 Garching,
Fed. Rep. of Germany

Scott, Gary W.
University of California
Dept. of Chemistry
Riverside, CA 92521, USA

Shank, C.V.
Bell Laboratories
Holmdel, NJ 07733, USA

Shapiro, Stanley L.
MS 567
Los Alamos Scientific Lab.
Los Alamos, NM 87545, USA

She, C.Y.
Colorado State University
Ft. Collins, CO 80523, USA

Shen, Y.R.
University of California
Berkeley, CA 94720, USA

Shimoda, Koichi
University of Tokyo
Physics Dept. Hongo
Tokyo 113, Japan

Shionoya, Shigeo
Inst. for Solid State Physics
The University of Tokyo
Roppongi, Minato-ku
Tokyo 106, Japan

Siegman, A.E.
Stanford University
Ginzton Laboratory
Stanford, CA 94305, USA

Sipp, Bernard
C.N.R.S.
Centre Nucléaire
F-67000 Strasbourg, France

Slayman, Charles
University of California
Berkeley, CA 94720, USA

Smirl, Arthur L.
North Texas State
Department of Physics
Denton, TX 76203, USA

Smith, Harry L.
EG and G Inc. MS N-22
P.O. Box 1912
Las Vegas, NV 89101, USA

Spears, Kenneth G.
Northwestern University
2145, Sheridan Road
Evanston, IL 60201, USA

Staerk, Hubert
Max-Planck-Institute
MPl Biophys. Chem.
D-3400 Göttingen
Fed. Rep. of Germany

Steinmetz, Lloyd
Lawrence Livermore Laboratory
Box 808, L-323
Livermore, CA 94550, USA

Svelto, Orazio
Politecnico Milano
P. Leonardo Da Vinci, 32
Milano, Italy

Swofford, Robert L.
Standard Oil Company
4440 Warrensville Ctr.
Cleveland, OH 44128, USA

Taggart, Eric S.
Science Applications
4126 Linden Avenue
Dayton, OH 45432, USA

Tanaka, Yuichi
Inst. for Solid State Physics
University of Tokyo
Roppongi, Minato-ku
Tokyo 106, Japan

Takayoshi, Kobayshi
Bell Laboratories
600, Mountain Avenue
Fanwood, NJ 07094, USA

Teschke, Omar
UNICAMP - Inst. Fisica
CID Universitaria
Campinas SP 13100, Brazil

Topp, Michael R.
University of Pennsylvania
Philadelphia, PA 19104, USA

Turner, John J.
Institute of Defense Analysis
400 Army-Navy Drive
Arlington, VA 22202, USA

Voorst van, Jon
University of Amsterdam
Nieuwe Prinsengracht 126
Amsterdam, The Netherlands

Vidal, Carl Rudolf
MPI für Extraterrestrische Physik
D-8046 Garching
Fed. Rep. of Germany

Von der Linde, Dietrich
Max-Planck-Institute
Heisenberg Str. 1
D-7000 Stuttgart
Fed. Rep. of Germany

Weber, Heinz P.
University of Berne
Sidlerstraße 2
CH-3122 Berne, Switzerland

Weisman, R. Bruce
Department of Chemistry
University of Pennsylvania
Philadelphia, PA 19104, USA

Wheeler, John Post
Coherent Inc.
3210, Porter Drive
Palo Alto, CA 94304, USA

Wiersma, Douwe A.
University of Groningen
Nyenborgh 16
Groningen, The Netherlands

Wiesenfeld, Jay
Bell Laboratories
4C-514
Holmdel, NJ 07733, USA

Williams, David J.
Xerox Corporation
800, Phillips Road
Webster, NY 14450, USA

Williams, Richard T.
Naval Research Laboratory
Code 5581
Washington, DC 20375, USA

Wilson, Kent R.
University of California
at San Diego
Department of Chemistry
La Jolla, CA 92093, USA

Windsor, Maurice W.
Washington State University
Department of Chemistry
Pullman, WA 99104, USA

Wirth, Mary J.
Purdue University
Chemistry Department
W. Lafayette, IN 47907, USA

Yablonovitch, Eli
Harvard University
Pierce Hall 234
Cambridge, MA 02138, USA

Yajima, Tatsuo
University of Tokyo
Roppongi Minatoku
Tokyo 106, Japan

Yamanaka T.
Osaka University
Suita Osaka, Japan

Yu, William
Hamamatsu Corporation
120, Wood Avenue
Middlesex, NJ 08846, USA

Zewail, Ahmed H.
Caltech
Pasadena, CA 91125, USA

Zych, Ludwig J.
Stanford University
Stanford, CA 94040, USA

Topics in Applied Physics

Founded by H. K. V. Lotsch

Volume 18

Ultrashort Light Pulses

Editor: S. L. Shapiro

1977. 173 figures. XI, 389 pages
ISBN 3-540-08103-8

With the advent of picosecond light pulses a decade ago numerous scientists recognized that new methods of prime importance for exploring molecular interactions were feasible and also that extremely rapid devices based on new principles were possible. Now that the basic fundamentals of picosecond technology are well understood, and many of the early exploration goals have been realized, it is important to present the first comprehensive treatment by distinguished experts on both pulse generation and pulse interactions with matter. Like the spectroscopy field (where phenomena are understood through measurements in the frequency domain), the picosecond pulse field (where phenomena are analyzed in the time domain) is here to stay. The book aims to summarize the state of the art now that picosecond technology is rapidly emerging as a general tool in many different disciplines. Students and professionals in the sciences and engineering will be fascinated by the new developments and will profit from understanding these new general techniques.

Contents:

Volume 13

High-Resolution Laser Spectroscopy

Editor: K. Shimoda

1976. 132 figures. XIII, 378 pages
ISBN 3-540-07719-7

Contents:

Springer-Verlag
Berlin
Heidelberg
New York

Springer Series in Optical Sciences

Editor: D. L. MacAdam

Volume 3:

Tunable Lasers and Applications

Proceedings of the Loen Conference, Norway, 1976
Editors: A. Mooradian, T. Jaeger, P. Stokseth

1976. 238 figures. VIII, 404 pages
ISBN 3-540-07968-8

Volume 4
V. S. Letokhov, V. P. Chebotayev

Nonlinear Laser Spectroscopy

1977. 193 figures, 22 tables. XVI, 466 pages
ISBN 3-540-08044-9

Volume 6
B. Saleh

Photoelectron Statistics

With Applications to Spectroscopy and Optical Communication

1978. 85 figures, 8 tables. XV, 441 pages
ISBN 3-540-08295-6

Volume 7

Laser Spectroscopy III

Proceedings of the Third International Conference, Jackson Lake Lodge, Wyoming, USA, July 4–8, 1977

Editors: J. L. Hall, J. L. Carlsten

1977. 296 figures. XII, 468 pages
ISBN 3-540-08543-2

Springer-Verlag
Berlin
Heidelberg
New York